Integrating Information in Built Environments

In an increasingly globalised built environment industry, achieving higher levels of integration across organisational and software boundaries can lead to improved economic, social and environmental outcomes. This book is the direct result of a collaborative global network of industry and academic researchers spread across nine countries as part of CIB's (International Council for Research and Innovation in Building and Construction) Task Group 90 (TG90) Information Integration in Construction (IICON).

The book provides a broad view of some of the opportunities and challenges brought by integrating information across organisational and system boundaries in the built environment industry. Chapters cover a large range of topics and are separated into three sections: resources, processes and added value. They provide a much-needed international perspective on a current global evolution in the industry and present leading original research and valuable lessons for researchers, industry practitioners, government clients and policy makers across the industry.

Key features include:

- a broad range of topics that are not covered elsewhere in the literature;
- contributions from a diverse group of industry research leaders from across the globe;
- exemplar case studies providing real-world examples of where information integration has been a key factor for success or lack thereof has been the root cause of failure;
- an analysis of future priority areas for research and development investment as well as their strategic implications for public and private decision-makers;
- the book will deliver innovation in best practice methodology for information sharing across disciplines and between the design, construction and asset management sectors.

Adriana X. Sanchez is currently a research associate at the Australian Sustainable Built Environment National Research Centre and a PhD Candidate at the University of New South Wales through the Cooperative Research Centre for Low Carbon Living. She is one of the coordinators of the CIB Task Group 90: Information Integration in Construction.

Keith D. Hampson is CEO of the Australian Sustainable Built Environment National Research Centre, and Professor of Construction Innovation at Curtin University, Australia. He has a Bachelor of Engineering with Honours and an MBA from Queensland University of Technology (QUT) and a PhD in Civil and Environmental Engineering from Stanford University, USA.

Geoffrey London currently works at the University of Western Australia. He served as Government Architect for the last 13 years in Victoria and Western Australia; this is a role that nourishes the production of well-produced sustainable buildings.

About CIB and the CIB series

CIB, the International Council for Research and Innovation in Building and Construction, was established in 1953 to stimulate and facilitate international cooperation and information exchange between governmental research institutes in the building and construction sector, with an emphasis on those institutes engaged in technical fields of research.

CIB has since developed into a worldwide network of over 5,000 experts from about 500 member organisations active in the research community, in industry or in education, who cooperate and exchange information in over 50 CIB Commissions and Task Groups covering all fields in building and construction related research and innovation.

http://www.cibworld.nl/

This series consists of a careful selection of state-of-the-art reports and conference proceedings from CIB activities.

R&D Investment and Impact in the Global Construction Industry
K.D. Hampson et al.
ISBN: 9780415859134. Published: 2014

Public Private Partnerships A. *Akintoye* et al.
ISBN: 9780415728966. Published: 2015

Court-Connected Construction Mediation Practice A. *Agapiou and D. Ilter*
ISBN: 9781138810105. Published: 2016

Advances in Construction ICT and e-Business S. *Perera* et al.
ISBN: 9781138914582. Published: 2017

Clients and Users in Construction K. *Haugbolle and D. Boyd*
ISBN: 9781138786868. Published: 2017

Integrating Information in Built Environments A. X. *Sanchez* et al.
ISBN: 9781138706323. Published: 2018

Integrating Information in Built Environments

From Concept to Practice

Edited by Adriana X. Sanchez,
Keith D. Hampson and
Geoffrey London

First published 2018
by Routledge
2 Park Square, Milton Park, Abingdon, Oxon OX14 4RN

and by Routledge
711 Third Avenue, New York, NY 10017

Routledge is an imprint of the Taylor & Francis Group, an informa business

© 2018 selection and editorial matter, Adriana X. Sanchez, Keith D. Hampson and Geoffrey London; individual chapters, the contributors

The right of Adriana X. Sanchez, Keith D. Hampson and Geoffrey London to be identified as the authors of the editorial material, and of the authors for their individual chapters, has been asserted in accordance with sections 77 and 78 of the Copyright, Designs and Patents Act 1988.

All rights reserved. No part of this book may be reprinted or reproduced or utilised in any form or by any electronic, mechanical, or other means, now known or hereafter invented, including photocopying and recording, or in any information storage or retrieval system, without permission in writing from the publishers.

Trademark notice: Product or corporate names may be trademarks or registered trademarks, and are used only for identification and explanation without intent to infringe.

British Library Cataloguing-in-Publication Data
A catalogue record for this book is available from the British Library

Library of Congress Cataloging-in-Publication Data
Names: Sanchez, Adriana, editor. | Hampson, Keith (Keith Douglas), editor. | London, Geoffrey, editor.
Title: Integrating information in the built environments : from concept to practice / edited by Adriana X. Sanchez, Keith D. Hampson and Geoffrey London.
Description: Abingdon, Oxon ; New York, NY : Routledge, 2018. | Series: CIB series | Includes bibliographical references and index.
Identifiers: LCCN 2017004102 | ISBN 9781138706323 (hardback : alk. paper) | ISBN 9781315201863 (ebook)
Subjects: LCSH: Construction industry--Information services. | Building--Data processing. | Building information modeling.
Classification: LCC TH216 .I59 2018 | DDC 690.028/5--dc23
LC record available at https://lccn.loc.gov/2017004102

ISBN: 9781138706323 (hbk)
ISBN: 9781315201863 (ebk)

Typeset in Goudy
by Saxon Graphics Ltd, Derby

Printed and bound by CPI Group (UK) Ltd, Croydon, CR0 4YY

Contents

List of figures	viii
List of tables	x
Notes on contributors	xi
Foreword	xxvii
Preface	xxix
Acknowledgements	xxxi
List of abbreviations	xxxii

1 Integrating information across built environment boundaries   1
ADRIANA X. SANCHEZ, KEITH D. HAMPSON AND GEOFFREY LONDON

PART 1
Resources   11

2 Integrating information for more productive social housing outcomes: an Australian perspective   13
JUDY A. KRAATZ, ANNIE MATAN AND PETER NEWMAN

3 Resilient policies for wicked problems: increasing resilience in a complex and uncertain world through information integration   35
ADRIANA X. SANCHEZ, PAUL OSMOND AND JEROEN VAN DER HEIJDEN

4 Internet of Things for urban sustainability   52
FONBEYIN H. ABANDA AND JOSEPH H. M. TAH

5 Digital technologies improving safety in the construction industry   70
WEN YI, PENG WU, XIANGYU WANG AND ALBERT P. C. CHAN

vi *Contents*

6 Information integration and interoperability for BIM-based
life-cycle assessment   91
RUBEN SANTOS AND ANTÓNIO AGUIAR COSTA

PART 2
Processes   109

7 Precinct Information Modelling: a new digital platform for
integrated design, assessment and management of the built
environment   111
PETER NEWTON, JIM PLUME, DAVID MARCHANT, JOHN MITCHELL
AND TUAN NGO

8 Information integration for asset and maintenance management   133
SONIA LUPICA SPAGNOLO

9 IT support for contractor monitoring of refurbishment projects   150
JAN BRÖCHNER AND AHMET A. SEZER

10 Experience with the use of commissioning advisers in design:
a Danish context   160
MARIANNE FORMAN

11 Turning energy data into actionable information: the case of
energy performance contracting   178
FRÉDÉRIC BOUGRAIN

12 Stakeholder perspectives and information exchange in AEC
projects   194
TORILL MEISTAD, MARIT STØRE-VALEN, VEGARD KNOTTEN,
ALI HOSSEINI, OLE JONNY KLAKEGG, ØYSTEIN MEJLÆNDER-LARSEN,
EILIF HJELSET, FREDRIK SVALESTUEN, OLA LÆDRE, GEIR K. HANSEN
AND JARDAR LOHNE

PART 3
Added Value   211

13 The concept of value of buildings in use   213
MARIT STØRE-VALEN, TORILL MEISTAD, KNUT BOGE,
MARGRETHE FOSS, LEIF D. HOUCK AND JARDAR LOHNE

Contents vii

**14 Information integration and public procurement: the role of
monitoring, benchmarking and client leadership** 231

ADRIANA X. SANCHEZ, JESSICA BROOKS AND KEITH D. HAMPSON

15 Four metaphors on knowledge and change in construction 245

KIM HAUGBØLLE

16 Contrasting aspects of information integration 265

ADRIANA X. SANCHEZ, GEOFFREY LONDON AND KEITH D. HAMPSON

Index 274

Figures

2.1	Overview of the housing system in Australia and data sources, at 30 June 2013	18
2.2	Main social housing assistance programmes in Australia and total number of households/clients assisted per programme, 30 June 2013	19
2.3	Integrating broad-based stakeholder perspectives	22
2.4	Strategic evaluation framework	24
2.5	Two-tiered approach to return on investment	28
5.1	Parametric-driving modular and mobile scaffold designs	75
5.2	Building scaffolds	77
5.3	Rule-based hole and edge detection	77
5.4	Guardrail systems at edge and cover access opening	78
5.5	Modular scaffold installation in BIM	78
5.6	Clash detection in BIM between scaffold, building and mobile devices	79
5.7	Site traffic simulation and verification	79
5.8	Unsafe distance warning in site traffic planning	80
5.9	Prototype of the early-warning system	83
5.10	Process map of the early-warning system	83
6.1	LCA methodology phases	93
6.2	Input/output of a product's life-cycle	94
6.3	BIM–LCA data	100
6.4	NATSPEC BIM object/element matrix – exterior wall	102
6.5	Example of NBS BIM object guide	103
6.6	BIM–LCA integration framework	104
7.1	The urban sustainability framework: products, buildings, infrastructure, neighbourhoods and cities	112
7.2	ISO model of multiple information layers required for precinct representation and performance assessment of smart sustainable built environment	112
7.3	Spatial information platforms for the built environment	113
7.4	Temporal information platform for the built environment: precinct life-cycle	114
7.5	Precinct performance assessment framework	115

Figures ix

7.6	Precinct Information Model for an urban retrofit project	119
7.7	Land use map of the 'Empowering Broadway' precinct	120
7.8	(a) Zonal functional typology – schematic master plan; (b) Built facility type – development proposal; (c) Elemental object typology – detailed asset model	121
7.9	PIMViewer application	123
7.10	BIM model derived from City of Sydney 2012 FES data	126
7.11	Alternatives to BAU energy scenario for new residential building at Fishermans Bend	129
8.1	Comparison between existing classification systems	137
8.2	INNOVance partnership	141
8.3	INNOVance hierarchical structure	142
8.4	INNOVance coding system	144
11.1	Modes of knowledge creation	182
12.1	Integration in early phases offers opportunities for large impact on performance	196
12.2	Design process in different trades	200
12.3	Parallelism between engineering, procurement and construction	202
12.4	Main stages in the scan-to-BIM process	203
13.1	Construction project main roles and their relationship when creating values	220
13.2	Complicity of stakeholders involved in the construction process	221
13.3	Factors emphasising choice of façade materials	226
15.1	Dominant focus of Danish policy studies	250
15.2	Dominant R&D stems from the industrial supply network	251
15.3	Absorptive capacity in construction firms	254
15.4	Multiple communities of practice in construction	257
15.5	Knowledge as formative	260

Tables

2.1	State-by-state transformation matrix	20
2.2	Layers of performance indicators	25
2.3	Some relevant datasets	30
4.1	Applications of IoT-based technologies for sustainability	57
8.1	List of noteworthy factors used in the factor method	139
10.1	List of existing objects from the design phase, new objects from the operating organisation and new Cx objects	169
10.2	Overview of dilemmas that the use of commissioning causes for the various parties	174
11.1	Results of the EPC project concerning 18 high schools	185
11.2	Characteristics of the call for tenders concerning the EPC of 14,000 dwellings	189
12.1	Outline of the Next Step framework	204
15.1	Technological frame: knowledge as flows	252
15.2	Technological frame: knowledge as capabilities	255
15.3	Technological frame: knowledge as practices	258
15.4	Technological frame: knowledge as formative	261

Contributors

Fonbeyin Henry Abanda
Senior Lecturer, Oxford Brookes University
Oxford, UK
Fonbeyin has a BSc (Hons) from the University of Buea-Cameroon and an Engineering Diploma in Mathematics/Physics and Civil Engineering from the Ecole Nationale Supérieure Polytechnique de Yaoundé-Cameroon respectively. He obtained his PhD from Oxford Brookes University in the UK in 2011. His PhD investigated the extent to which Semantic Web technologies can be used in managing sustainable building technology knowledge for use in different building projects. Currently, Fonbeyin is a senior lecturer in the School of the Built Environment at Oxford Brookes University. He teaches Construction Project Management/Building Information Modelling at undergraduate and postgraduate levels. He supervises PhD students in the area of Construction Information Technologies. Fonbeyin is a Chartered Engineer, an assessor at the UK Institution of Engineering and Technology and a fellow of the UK Higher Education Academy.

Knut Boge
Associate Professor, Oslo and Akershus University College of Applied Sciences
Oslo, Norway
Knut is an associate professor and coordinator of Oslo Business School's Bachelor's programme in Facility Management. Knut has also been employed at Oslo and Akershus University College of Applied Sciences since 2006, and his research interests mainly focus on facility management, real estate, and innovation and entrepreneurship. He is particularly interested in service innovations. Prior to his current role, Knut was an assistant professor at BI Norwegian School of Management doing research and teaching about the intercept areas of political science, economic history, and innovation and entrepreneurship.

xii *Contributors*

Frédéric Bougrain
Researcher, Economics and Human Sciences Direction, Centre Scientifique et Technique du Bâtiment (CSTB)
Champs-sur-Marne, France
Frédéric works as a researcher for the Economics and Human Sciences Department at Technical Centre for Building (CSTB, Centre Scientifique et Technique du Bâtiment), a state-owned industrial and commercial research centre under the wing of the Ministry of Housing in France. His research concentrates on societal issues with a particular focus on innovations in the building and construction industry and energy saving performance contracts. Frédéric previously lectured at the University of Orléans, France, where he defended a thesis on innovation, small and medium-sized enterprises and the consequences for regional technology policy. Frédéric has published works on public–private partnerships, energy-saving performance contracts, innovation in small and medium-sized enterprises and the social housing sector.

Jan Bröchner
Professor and Chair of Organization of Construction, Division of Service Management, Chalmers University of Technology
Gothenburg, Sweden
Jan holds the Chair of Organization of Construction at Chalmers University of Technology in Gothenburg, Sweden. He is an adjunct member, representing the Swedish universities of technology, of the Research Committee of the Development Fund of the Swedish Construction Industry (SBUF). Currently, he leads the productivity and innovation effect analysis project within the 'Bygginnovationen' Programme, supported by Swedish Governmental Agency for Innovation Systems (VINNOVA) and an industry consortium.

Jessica Brooks
Research Assistant at Sustainable Built Environment National Research Centre (SBEnrc)
Sydney, Australia
Jessica has a Bachelor's in Psychology from University of Wollongong, Wollongong, and has been working at the Australian Sustainable Built Environment National Research Centre (SBEnrc) headquartered at Curtin University, Perth since August 2015. In this role, she has been involved in the development of Australia's first BIM decision-support tool, BIM Value, and is currently developing Australia's first BIM benefits benchmarking system: BIM Value Benchmark. This system aims to overcome barriers traditionally found in industry benchmark development by applying crowdsourcing principles and building on a collaborative effort between industry, government and research.

Albert PC Chan
Head of Department and Associate Director of the Research Institute for Sustainable Urban Development, The Hong Kong Polytechnic University
Hong Kong, China

Albert is Head of Department of Building and Real Estate and Chair Professor of Construction Engineering and Management as well as a chartered construction manager, engineer, project manager and surveyor by profession. He has worked in a number of tertiary institutions both in Hong Kong and internationally. Albert's research and teaching interests include project management and project success, construction procurement and relational contracting, construction management and economics, construction health and safety, and construction industry development. Albert holds an MSc in Construction Management and Economics from the University of Aston in Birmingham and a PhD in Project Management from the University of South Australia.

Antonio Aguiar Costa
Assistant Professor, University of Lisbon
Lisbon, Portugal

António is an assistant professor at the Department of Civil Engineering and Architecture of the Superior Technical Institute (IST, Instituto Superior Técnico), University of Lisbon. His research interests are information management and integration, interoperability, Building Information Modelling (BIM) implementation and standardisation and BIM-based e-procurement. His other areas of research include intelligent buildings, and Internet of Things for buildings and smart cities. At the national level, Antonio is Chairman of CT197-BIM, the Portuguese Technical Committee for BIM standardisation, and Coordinator of the Information and Design Support Systems Research Group of Civil Engineering Research and Innovation for Sustainability (CEris). Internationally, he is the representative of Portugal at CEN/TC442 (BIM European Standards) and member of the European Union BIM Task Group supported by European Commission. Antonio is also a member of CIB TG90: Information Integration in Construction and CIB W98: Intelligent and Responsive Buildings and works as an independent expert for the European Commission; participating in evaluation of Horizon 2020 projects.

Marianne Forman
Senior Researcher, Danish Building Research Institute, Aalborg University
Copenhagen, Denmark

Marianne is a senior researcher at the Danish Building Research Institute, Aalborg University. Her research area encompasses innovation, user-driven innovation, sustainable transition, environmental management in companies and product chains, project management and change processes. Since 2006, her research has been related to challenges within the construction industry. She has an MSc in Civil Engineering from the Danish Technical University, where she

xiv *Contributors*

also defended her PhD thesis on 'Processes of Change and Participation Forms in Preventive Environmental Work'.

Margrethe Foss
Senior Advisor, Multiconsult / Lecturer, Norwegian University of Science and Technology (NTNU)
Trondheim, Norway
Margrethe is currently holding three positions in Norway: Senior Advisor Multiconsult (since 2009), Benchmarking Officer for NfN (Norwegian Facilities Management Network), Course Coordinator and Lecturer in Facility Management (FM) at the Master's programme for practitioners at NTNU (Norwegian Institute of Technology). Margrethe's main areas of competence are developing FM strategies and organisation, early phase building projects, workplace strategies, business development, benchmarking (FM and RE), procurement processes, localisation studies, life-cycle evaluations, organisation and competence development. She has often had roles as advisor, lecturer or in project/process management. She has always strived towards optimising the cost–benefit factor for the customers so real estate and FM activities bring more value to the core business. Margrethe strongly believes that a good process is often the key factor in a good product/result. Margrethe's previous experience includes Managing Director for Analyse & Strategi AS (2005–2009), Concept Developer for Veidekke Bostad AB (Stockholm 2005–2007), and Project Manager for Ericsson Real Estate and Ericsson Telecom (Stockholm, 1995–2005).

Keith D. Hampson
CEO, Sustainable Built Environment National Research Centre, Curtin University
Perth, Australia
Keith has over thirty years of industry, government and research leadership. He has a Bachelor of Engineering with Honours and an MBA from Queensland University of Technology (QUT), and a PhD from Stanford University focusing on innovation and business performance. He is a Fellow, Institution of Engineers Australia; Fellow, Australian Institute of Company Directors and Fellow, Australian Institute of Management. Keith currently co-coordinates the CIB TG90: Information Integration in Construction. Keith serves as CEO of the Sustainable Built Environment National Research Centre, successor to the Australian CRC for Construction Innovation, for which he led the bid team in 2000 and was CEO for its nine years of operation. As Professor of Construction Innovation at Curtin University, he continues to work collaboratively with colleagues across Australia and globally to transform industry performance in sustainability, safety and productivity for a stronger and more competitive industry.

Geir K. Hansen
Professor, Norwegian University of Science and Technology (NTNU)
Oslo, Norway
Geir is a professor in Architectural and Facilities Management at the Department of Architectural Design and Management, Faculty of Architecture and Fine Art, NTNU. He is Head of the Centre for Real Estate and Facilities Management and leads two Master's programs in Real Estate and Facilities Management at NTNU. Geir graduated as an architect from the Faculty of Architecture at the Norwegian Technical University. His main fields of research are programming, evaluation of processes and buildings, and usability of buildings related to the user and organisational perspective. Geir has previously worked with topics such as adaptability and flexibility, new use and transformation of existing buildings, and run several architectural design courses on this. Geir has been an active member of International Council for Building (CIB) working commissions: W96: Architectural Management; W70: Facilities Management; and is the coordinator of W111: Usability of Buildings.

Kim Haugbølle
Senior Researcher, Danish Building Research Institute (SBi), Aalborg University
Copenhagen, Denmark
Kim conducts advisory services to the Danish Government, undertakes teaching, provides training of professionals and develops research-based knowledge to improve the built environment. He has authored or co-authored more than 200 publications on innovation and socio-technical change in the construction industry with a special focus on the role of the construction client, life-cycle economics and sustainability. Kim has been coordinating and managing several national and international R&D projects. Previously, he headed the secretariat of the think tank Danish Building Development Council and, later, a research department at the national building research institute. Kim is the international co-coordinator of the CIB Working Commission W118: Clients and Users in Construction as well as a board member of the Nordic researchers' network on construction economics and organisation, CREON.

Jeroen van der Heijden
Associate Professor, Australian National University (ANU)
Canberra, Australia
Jeroen is an associate professor at the ANU's Regulatory Institutions Network (RegNet) and at the University of Amsterdam's Law School. He received his PhD in Public Administration (highest honours) in 2009 and his MSc in Architecture (high distinction) in 2002, both from Delft University of Technology, the Netherlands. Jeroen works at the intersections of regulation and governance, policy change, and urban development and transformation. His research aims to improve local, national and international outcomes of urban governance on some of the most pressing challenges of our time: climate change, energy and water use, and a growing and increasingly urbanising world population.

xvi *Contributors*

Through his work, Jeroen seeks to inform ongoing academic debates on these challenges as well as provide hands-on lessons to policy-makers and practitioners on how to govern urban sustainability and resilience on a day-to-day basis.

Eilif Hjelset
Associate Professor, Oslo and Akershus University College of Applied Sciences
Oslo, Norway

Eilif is an associate professor at Oslo and Akershus University College of Applied Sciences, where he focuses on integrating Building Information Modelling (BIM) into the existing curriculum. His research interest is in BIM-based model checking, specifications for information flow in the life-cycle of buildings, in addition to processes for implementing BIM in the industry. He is Norwegian representative for European Conferences of Product and Process Modeling (ECCPM) and member of the Scientific Committee of IT in Construction (CIB W78). Eilif also serves as Educational Coordinator at buildingSMART Norway, connecting students with industry. He is active as ISO expert in development of BIM-related standards. Prior to these roles, he worked for the Norwegian Building Authority (DIBK), Standards Norway and the Norwegian University of Life Sciences.

Ali Hosseini
PhD Candidate, Norwegian University of Science and Technology (NTNU)
Oslo, Norway

Ali works as a researcher in the Coastal Highway Route E39 Project, funded by Norwegian Public Road Administration (NPRA). He is also carrying out his PhD candidacy at the Department of Civil and Transport Engineering, NTNU, while supervising Master's students working on implementation strategies and types of contract. His research currently focuses on innovative and relational delivery methods for infrastructure projects. Prior to this role, Ali worked as a project planner (detailed planning in engineering fabrication and procurement) in the Norwegian oil sector while he was involved in educating and training new employees for ÅF Reinertsen AS. He started his career as an industrial engineer working in the construction industry and later completed a Master of Engineering in manufacturing system engineering.

Leif D. Houck
Head of Department of Building and Environmental Technology, University of Life Sciences (NMBU) and Partner at SPINN architects AS
Oslo, Norway

Leif has been teaching architectural design and project management at NMBU since 2009. As a researcher, he has specialised in school buildings, daylight and universal design. He is internationally recognised for his research on school architecture and has published several papers on school design, especially linked to daylight, sustainability and universal design. As an architect, Leif has worked with several school buildings in Snøhetta, Kristin Jarmund Architects (former partner), and currently SPINN Architects (founding partner). SPINN Architects is specialised in the programming and designing of school buildings. Leif has also participated in several school competition juries and has worked as an advisor in procurement and evaluation methods.

Ole Jonny Klakegg
Professor, Norwegian University of Science and Technology (NTNU)
Trondheim, Norway
Through his twenty-six years of work experience, Ole Jonny has alternated between teaching and research at the university and working as a consultant in project management, building substantial experience as well as theoretical and practical insight. Ole Jonny shares his time between his current position as Professor of Project Management and his role as R&D Director of WSP Norway. He has been involved in a large number of major projects in Norway in both public and private sectors, including building, civil engineering, transport, health, defence and organisational development. He is currently involved in research on project governance, project risk management, target value delivery and digitalisation of the building process. He also serves as leader for the construction programme in Project Norway.

Vegard Knotten
PhD Candidate, Norwegian University of Science and Technology (NTNU)
Oslo, Norway
Vegard's research focuses on a new integrated methodology for design management. This project is the result of a collaborative effort between NTNU, University of Agder and several industrial partners. The project compares the architectural, engineering and construction industry with offshore and shipping industries, while focusing on the early design phases of building design management. Vegard has extensive industry experience accrued through his work as Project Manager for Veidekke Entreprenør AS, Project and Construction Manager for Helsebygg Midt-Norge and Senior Engineer at Interconsult ASA.

Judy A. Kraatz
Senior Research Fellow, Griffith University
Brisbane, Australia
Judy is a senior research fellow with the Cities Research Centre at Griffith University. Judy has over twenty-five years of professional activity in the built environment as a design architect; leading a team of professionals delivering city-wide solutions for public buildings and parklands; and integrating sustainability into curriculum, design practice and business solutions. Judy's research addresses issues of corporate and social responsibility in the delivery of urban and social infrastructure. Judy brings a focus on meta-research and evaluation frameworks to better leverage research and achieve practical outcomes for both the urban environment as well as its residents. Current research focuses on the need for an efficient, effective and equitable social housing sector in Australia.

xviii *Contributors*

Ola Lædre
Associate Professor, Norwegian University of Science and Technology (NTNU)
Oslo, Norway
Ola is an associate professor at NTNU's Department of Civil and Transport Engineering. His most recent research publications deal with contracts and contract strategies, viability of large investment projects, internal rent schemes and building design management. Ola completed his PhD with the thesis 'Selection of contract strategy in construction projects' in 2006 and has working experience in research, private industry and local government.

Jardar Lohne
Research Scientist, Norwegian University of Science and Technology (NTNU)
Oslo, Norway
Jardar presently works as a research scientist/post-doctoral fellow at the Department of Civil and Transport engineering and at Klima 2050, a centre for research-based innovation. Over the years, he has published on a wide range of areas of research, including facilities management, project management, contract strategies, process innovation, sustainability analysis and formal frameworks for the construction sector. He is currently working on several publications within the field of climate change adaptation in the built environment.

Geoffrey London
Professor, The University of Western Australia
Perth, Australia
Geoffrey is a Professor of Architecture at The University of Western Australia where he is a past Dean and Head of School. He is a Professorial Fellow at The University of Melbourne, a Life Fellow of the Australian Institute of Architects (AIA) and an Honorary Fellow of the New Zealand Institute of Architects. He previously held the positions of Victorian Government Architect (2008–2014) and Western Australian Government Architect (2004–2008). He has been involved in advising those state governments on a wide range of projects, from the scale of individual houses to the complexity of major new tertiary hospitals. He has advised on issues that include design quality, project procurement, heritage, master planning, sustainability and development strategies, and been responsible for setting up design workshops on key, large-scale projects. He maintains a role as a consultant on urban design, architecture, design review and architectural competitions. He is an active researcher and program director in the Cooperative Research Centre for Water Sensitive Cities and has a long-term professional and research interest in medium density housing and forms of delivery that provide more affordable and better design.

David Marchant
Senior Research Fellow, University of New South Wales
Sydney, Australia

David is a senior research fellow at University of New South Wales in Sydney, Australia. He is currently a member of the team working on the Precinct Information Modelling (PIM) research project within the Cooperative Research Centre (CRC) for Low Carbon Living. His doctorate research addressed integration of design briefing within building information models. This research is also applicable to record design intent for data models addressing larger scales of planning such as precincts. David is a registered architect in NSW. He has also been an adjunct associate professor in the Key Centre for Design Computing and Cognition, Faculty of Architecture, University of Sydney. Prior to the current CRC, he participated as an active industry member of the CRC for Construction Innovation, particularly focused on research projects addressing use cases around Building Information Modelling and team collaboration.

Annie Matan
Adjunct Senior Research Fellow, Curtin University
Perth, Australia

Annie is Adjunct Senior Research Fellow at Curtin University Sustainability Policy (CUSP) Institute in Perth, Western Australia. She is interested in creating sustainable, vibrant and people-focused urban places. Her research focus is on social housing and active transport, particularly walking and cycling, pedestrian planning and urban design, focusing on how people interact with the built environment and human health outcomes of planning decisions. She worked for state and local government before joining Curtin University in 2011.

Torill Meistad
Senior Advisor, Nordic Energy Research
Oslo, Norway

Torill's work currently focuses on renewable energy systems in Nordic countries. She coordinates collaborative studies among Nordic universities and institutes about sustainable and reliable energy for transport, heating and industry production. Her work has especially focused on energy-efficient buildings, leading to publications on the issues of early involvement and integration in planning and construction projects. She completed her PhD at the Norwegian University of Science and Technology (NTNU). She has also worked at the Department of Civil and Transport Engineering, the Research Centre for Zero Emission Buildings (ZEB) and the Department of Architectural Design and Management. Torill has previously been involved with applied research projects related to industrial transformation and local community development at Centre for Rural Research, Trøndelag R&D Institute and NORUT Finnmark AS.

xx *Contributors*

Øystein Mejlænder-Larsen
Technology Manager, Multiconsult/ PhD Candidatee, Norwegian University of Science and Technology (NTNU)
Oslo, Norway

Øystein is an industrial PhD candidate, employed at Multiconsult and associated with NTNU. The PhD is part of the research project 'Collaboration in the building process – with Building Information Modelling (BIM) as a catalyst' ('SamBIM'), with Statsbygg, Skanska, Link Arkitektur and Multiconsult as industry partners, and NTNU, SINTEF and FAFO as research partners. The focus of his PhD is to explore how the use of project execution models and BIM can increase the efficiency of the building process, based on experiences from the execution of major oil and gas projects. Øystein currently has a leave of absence as a technology manager in Multiconsult, one of the leading firms of consulting engineers and designers in Norway, to pursue his industrial PhD. He holds a Master of Science in Civil and Environmental Engineering (NTNU) and Master of Technology Management (NTNU, NHH, MIT Sloan). Prior to joining Multiconsult in 2009, Øystein worked as a technology manager at Selvaag Bluethink (2000–2009).

John Mitchell
Senior Research Fellow, University of New South Wales
Sydney, Australia

John is Chairman of the buildingSMART Australasia Chapter and committed to the adoption of digital technologies for the design and construction professions. He takes an active role in research that provides new ways of working, increased productivity and extending building performance. One of his key interests is the Industry Foundation Classes (IFC) openBIM standard and the use of IFC model-servers that manipulate large datasets to support precinct-scale integrated built asset models being developed by the Precinct Information Model project at the Australian Cooperative Research Centre for Low Carbon Living. This development and the extension of the standard into all infrastructure types provides an innovative toolkit for the management of urban development by local government.

Peter Newman
Professor, Curtin University
Perth, Australia

Peter is a Professor of Sustainability at Curtin University. He has authored sixteen books and over 300 papers. His books include *The End of Automobile Dependence* (2015), *Green Urbanism in Asia* (2013) and *Sustainability and Cities: Overcoming Automobile Dependence* which was launched in the White House in 1999. Peter was a Fulbright Senior Scholar at the University of Virginia Charlottesville and was on the Inter-governmental Panel on Climate Change (IPCC) for their Fifth Assessment Report. In 2014, he was awarded an Order of Australia for his contributions to urban design and sustainable transport. He is a Fellow of the

Academy of Technological and Engineering Sciences Australia. Peter has worked in local government as an elected councillor, in state government as an advisor to three Premiers and in the Australian Government on the Board of Infrastructure Australia.

Peter Newton
Research Professor, Swinburne University of Technology/Research Program Leader, Cooperative Research Centre for Low Carbon Living
Melbourne, Australia
Peter is a Research Professor in Sustainable Urbanism at Swinburne University of Technology in Melbourne, Australia. His research is focused on pathways capable of achieving a sustainability transition for cities and their residents. They include technological innovation in urban infrastructure, innovation in urban planning and design, and in understanding household consumption behaviour. Before joining Swinburne in 2007 Peter was Chief Research Scientist at the Australian Commonwealth Scientific and Industrial Research Organisation (CSIRO). He is a member of Australia's major urban research networks: CRC for Low Carbon Living, CRC for Spatial Information, CRC for Water Sensitive Cities, AHURI, AURIN and SBEnrc. His research on low carbon living includes studies on zero carbon housing, quantifying the significance of alternative urban forms for energy use in cities, a study identifying the determinants of household energy use and an Australian Research Council (ARC) Discovery project on the green economy.

Tuan Ngo
Associate Professor, the University of Melbourne/Research Director, Australian Research Council (ARC) Centre for Advanced Manufacturing of Prefabricated Housing
Melbourne, Australia
Tuan is the leader of the MUtopia Platform, a simulation and visualisation system for designing and assessing sustainable precincts as well as sustainable cities. The MUtopia platform has been used in a range of urban development projects in Australia.

Paul Osmond
Director of the Sustainable Built Environment Program, University of New South Wales
Sydney, Australia
Paul has been engaged with sustainable development since the 1980s in practice and more recently through teaching and research. Prior to taking on an academic position with the University of New South Wales Built Environment Faculty, he managed the former UNSW Environment Unit. Paul has also worked in consultancy and local government roles, where he was responsible for the delivery of a variety of pioneering environmental management, landscape and urban design programs and projects. His previous professional background includes experience in forestry, freelance technical journalism and

xxii *Contributors*

the metal industry. Paul has qualifications in applied science, environmental management and landscape design. His PhD research focused on methods for evaluation and design of sustainable urban form. He is a Certified Environmental Practitioner, Associate of the Institute of Environmental Management and Assessment, Registered Environmental Auditor and Green Star Accredited Professional.

Jim Plume
Senior Research Fellow, University of New South Wales
Sydney, Australia

Following an academic career that spanned over thirty years, Jim now holds a half-time position as a senior research fellow at the University of New South Wales in Sydney, leading a research project on Precinct Information Modelling. His current research focus is concerned with extending the concept of BIM to the scale of urban precincts, specifically to facilitate the measurement, assessment and management of carbon impact to achieve sustainable built environments. Jim is on the Board of buildingSMART Australasia and is a member of the Infrastructure Committee for buildingSMART International, contributing to the development of international standards for information modelling of the built environment. Recently, he was lead author on a position paper for the Australian Commonwealth Government, written in collaboration with Spatial Industry Business Association (SIBA) and examining the relationship between construction and spatial modelling. He also co-chairs an international buildingSMART Working Group, working closely with Open Geospatial Consortium (OGC) to examine the role of standards across the building and spatial domains.

Adriana X. Sanchez
Research Associate, Sustainable Built Environment National Research Centre, Curtin University and PhD Candidate, University of New South Wales
Sydney, Australia

Adriana has ten years of research experience accrued across four continents and resulting in a growing list of academic and industry publications. These include the edited Routledge books *R&D Investment and Impact in the Global Construction Industry* and *Delivering Value with BIM: A Whole-of-life Approach*. Adriana also teaches at UNSW and University of Technology Sydney (UTS) on topics related to sustainable, resilient and smart cities. She is one of the coordinators of the International Council for Research and Innovation in Building and Construction (CIB) Task Group 90: Information Integration in Construction. Adriana's current interests centre on increasing urban resilience and leveraging information across organisational and system boundaries. Her research activities focus on how to translate policies into tangible outcomes and research into practice. Her most recent experience has been in sustainable infrastructure, urban resilience, developing national strategies for the adoption of new technologies and how to maximise and monitor benefits from the implementation of these technologies.

Ruben Santos
PhD Candidate, University of Lisbon
Lisbon, Portugal

Ruben is a PhD Candidate in Civil Engineering at Superior Technical Institute (IST, Instituto Superior Técnico) at the University of Lisbon. He specialises in Building Information Modelling (BIM) and has experience in the fields of sustainable construction, energy efficiency, life-cycle assessment (LCA). Ruben has been doing research at the Civil Engineering Research and Innovation for Sustainability (CEris) (IST, University of Lisbon) on BIM, multi-objective optimisation models for energy efficiency based on BIM models and IFC as well as about the integration of smart technology with BIM-based objects. Ruben is also engaged in the promotion of BIM in Portugal, particularly in the field of normalisation, through his work as secretary of the Portuguese Technical Committee for BIM Standardisation (CT197-BIM). He also has experience in the field of structural design, having practised in Portugal and the United Kingdom.

Ahmet Anıl Sezer
PhD Candidate, Chalmers University of Technology
Gothenburg, Sweden

Ahmet is a PhD candidate at the Division of Service Management and Logistics of Chalmers University of Technology. His research focuses on performance measurement and resource-use management at refurbishment sites, including information and communication technologies (ICT) support for these activities. He is interested in the links between productivity and sustainability in housing and office refurbishment projects. Ahmet has a Master's in Design and Construction Project Management from Chalmers University of Technology and a Bachelor's in Civil Engineering from Karadeniz Technical University.

Sonia Lupica Spagnolo
Adjunct Professor and Research Fellow, Polytechnic of Milan (Politecnico di Milano)
Milan, Italy

Sonia is Adjunct Professor and Research Fellow at the Department of Architecture, Built environment and Construction Engineering (ABC), Politecnico di Milano. She holds a PhD in Building Systems and Processes, and a Master of Science in Building Engineering. Her research interests and expertise lie in the areas of durability, maintenance management, information integration in construction, performance decay over time and energy efficiency. She was coordinator of three experimental programmes at Politecnico di Milano, two on photocatalytic materials and one on External Thermal Insulation Composite Systems (ETICS). She also collaborated with the Italian Institute for Building Technologies – National Council for Research (ITC-CNR) and with the French Scientific and Technical Centre for Building (CSTB, Centre Scientifique et Technique du Bâtiment) developing new methods and tools for service-life prediction. Sonia is xxiv *Contributors*

a member of the CIB Working Commission 80 (W080): Prediction of Service Life of Building Materials and Components, and of CIB Task Group 90 (TG90): Information integration in Construction. She has authored over sixty academic publications, including books, peer-reviewed journal papers, conference papers and book chapters.

Marit Støre-Valen
Associate Professor, Norwegian University of Science and Technology (NTNU)
Trondheim, Norway
Marit leads the research on Property Management at NTNU and teaches Master's courses within the Civil Engineering and Environment Master's programme at the Department of Civil and Transport Engineering, Faculty of Engineering and Technology Science, NTNU. She also participates in the Master's programme in Real Estate and Property Development and Management at the Faculty of Architecture and Fine Art. Her field of research is real estate and property management of public and private building portfolios, refurbishment and development of buildings, strategic facilities management and maintenance planning, assessment tools for whole-of-life sustainable buildings and building processes, and use of smart technology. Marit was Head of Department at NTNU (2009-2013) and has since served as a Senior Consultant/Project Manager for the Norwegian Building Authority (DIBK), where she led 'Bygg21' (2013–2014), an initiative to develop a policy instrument and strategies for the Norwegian construction industry. Marit is an active member of the CIB, participating in task groups, contributing to the development of roadmaps for research, and publishing articles and book chapters. She is an active networker among Norwegian researchers with a special interest in improvements within the construction industry.

Fredrik Svalestuen
PhD Candidate, Norwegian University of Science and Technology (NTNU)
Oslo, Norway
Fredrik is doing his PhD on the communication between engineering and production in construction projects at NTNU's Department of Civil and Transport Engineering. Prior to this, Fredrik worked as Site Manager for a subsidiary of Veidekke Group involved in asphalt/aggregate operations and public road maintenance in Norway. Fredrik has a growing list of publications ranging in topic from performance measurements, improving design management, using mobile devices to improve communication in construction projects, to Virtual Design and Construction.

Joseph Handibry Mbatu Tah
Head of School and Professor in Project Management, Oxford Brookes University
Oxford, UK
Joseph is Professor in Project Management and Head of School of the Built Environment at Oxford Brookes University in the UK. He has extensive

experience in the application of artificial intelligence, distributed computing and Building Information Modelling techniques to systems for managing large-scale projects and extended enterprises in the construction and related industries. He has published widely in these areas and provided consultancy and advisory services to national and international companies and governments. He is a Fellow of the Royal Institution of Chartered Surveyors (FRICS) and a member of the Chartered Institute of Building (MCIOB).

Xiangyu Wang
Professor and Director, Australasian Joint Research Centre for Building Information Modelling (BIM), Curtin University
Perth, Australia

Xiangyu is the Director of the Australasian Joint Research Centre for Building Information Modelling at Curtin University. He is also a project leader at the Sustainable Built Environment National Research Centre. Xiangyu serves as Curtin-Woodside Chair Professor for Oil, Gas & LNG Construction and Project Management and Chair of Curtin Advanced Technology Research and Innovation Alliance (CATRINA). His research interests include Building Information Modelling, information technology in construction, virtual, augmented and mixed reality, computer-supported cooperative design/work, mobile, pervasive and ubiquitous computing in design and construction, computer-aided design and e-learning. He holds a PhD from Purdue University and has published over a hundred academic papers, chapters and books.

Peng Wu
Senior Lecturer, Curtin University
Perth, Australia

Peng is a senior lecturer at Department of Construction Management, Curtin University. He did his PhD in Project Management at the National University of Singapore. He also holds a Master of Science in Construction Management from Loughborough University, UK and a Bachelor of Science in Project Management from Tsinghua University, China. His research areas include sustainable construction, lean production and construction, production and operations management and life-cycle assessment. He has provided consultancy services to many clients in the construction sector, including the Housing and Development Board, Singapore.

Wen Yi
Research Fellow, Australasian Joint Research Centre for Building Information Modelling, Curtin University
Perth, Australia

Wen is a research fellow at the Australasian Joint Research Centre for Building Information Modelling at Curtin University and a Post-Doctoral Fellow at the Department of Building and Real Estate at the Hong Kong Polytechnic University in Hong Kong, China. Wen's research interests include occupational

xxvi *Contributors*

health and safety, information technology in construction management, construction labour productivity and ergonomics. Wen holds an MSc in Construction Management and Economics from the Chongqing University China and a PhD in Construction Management from the Hong Kong Polytechnic University.

Foreword

Business leaders globally are seeking solutions to address economic, social and environmental challenges. They are now increasingly looking to take advantage of the emerging digital era.

In an increasingly globalised built environment industry, achieving higher levels of integration across organisational and software boundaries is now demonstrably identifying opportunities to find new solutions to these complex challenges. The rise of Building Information Modelling (BIM)/Digital Engineering (DE), Geographic Information Systems (GIS) and the Internet of Things (IoT) are specific examples we see in the built environment.

New digital technologies can help projects, firms, industries and countries deliver built environments that are more productive, managed more effectively and deliver better environmental and social outcomes. This book *Integrating Information in Built Environments: From Concept to Practice* makes a valuable global contribution to this effort.

My experience over the past decade in leading the shaping of the United Kingdom Government's response to digital advances in construction has highlighted significant economic, social and environmental opportunities.

I commend the authors for their insightful distillation of a range of international perspectives on the current global evolution of information integration in the built environment industry. This book provides much needed original research and valuable lessons for researchers, industry practitioners, public and private clients and policy makers across the international built environment industry.

Mark Bew MBE
Chair, Professional Construction Strategies Group (PCSG) Ltd
Chair, UK Government Construction BIM Task Group
Chair, buildingSMART (UK)

xxviii *Foreword*

Networks rule – but only if they're integrated …

The global built environment industry has recognised the vast potential for improved performance of its value chains from integrating planning, design, construction and asset management processes. Such process integration depends on integrating information – the basis of work for any project team – across project phases and disciplines.

Ultimately, the value of any such efforts will be judged by the end-user and the asset owner. Does it provide a better user-experience and is the return on investment worth it? Again, such assessments depend on the quality, timeliness and integration of information.

In over a decade of leadership of Stanford University's Center for Integrated Facility Engineering (CIFE), I have witnessed many exemplary international research initiatives that have advanced the field of information integration in construction. This publication is one such exemplar.

Integrating Information in Built Environments: From Concept to Practice documents innovation in information sharing across disciplines with contributions across nine countries as part of the International Council for Research and Innovation in Building and Construction (CIB) network. It provides a global perspective on current research in this critical space and an analysis of future priority areas for research and development as well as strategic implications for public and private decision-makers.

I commend this publication as a leading international academic reference.

Martin Fischer PhD
Professor of Civil and Environmental Engineering
Director, Center for Integrated Facility Engineering (CIFE)
Stanford University

Preface

Globally, we are facing challenges due to political and economic unpredictability, changing climate and technological advances. Traditional delivery and management models in the built environment industry additionally often promote adversarial and inefficient behaviours across the supply chain, which are reinforced throughout the delivery process. This realisation is leading to a move towards more integrated approaches that may open the gate to a more productive and sustainable industry.

Ensuring timely access to relevant, reliable and actionable information that is stored in a standardised and interoperable format is at the centre of this movement. However, while technological approaches to this issue are important, process and governance considerations can prove to be at least as important. This book brings together input from industry researchers globally to provide insight into challenges and opportunities related to achieving higher levels of information integration across life-cycle phases, stakeholder groups and portfolios of built assets. It does so by focusing on different goals and sectors through case studies and strategies that build on international experience from nine countries.

With contributions from 43 authors, we aimed to provide a much-needed international perspective on topics relevant to information integration in the built environment. The book presents leading original research and valuable lessons for researchers, industry practitioners, government clients and policy-makers.

This publication is the direct outcome of the close collaboration fostered by the International Council for Research and Innovation in Building and Construction (CIB) Task Group 90 (TG90): Information Integration in Construction. This Task Group has built a vigorous international network which now benefits from the active involvement of 42 members from 14 countries. It was formed in 2014 in response to the desire of industry, policy-makers and analysts, government clients and researchers to address issues such as efficient knowledge creation, preservation and integration across the life-cycle of constructed facilities, and relevant, reliable, interoperable and long-lasting data and information gathering and analysis. This book aims to help achieve the group's objective of creating a more effective and reflective industry.

xxx *Preface*

Contributions from individuals who strive to make a difference are essential if we are to deliver the benefits of research into practice. We look forward to continuing to work together to better align industry research policies, funding and collaborative teams for a stronger, more sustainable and productive built environment industry.

Adriana X. Sanchez, Keith D. Hampson and Geoffrey London

Acknowledgements

The editors wish to thank all those who have made this publication possible through their contributions and support.

We firstly wish to thank our national and international group of authors who have contributed to the chapters and generously shared the outcomes of many years of industry research experience in this field. These contributions have been crucial, and without them this book would not have been possible. We would also like to expressly thank those who provided an early review of this book's intent.

This publication is the outcome of the International Council for Research and Innovation in Building and Construction (CIB) Task Group 90: Information Integration in Construction, formed in 2014. We therefore extend our sincere thanks to those in the CIB Secretariat who have facilitated and encouraged the interactions of this Task Group, of which this publication is a major outcome. Acknowledgment is also made of the work of the former Task Groups – TG85: Research Investment and Impact, TG58: Clients and Construction Innovation and TG47: Innovation Brokerage in Construction, upon which many personal contacts and friendships have been founded.

The editors received both encouragement and financial support from the Australian Sustainable Built Environment National Research Centre (SBEnrc), Curtin University, University of New South Wales and University of Western Australia. Without support from these organisations, together with our global innovation networks, this publication would not have been realised.

Finally, we would also like to acknowledge those who have granted permission to reproduce their material in this book.

Abbreviations

Acronym	Term
ACGIH	American Conference of Governmental Industrial Hygienists
AEC	architecture, engineering and construction
AFO	Aree funzionali omogenee (homogeneous functional environments)
AIA	American Institute of Architects
AoT	Array of Things
ASCE	American Society of Civil Engineers
ASO	aree spaziali omogenee (homogeneous spatial environments)
ATSI	Aboriginal and Torres Strait Islander
B2B	business-to-business
BAS	Building Automation Systems
BAU	business as usual
BI	Swedish Construction Federation (Sveriges Byggindustrier)
BIM	Building Information Modelling
BIM&M	Building Information Modelling and Management
BIMCS	BIM Cloud Score
BIoT	Building Internet of Things
BLS	Bureau of Labour Statistics
bpm	beats per minute
BR06	Danish Building Regulations 2006
BREEAM	Building Research Establishment Environment Assessment Method
BtO	From Build to Operation
CAD	computer-aided design
CASBEE	Comprehensive Assessment System for Built Environment Efficiency
CBD	central business district
CCTV	closed-circuit television
CIB	International Council for Research and Innovation in Building and Construction
CIC	Construction Industry Council
CII	Construction Industry Institute
CIM	City Information Modelling

CIM	construction information management
CMB	Centre for Management of the Built Environment
CO_2	carbon dioxide
COBIE	Construction-Operations Building Information Exchange
COP	Conference of the Parties
CPIC	Construction Project Information Committee
CRC	Cooperative Research Centre
CRC LCL	Cooperative Research Centre for Low Carbon Living
CSI	Construction Specification Institute
CSTB	Centre Scientifique et Technique du Bâtiment (Scientific and Technical Centre for Building)
DBE	digital built environment
DBO	design, build and operate
DIKW	data–information–knowledge–wisdom
DPSEEA	driving forces, pressures, state, exposures, effects, actions
DPSIR	driving forces, pressures, state, impacts, response
EOL	end of life
EPC	energy performance contracting
EPC	engineering, procurement and construction
EPD	Environmental Product Declarations
ERP	enterprise resource planning
ESCO	energy service company
ESL	estimated service-life
EU	European Union
ESP	Envision Scenario Planner
EUR	euro
FM	facilities management
GDP	gross domestic product
GFC	global financial crisis
GIS	Geographic Information System
GML	Geography Markup Language
GNSS	global navigation satellite system
GSA	US General Services Administration
GSM	Global System for Mobile Communications
HACT	Housing Associations' Charitable Trust
HiB	College School of Bergen
HRI	heat-related illness
HSE	Health and Safety Executive
HVAC	heating, ventilation and air-conditioning
ICT	information and communications technology/ies
IDDS	integrated design and delivery solutions
IDP	integrated design process
IEA	International Energy Agency
IED	integrated energy design
IFC	Industry Foundation Classes

xxxiv *Abbreviations*

IICON	Information Integration in Construction
IoE	Internet of Everything
IoP	Internet of People
IoT	Internet of Things
IP	Internet Protocol
IPCC	Intergovernmental Panel on Climate Change
ISO	International Organization for Standardization
IT	information technology/ies
ITI	US Information Technology Industry Council
KPI	key performance indicator
LCA	life-cycle assessment
LCC	life-cycle costs
LCCA	life-cycle cost analysis
LCEA	life-cycle energy analysis
LCI	life-cycle inventory
LCIA	life-cycle impact assessment
LED	light-emitting diode
LEED	Leadership in Energy and Environmental Design
LiDAR	Light Detection and Ranging
LOD	level of development
M&V	measurement and verification
MW	megawatt
NatHERS	Nationwide House Energy Rating Scheme
NCC	National Construction Code
NCCARF	Australian National Climate Change Adaptation Research Facility
NGR	New Generation Rollingstock
NIHL	noise-induced hearing loss
NIST	US National Institute of Standards and Technology
NRC	National Research Council
NSW	New South Wales
NYC	New York City
OC	offshore construction
OCR	optical character recognition
OGC	Open Geospatial Consortium
OSM	OpenStreetMap
PCH	Perth Children's Hospital
PCS	personal cooling system
PEM	project execution model
PIM	Precinct Information Modelling
PPP	public-private partnership
PV	photovoltaic
QA	quality assurance
QTMR	Queensland Department of Transport and Main Roads
R&D	research and development
RFID	radio frequency identification

RIBA	Royal Institute of British Architects
ROI	return on investment
RPE	rating of perceived exertion
RSL	reference service-life
SAP	System, Applications and Products in Data Processing
SARIG	South Australian Resource Information Geoserver
SB	shipbuilding
SBEnrc	Australian Sustainable Built Environment National Research Centre
SBHF	Sykehusbygg (Hospital Construction) HF trust
SCOT	social construction of technology
SECI	socialisation, externalisation, combination and internalisation
SfB	Samarbetskommittén for Byggnadsfrägor (Cooperative Committee for Construction Issues)
SLIP	Shared Location Information Platform
SLP	service-life planning
SME	small and medium-sized enterprise
SOH	Sydney Opera House
SROI	social return on investment
SSK	sociology of scientific knowledge
STS	science and technology studies
TG90	Task Group 90
THFC	The Housing Finance Corporation
TLV	threshold limit value
UDC	Universal Decimal Classification
UEEE	used electrical and electronic equipment
UK	United Kingdom
UN	United Nations
UNECE	United Nations Economic Commission for Europe
UNFCCC	United Nations Framework Convention on Climate Change
UNI	Ente Italiano Di Normazione (Italian Organisation for Standardisation)
Uniclass	Unified Classification for the Construction Industry
URI	uniform resource identifier
URL	uniform resource locator
US	United States of America
USD	United States dollar
UWB	ultra wideband
VDC	virtual design and construction
VWHA	Victorian Women's Housing Association
W3C	World Wide Web Consortium
WA	Western Australia
WAG	Western Australian Government
WLAN	wireless local area network
WVA	well-being valuation analysis
XML	Extensible Markup Language

1 Integrating information across built environment boundaries

Adriana X. Sanchez, Keith D. Hampson and Geoffrey London

Introduction

> We are living in a world of networks, and these networks are becoming more interdependent every day.
>
> (Barrett et al., 2011)

Globally, governments are increasingly showing concern about not only coping with but taking advantage of the new digital era. The rise of Building Information Modelling (BIM) and Geographic Information Systems (GIS) are but the tip of the iceberg. The UK Construction Industry Council has for example predicted that the adoption of digital built environments will be the beginning of a complete shift from the way that societies and industries operate. They predict that this digital transformation will eventually lead to the use of advanced robotics, autonomous vehicles, industrial 3D printing and self-healing materials, among other innovations (Philp and Thompson, 2014). These new technologies also have the potential to help countries deliver more resilient, liveable and sustainable built environments that in addition are more productive and managed more effectively.

In the early 2000s Clayton Christensen, a Harvard University professor, warned that the accelerating pace of change was causing a dramatic surge in the amount of information available to managers (Christensen, 2001). This trend has continued to prevail across the built environment together with the level of complexity that plans, designs, constructs and maintains the systems that form it. However, the information required for well-informed and effective decision-making is often scattered across disparate, uncoordinated and isolated sources. This, together with the complexity brought by the variety of stakeholders with different interests across the supply chain, and the sheer volume of data, makes information integration a key challenge for the built environment industry.

Another issue that makes integrating information a complex problem is that large amounts of data are already being generated and collected in parallel by specific sectors without taking into consideration the present or future needs of other parts of the system or life-cycle phases of that same asset. This often leads to decision-making that is based on the available but sometimes incomplete or

2   *Sanchez et al.*

imperfect information that was not collated or structured to answer that particular question. This in turn results in much of the research into information integration focusing on developing processes 'to fuse various sources of data in statistically rigorous ways, and in recognizing gaps in the data and developing well-founded models to fill these gaps' (Barrett et al., 2011). This has also led to a large body of literature about systems and approaches to integrate data and information (Caragea et al., 2005).

The following chapters will discuss two types of information integration: vertical and horizontal. Extrapolating from Marzouk et al.'s definition (2010), the former refers to sharing information and knowledge across value networks[1] or phases of the project life-cycle. The latter refers to sharing across entities within the same level of the value network or life-cycle phase. Within each of these, the chapters will mainly focus on integration across digital system boundaries but will also cover integration across organisational boundaries and how one can support the other.

Integration of information across organisational boundaries

The built environment industry is becoming more globalised and highly competitive, with clients also demanding faster response times, shorter production cycles and greater customisation. Organisations are entering new markets 'through the use of information and communication technologies and the development of partnerships with other organisations' (Barlow and Li, 2005). This is leading supply and value chain management to become a field of growing strategic value. Within this context, the level of integration across the chain can be a competitive advantage that enhances performance of individual firms and the industry as a whole. Supply chain integration can be defined as:

> The degree to which SC [supply chain] members achieve collaborative inter- and intra- organisational management on the strategic, tactical and operational levels of activities (and their corresponding physical and information flows) that, starting with raw materials suppliers, add value to the product to satisfy the needs of the final customer at the lowest cost and the greatest speed.
>
> (Alfalla-Luque et al., 2015)

Here the supply chain can be defined as 'The network of organizations that are involved, through upstream and downstream linkages, in the different processes and activities that produce value in the form of products and services delivered to the ultimate consumer' (Mentzer et al., 2001).

Some models also distinguish two types of integration: logistical and technological. The former relates to cooperating to manage basic information and material flows. This type of integration is based on information sharing about planning systems, production plans, inventory and interoperability. An example of this is showcased in Chapter 2 (Kraatz et al.), which explores the need for horizontal and vertical integration of data, information and knowledge within and across organisations involved in the social housing sector. Technological integration relates to sharing tacit knowledge in strategic areas along the chain and includes not only structural aspects but also infrastructural issues (methods and management systems). Examples of such an integration can be found in Chapter 11 (Bougrain), which explores this issue within the context of energy performance contracting in France.

Within this kind of model, low levels of logistical integration often manifest themselves in detailed contracts and information sharing being limited to minimising risk and performance monitoring. Greater levels of integration, on the other hand, have been linked to better environmental management during operations, higher efficiency of the supply chain and improved contract performance. The level of technological integration is then defined by the 'extent of sharing of technical and tacit knowledge, and the extent of interaction on new product and process design' (Vachon and Klassen, 2006). Haugbølle (Chapter 15) goes a step further by exploring the hurdles of turning information and knowledge into actions that change practices.

A value chain is 'a well-established concept for considering key activities that an organization can perform or manage with the intention of adding value for the customer as products and services move from conception to delivery to the customer' (Barlow and Li, 2005). The value chain concept is closely related to the supply chain but focuses on the activities or processes that add value from concept to delivery, rather than on the organisations themselves. Within this field of value chain integration, one of the focus areas is on enhancing the inter-organisational linkages and information sharing through the implementation of new technological solutions (Barlow and Li, 2005). Chapter 13 (Støre-Valen et al.) will exemplify this point through case studies in Norway that show how value is realised differently by different stakeholders. Vertical integration provides a competitive advantage, especially in sectors where the clients and end-users are highly demanding and are perceived as under-served by what is available in the market (Christensen et al., 2009). Chapter 12 (Meistad et al.) illustrates this issue through four case studies from Norway that explore different approaches to integrating information while balancing the needs and drivers of different stakeholders.

The level of success of the integration effort is partly dependent on coordination processes across agencies and the digitisation of that information so it can be shared across boundaries. It is also about standards and aligning technical requirements as well as understanding the information needs of other stakeholders. This point is perfectly illustrated by Chapters 6 (Santos and Aguiar), 7 (Newton et al.) and 8 (Lupica). Santos and Aguiar discuss the integration of BIM with Life-cycle Assessment methodologies, highlighting the importance of standards availability to reduce the investment and skills level requirements to succeed. Newton et al. on the other hand, explore the potential behind a digital built environment and open standards facilitating the road to this goal. Lupica focuses on information and communication technology (ICT) tools that enable more effective life-cycle management and can reshape the way companies interact

4 *Sanchez et al.*

with their customers. Sanchez et al. (Chapter 14) additionally discuss the challenges associated with benchmarking benefits from different stakeholders and propose a new system that builds on crowdsourcing principles.

Integrating information across software boundaries

Digital information and data is being generated and stored at an unprecedented pace. While the issue of integration may seem like just about organising this data and extracting the information required, the real challenge is about having the correct data in the right accessible format so it can be used (Barrett et al., 2011). This issue is exemplified in Chapters 3 (Sanchez et al.), 4 (Abanda and Tah) and 5 (Yi et al). Sanchez et al. explore the issues and potential gains behind open data portals and urban information models as ways to improve the effectiveness of urban resilience policy action. Abanda and Tah instead discuss the paradigm change and challenges associated with growing volumes of data through the Internet of Things (IoT) applications. Yi et al. explore how advanced ICTs can help bridge information gaps to improve safety in the construction industry.

Integrating disparate datasets is an issue that has been studied extensively over the years, with recent research focusing on data integration. This is 'the problem of combining the data residing at different sources, and providing the user with a unified view of these data, a common issue in today's industry' (Calì et al., 2002). Information system integration adds linkages to organisational process accounting, aiming to implement systematic, coordinated and standardised structures for the generation, management and storage of information (Chapman and Kihn, 2009). However, it also brings special challenges for certain sectors of the industry. Bröchner and Sezer for example highlight in Chapter 9 the important issue of downscaling such systems from large new construction project settings to small refurbishment projects.

In parallel to this growing field of research, 'information technologies are evolving into socio-technical systems and slowly becoming integral to every aspect of urban asset management and governance' (Sanchez et al., 2016b). Countries across the globe have started to move towards implementing new approaches such as BIM to cope with mounting external pressures and leverage emerging technologies to increase their industry's efficiency and productivity. However, issues such as lack of interoperability, complicated data ownership and still-emerging standards continue to be a major barrier for more efficient implementation of these systems (Sanchez et al., 2016a). These challenges are raised throughout this book from planning and design to operations across a number of sectors of the built environment.

About this book

In an increasingly globalised built environment industry, creating stronger links between international researchers can lead to improved economic, social and environmental outcomes by leveraging international knowledge and innovation networks. This book is the direct result of a collaborative global network of industry and academic researchers spread across 9 countries. This network is part of the International Council for Research and Innovation in Building and Construction (CIB).[2] The CIB was established in 1953 and is an international network that promotes knowledge exchange and cooperation for research and innovation in the building and construction industry. The CIB consists of over 5,000 experts from about 500 member organisations active in academia, research, government, industry and education. This organisation provides support to improve processes and performance in the built environment through international task groups and working commissions.

CIB's Task Group 90 (TG90) Information Integration in Construction (IICON) was established in Barcelona, Spain in October 2014. The Task Group focuses on addressing the need for:

- efficient knowledge creation, preservation and integration across the life-cycle of constructed built assets;
- relevant, reliable, interoperable and long-lasting data and information gathering and analysis;
- monitoring and feedback from end-users into the different stages of planning, design, construction and asset management of buildings and infrastructure.

TG90 coordinators are Adriana X. Sanchez and Keith D. Hampson from the Australian Sustainable Built Environment National Research Centre (SBEnrc) headquartered at Curtin University in Perth Australia, and Rasmus Rempling from Chalmers University and NCC in Sweden. At the time of writing, the group comprises 42 members from 14 countries and reflects a range of industry research backgrounds. The main objective of the group is to help create a more effective and reflective construction industry and deliver benefits to public and private sectors. This book is intended to contribute to this objective by providing a wide view of some of the opportunities and challenges brought by integrating information across organisational boundaries and system domains in the built environment industry.

The built environment

Within the context of this publication, the built environment refers 'to the man-made landscapes that provide the setting for human activity, ranging from the large-scale urban entities to personal dwelling places'. This includes buildings, infrastructure and cultural landscape that form the urban fabric which can be considered as an artefact 'in an overlapping zone between culture and nature, with causation occurring in both directions' (Hassler and Kohler, 2014). When discussing the built environment industry and value chain, this includes those involved in the planning, financing, design, construction and management of built assets.

6   *Sanchez et al.*

This includes the following stakeholders:

- client/owner: includes the person/organisation for whom a structure is constructed; the person or organisation that took the initiative of the construction, or that owns the built structure;
- designer: includes architectural, industrial, structural and civil designers (architects or engineers);
- contractor: refers to the head contractor; that is the main contractor responsible for the majority of work on a construction site, including sub-contract work, materials, and labour supplies;
- sub-contractor: refers to a person/organisation who carries out construction work or supply related goods and services under a construction contract otherwise than as head contractor;
- fabricator/manufacturer: refers to manufacturing firms who produce construction elements off-site at a factory or manufacturing plant;
- asset manager: includes facility, operations and maintenance managers of buildings and infrastructure;
- supplier: refers to individuals/organisations who provide physical supplies such as goods, materials and plant for construction project or the management of a built asset to the client/owner, asset manager or contractors and sub-contractors.

In addition to this, some chapters include 'end-users' which are in some cases the operations managers but most commonly refer to the inhabitants and other users of the built structure.

Book structure

This book is formed by 14 chapters in addition to this introductory chapter and a concluding chapter. These chapters cover a wide range of topics related to the role, relevance and technical issues linked to integrating information across organisational boundaries and software domains. The chapters have been arranged in three parts inspired by Clayton Christensen's innovation capabilities framework. The original framework includes three classes of factors that affect the capabilities of an organisation: resources, processes and values. These, Christensen argues, limit or enable the types of innovations that can be implemented by a given organisation (Christensen, 2001).

This framework was chosen because, on the one hand, as highlighted earlier, integrating information is a key challenge being faced globally by most industries, including the built environment industry. Achieving higher levels of integration will require the implementation of innovative processes and tools that help organisations overcome the challenges outlined previously. On the other hand, the topics discussed in Christensen's framework strongly resonated with the themes that the experts from the CIB TG90 group identified as relevant to the objectives of the group. However, the framework is not apt to be directly applied to the topics discussed here. This has led the editors to use the framework as inspiration and adapt it to suit the scope of the book. Accordingly, chapters have been organised in the following sections:

- *resources:* this section of the book discusses topics related to the need for information integration for allocation, management and optimisation of resources. These resources can be human, economic or environmental. It provides arguments for increasing the level of integration across organisational boundaries and software domains by discussing the potential improvements that can be achieved in built environment resource management. The chapters also discuss potential avenues to achieve these higher levels of integration. The reader will find in this section topics such as social housing, sustainability, safety and resilience;
- *processes:* this section covers areas related to processes that are involved in the vertical and horizontal integration of information across the built environment industry. These relate to 'the patterns of interaction, coordination, communication, and decision making' through which transformation can be achieved (Christensen, 2001). Chapters in this part of the book discuss more technical aspects of achieving higher levels of information integration for specific industry sectors and stakeholder groups. Topics include information technology (IT) support, delivery models and information modelling;
- *added value:* chapters in this section discuss cultural and 'less tangible' issues that relate to information integration. They explore the potential added-value of achieving higher levels of information integration. Topics covered in this section include knowledge, the role of clients and value of built assets.

Each chapter provides insight about the relevance of integrating information within their specific topic of discussion as well as strategic implications for decision-makers and managers. They additionally discuss priority areas for future research and investment related to their topics that have the potential to significantly improve the level of integration across the built environment industry and value chain.

Conclusions

The substance of this book is directed towards an implicitly agreed outcome: a more efficient and sustainable built environment industry that is able to produce higher-quality built assets for global societies. With the growing complexity in methods of procurement, financing, ownership and maintenance, combined with the shifting roles of consultants in the industry and developments in construction techniques, it is easy to lose sight of the bigger picture: to enable the best possible design to be delivered in a way that is affordable by its society, that provides high functionality, that does not damage the environment and that engenders community pride and identity. Achieving higher levels of

8   *Sanchez et al.*

information integration across organisations, individuals, and life-cycle phases and portfolios of built assets, will bring a number of challenges to the built environment industry but can also help achieve this vision. This book aims to provide insight about some of these challenges and potential benefits across a wide range of areas, and thereby contribute to a more effective and reflective built environment industry.

Notes

1 The context within which an organisation identifies and responds to stakeholder needs, procures inputs and reacts to competitors (Christensen and Rosenbloom, 1995).
2 For more information refer to www.cibworld.nl/site/about_cib/index.html.

References

Alfalla-Luque, R., Marin-Garcia, J. A. and Medina-Lopez, C., 2015. An analysis of the direct and mediated effects of employee commitment and supply chain integration on organisational performance. *International Journal of Production Economics*, 162, 242–257.

Barlow, A. and Li, F., 2005. Online value network linkages: Integration, information sharing and flexibility. *Electronic Commerce Research and Applications*, 4, 100–112.

Barrett, C., Eubank, S., Marathe, A., Marathe, M. V., Pan, Z. and Swarup, S., 2011. Information integration to support model-based policy. *Innovation Journal*, 16(1), 1–16.

Calì, A., Calvanese, D., De Giacomo, G. and Lenzerini, M., 2002. Data Integration under Integrity Constraints. In: A. Banks Pidduck, M. Tamer Ozsu, J. Mylopoulos and C. C. Woo (eds) *CAISE 2002*. Berlin: Springer-Verlag, pp. 262–279.

Caragea, D., Pathak, J., Bao, J., Silvescu, A., Andorf, C., Dobbs, D. and Honavar, V., 2005. Information integration and knowledge acquisition from semantically heterogeneous biological data sources. In: B. Ludascher and L. Raschid (eds) *DILS 2005*. Berlin: Springer-Verlag, pp. 175–190.

Chapman, C. S. and Kihn, L.-A., 2009. Information system integration, enabling control and performance. *Accounting, Organizations and Society*, 34, 151–169.

Christensen, C. M., 2001. Assessing your organization's innovation capabilities. *Leader to Leader*, Summer, 27–37.

Christensen, C. M. and Rosenbloom, R. S., 1995. Explaining the attacker's advantage: Technological paradigms, organizational dynamics, and the value network. *Research Policy*, 24, 233–257.

Christensen, C. M., Verlinden, M. and Westerman, G., 2009. Disruption, disintegration and the dissipation of differentiability. In: R. A. Burgelman, C. M. Christensen and S. C. Wheelwright (eds) *Strategic Management of Technology and Innovation* 5th edn. New York: McGraw-Hill Irwin, pp. 363–387.

Hassler, U. and Kohler, N., 2014. The ideal of resilient systems and questions of continuity. *Building Research & Information*, 42(2), 158–167.

Marzouk, M., Hisham, M., Ismail, S., Youssef, M. and Seif, O., 2010. *On the Use of Building Information Modeling in Infrastructure Bridges*, paper presented at 27th International Conference–Applications of IT in the AEC Industry (CIB W78), Cairo, Egypt, 16–19 November, pp. 1–10.

Mentzer, J. T., DeWitt, W., Keebler, J. S., Min, S., Nix, N. W., Smith, D, C. and Zacharia, Z. G., 2001. Defining supply chain management. *Journal of Business logistics*, 22(2), 1–25.

Philp, D. and Thompson, N., 2014. *Built Environment 2050: A Report on Our Digital Future*, London: Construction Industry Council.

Sanchez, A. X., Hampson, K. D. and Vaux, S., 2016a. *Delivering Value with BIM: A Whole-of-life Approach*. London: Routledge.

Sanchez, A. X., van der Heijden, J., Osmond, P. and Prasad, D., 2016b. *Urban Sustainable Resilience Values: Driving Resilience Policy that Endures*, paper presented at CIB World Building Congress, Tampere, Finland, 30 May–2 June.

Vachon, S. and Klassen, R. D., 2006. Extending green practices across the supply chain: The impact of upstream and downstream integration. *International Journal of Operations and Production Management*, 26(7), 795–821.

Part 1
Resources

2 Integrating information for more productive social housing outcomes

An Australian perspective

Judy A. Kraatz, Annie Matan and Peter Newman

Introduction

The United Nations Economic Commission for Europe (UNECE) broadly states that social housing is housing 'supplied at prices that are lower than the general housing market and ... distributed through administrative procedures ... some form of state support and subsidy are inevitably involved with this tenure' (Rosenfeld, 2015). Based on the Australian Productivity Commission's definition, social housing can be described as 'below-market rental housing for people on low incomes and for those with special needs', most of which is 'highly subsidised and rent is determined by tenant income (generally set at 25 or 30% of household income)' (Yates, 2013). In Canada, social housing is

> an umbrella term to refer to all forms of housing developed under various government subsidy programs in both the private and public sectors. It includes housing now discontinued under the public housing program, all housing that is owned and operated by the federal, provincial, territorial and municipal governments, and housing that has been subsidized by the government and developed by a private and/or non-profit organization.
> (Moskalyk, 2008)

In the Netherlands, 'the principal target group are low-income households (e.g. families with an income below 29,000 Euros/year).[1] The cheaper rented housing is intended primarily for this group, which gets housing benefits from the government' (Aedes, 2016). In the United Kingdom (UK),

> Social rented housing is owned by local authorities and private registered providers (as defined in section 80 of the Housing and Regeneration Act 2008), for which guideline target rents are determined through the national rent regime. It may also be owned by other persons and provided under equivalent rental arrangements to the above, as agreed with the local authority or with the Homes and Communities Agency.
> (UK Department of Communities and Local Government, 2012)

14 *Kraatz et al.*

And in the United States (US),

> Public housing was established to provide decent and safe rental housing for eligible low-income families, the elderly and persons with disabilities. Public housing comes in all sizes and types, from scattered single family houses to high rise apartments for elderly families.
>
> (US Department of Housing and Urban Development, 2016)

It should be noted that national terms for this provision vary, including, for example, social housing in Canada, Australia, the Netherlands and the UK; public housing in the US and Israel; common housing in Denmark; and housing promotion in Germany.

The effective and appropriate provision of social housing, as an integral part of the housing continuum, is increasingly difficult in light of current fiscal constraints and increasing housing needs being experienced globally, especially since the global financial crisis (GFC) of 2007–09. Achieving an economically and socially sustainable framework for the provision of social housing, as part of addressing the pressing need for affordable housing, is vital. To meet this challenge, many innovative models are being explored internationally, including partnerships and financing arrangements involving a mix of public, private and third-sector community provider funds.

The case study that will be presented in later sections of this chapter focuses on establishing clear links between social housing and improving economic productivity as well as non-economic benefits for the tenant, the government and the country as a whole. The Australian Council of Social Services notes this link:

> Housing, affordability and location are integral to enabling population growth, and labour mobility, supporting improvements in participation rates and improving productivity. The housing and construction industries are also key drivers of economic activity, and associated jobs growth.
>
> (Australian Council of Social Services, 2014)

The integral role of housing in broader social and economic outcomes is again highlighted by the UNECE.

> Housing is an integrative good, it is linked to many other sectors such as: health, economic security, energy security, transportation, education, employment. Housing also influences issues such as social cohesion and neighbourhood security. As an aggregate part of development efforts, housing is a key element in delivering sustainable urban development. The integrative nature of housing requires the social, cultural, environmental and economic facets of housing to be addressed in an integrated way.
>
> (Rosenfeld, 2015)

More productive social housing outcomes   15

This chapter will explore the need for horizontal and vertical integration of data, information and knowledge within and across departments and organisations (both for-profit and not-for-profit) to provide a more comprehensive rationale for a sustainable social housing system. It will do so from an Australian perspective, exploring and discussing the integration required to determine the broad spectrum of social housing benefits and outcomes from this productivity perspective. It will present a case study of a multi-agency collaboration to better portray the whole-of-society and whole-of-government productivity benefits of providing safe and secure social housing.

The social housing sector

The scale of social housing provision and need is substantial. In Canada, 'the social and affordable housing sector comprises about 4–5 per cent of the total' housing sector in that country (Carlson, 2014), while in the UK this figure is 19 per cent (Bourne, 2016). In the US, there are 1.2 million households living in public housing units (US Department of Housing and Urban Development, 2016). In the Netherlands, social housing organisations provide 2.4 million houses for four million people; out of a total national population of 16.8 million (Aedes, 2016). In 2010 in France, about ten million people or 17.3 per cent of the French population were tenants of social housing units (Wong and Goldblum, 2016). As of June 2013, around 414,000 households in Australia, from a total population of approximately 23.1 million, were living in social housing (Department of the Prime Minister and Cabinet, 2014).

It should be noted that some of the variation between the level of provision in various countries can be attributed to the approach taken, both historically and currently, to the provision of housing. The UNECE identifies three categories of social housing, namely: 'universal' which provides social housing to anyone, regardless of their income; 'targeted', where social housing is allocated based on pre-defined income levels; and 'residual', which allocates housing only to those of greatest need (Rosenfeld, 2015). Rosenfeld notes that residual allocation is dominant in many UNECE member states as well as in other countries such as Australia. This 2015 report draws upon an extensive review of the literature from over 50 countries and 30 interviews from a cross-section of representatives. As such it provides a credible backbone for much of the following discussion.

Beyond what is currently provided, significant waiting lists exist for social housing. In the UK, the waiting list was 1.8 million in 2014, up 81 per cent since 1997; in France, there were 1.7 million applications for social housing in 2014; in the US, there was a shortfall of 5.3 million affordable housing units in 2013 (Rosenfeld, 2015). In Australia in June 2013, there were 158,971 applicants on the public rental housing waiting list (Australian Institute of Health and Welfare, 2014). While in Canada, as reported prior to the 2015 national election in that country, there were 140,000 families awaiting rent-subsidised housing (Young, 2015).

16 *Kraatz et al.*

Adequate housing is also a basic necessity and human right which impacts on education, health and employment outcomes, as well as the overall well-being of the population. Having a private place to be which is decent and over which we have some real control is fundamental to the well-being of every one of us as individuals and communities. In this sense, affordable housing is both vital economic and social infrastructure.

(Australian Council of Social Service, 2014)

The degree of subsidisation of social housing rent is typically determined by tenant income. In Australia, this is generally set at 25 or 30 per cent of household income (Yates, 2013), while in the US this is 10–30 per cent (US Department of Housing and Urban Development, 2016). In the Netherlands, social housing organisations 'set their own rent policy within the limits of the national rent regulations. On average, social rent levels are approximately 30 percent below the maximum permitted rent' (Aedes, 2016).

The social housing sector in Australia

This sector in Australia includes public housing, community housing, as well as state-owned and managed Aboriginal and Torres Strait Islander (ATSI) housing (Romans, 2014). As of June 2013, around 414,000 households across the country were living in social housing. General housing affordability in Australia continues to decline, with large increases in residential property values and slow development of well-targeted affordable housing. Therefore, the social housing sector is increasingly under pressure to assist households to access appropriate, secure dwellings. As a result, current demand for social housing is much higher than supply, and waiting lists and times are extensive (Department of the Prime Minister and Cabinet, 2014).

As a result of limited investment in the sector over many years, social housing in Australia has become 'residualised': increasingly targeted to those with the greatest and most complex needs. This has led to falling rent revenue as the client's capacity to pay has declined, and has created a cycle of stock deterioration and reduction through an ongoing lack of funds for maintenance and new supply of public housing (Queensland Department of Housing and Public Works, 2014). For example, National Shelter has identified a 3.1 per cent decrease in public housing supply between 2006 and 2012 in Australia (National Shelter, 2014). Much of the public rental housing stock is now at the end of its economic life or does not meet current needs. The poor maintenance of dwelling stock can then create stigma and negative stereotyping of social housing tenants (Jacobs et al, 2010). Underutilisation of housing stock has also become a challenge as typical household sizes have decreased and tenants are living in social housing that does not match their household size or needs (New South Wales Auditor-General's Office, 2013).

Around Australia, more social housing is being provided by the community housing sector, with governments increasingly partnering with not-for-profit housing providers to finance, supply and manage affordable housing stock. Figure 2.1 provides an overview of the Australian system in 2014, and the position of social housing within this system. Alternate sources of financing for this sector in Australia are, however, still limited when compared to other countries. In the UK

> The European Investment Bank has agreed to provide GBP 1 billion for new social housing investment across the UK in partnership with the Housing Finance Corporation (THFC). The expanded Affordable Housing Finance programme will help to alleviate shortages in affordable housing and accelerate construction of new build social housing.
>
> (European Investment Bank, 2016)

On the other hand, in the US, 'Google Inc., Kroger Co. and Waste Management Inc. are investing in low-income rental housing as companies are lured to a field long dominated by financial firms with returns that have doubled to almost 10 percent since 2006' (Gopal, 2010).

Current research at Griffith University School of Business, funded by the National Affordable Housing Consortium, is seeking to develop innovative financial instruments to attract institutional investments into the Australian social housing sector. This project is evaluating the risks and returns of social housing based on advanced and recently developed models and theories in finance, such as the real options model, in order to develop the incomplete financial market for social housing in this country.

Figure 2.2 provides a further breakdown of the social housing sector in Australia.

Australia has a three-tiered system; federal, state and local governments all have a role to play in the availability of social housing. State jurisdictions have traditionally taken a lead role in social housing delivery, for example:

- in Queensland, 75 per cent (54,394 out of 72,329) of social housing stock was government-owned in 2012. This stock was managed through a state-wide network involving 23 Housing Service Centres, with the further 25 per cent (17,935) being owned and managed by community housing providers (Queensland Department of Housing and Public Works, 2014). In an ambitious and transformative move, as part of the Housing 2020 strategy, the Queensland Government is aiming to transfer 90 per cent of all state-managed dwellings to the community housing sector by 2020 (Queensland Department of Housing and Public Works, 2014);
- in Western Australia (WA), the majority of social housing continues to be managed by the state government. The WA Housing Authority manages approximately 36,000 of the 44,700 social housing properties, with community housing associations in charge of approximately 7,700 additional properties. They also offer an array of products through 'Keystart Home Loans'. 'These loan products help eligible people to buy their own homes through low deposit loans and shared equity schemes. Specific loan assistance

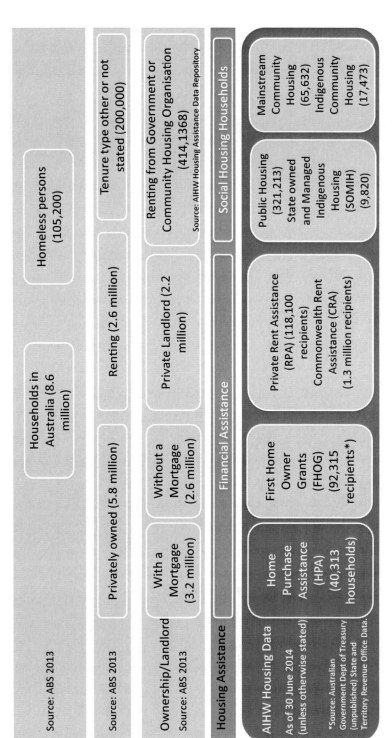

Figure 2.1 Overview of the housing system in Australia and data sources, at 30 June 2013 (Australian Institute of Health and Welfare, 2014).

More productive social housing outcomes 19

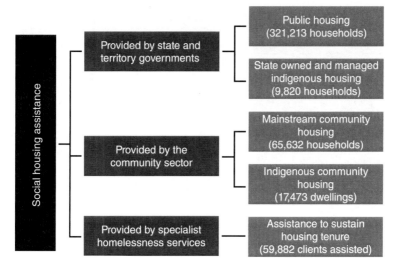

Figure 2.2 Main social housing assistance programmes in Australia and total number of households/clients assisted per programme, 30 June 2013 (Australian Institute of Health and Welfare, 2014).

Notes:
1 This figure does not include social housing dwellings provided to indigenous households in remote areas of the Northern Territory that are not captured in the social housing administrative collections. As of 30 June 2012, an estimated 4,965 dwellings fell into this category.
2 The number of households living in mainstream community housing excludes those in the Northern Territory since data were not available.
3 Data for indigenous housing are from 30 June 2012 since 2013 data were not available. The number of dwellings pertains to permanent dwellings.

is available for public housing tenants, sole parents, people living with a disability and Aboriginal borrowers' (WA Housing Authority, 2015). Access Housing and Foundation Housing are the two largest community housing providers with sufficient assets to leverage funds for growth. Although many smaller players exist, they remain limited in capacity;

- in South Australia, 'community providers manage approximately 13 per cent of South Australia's social housing supply. Existing government commitments will take this to 27 per cent over the next 5 years and initiatives by the community housing sector alongside government support are expected to increase it even further' (Renewal SA, 2013a). As of 2013 there were 48,780 social housing dwellings in South Australia (Renewal SA, 2013b).

Table 2.1 is based on an initial review of the various state-based transformations underway in 2015 to address the provision of social housing.

20   *Kraatz et al.*

Table 2.1 State-by-state transformation matrix

Principle/Policy/Practice[1]	SA	QLD	WA	NSW	VIC	TAS
Housing continuum – crisis – social – private rental to home ownership & 'optimum point'	Y	Y	Y	Y	Y	Y
Common register (integrated system multi provider, multi option)	Y	Y	Y	Y	Y	Y
Duration of need in social housing		Y	Y			Y
Reducing under-occupancy		Y	Y			
Private rental brokerage/assistance	Y	Y	Y	Y	Y	Y
Re-alignment of housing portfolios	Y	Y	Y	Y	Y	Y
Asset transfer	Y	?	Y	Y	Y	?
Management transfer[2]	Y	Y[3]	Y	Y	Y	Y
Urban renewal	Y	Y	?	Y	Y	
Public/Private Partnerships	?	Y	Y	Y	Y	Y
Housing impact statements – major developments			Y			
Model planning policies			Y			
Inclusionary zoning	Y		Y[4]			
Not for Profit/housing organisations – viable partners[5] plus companion/support	Y	Y	Y	Y	Y	Y

Compiled by the research team in 2015
Notes:
1 List based on terminology from WA Government Affordable Housing Strategy 2010 document.
2 Housing ministers' 2009 agreement for up to 35% transfer of management by 2014. Indigenous housing included. Only Tasmania has reached target thus far.
3 Qld 90% transfer of management by 2020.
4 15% minimum affordable housing only on government land and housing developments.
5 The impact of the 2014 federal Budget, which cut the final round ($1 Billion) of NRAS grants, remains unclear.

The need for information integration

> Housing is a complex good that brings together social, economic and environmental concerns.
>
> (Rosenfeld, 2015)

Rosenfeld highlights the need for clarity around the vision and purpose of social housing, and stakeholders' roles and responsibilities, in light of the continuing trend for the decentralisation of state responsibilities. In addition, she notes the integrative nature of the sector, which requires both the horizontal and vertical integration of efforts and funding across ministries, departments and levels of government with responsibilities such as social services, health and infrastructure. Beyond this, there is also the role of the private and not-for-profit sectors in provision, financing, maintenance and the like. In many countries, this extends to the role of international organisations who are seen 'as the hubs of knowledge exchange' (Rosenfeld, 2015). A further key challenge is to better understand and quantify the broader productivity benefits from investment in social housing, for tenants, the community and the wider economy.

> Housing matters bring together numerous sectors and disciplines. If treated as a mere piece of other disciplines, the solutions are unlikely to bring impactful results. Instead, housing should be treated as a sovereign professional discipline that convenes disparate efforts for integrated solutions.
>
> (Rosenfeld, 2015)

The Australian case study presented in this chapter discusses research exploring the impact of social housing across these various layers, through a productivity lens. This is examined by looking at an array of benefits from tenant, macro-economic, fiscal and non-economic perspectives. This is a broad-based approach with a focus on practical outcomes requiring an integrated approach across the various levels of government from national to local, to private sector and not-for-profit providers (Figure 2.3).

To achieve this broader understanding, government and industry knowledge, information and data will need to be used in a connected manner to identify productivity benefits and impacts. Tracking the long-term benefits of providing safe and secure housing will be crucial. Rosenthal notes the following initiatives communicated to the UNECE by national and sector leaders in housing include: the development of topic-specific think tanks; capacity building for local authorities; knowledge exchange platforms across the private and public sector; and a commitment to 'housing as an integrative field' (Rosenthal, 2015).

This will require innovative thinking around how information, knowledge and data are gathered to articulate the social along with the economic impact of providing social housing. McCreless and Trelstad (2012) highlight the need to combine 'information from disparate methodologies into a coherent, internally consistent and accurate categorisation of investments by level of social impact' (McCreless and Trelstad, 2012). These authors adopt the triangulation approach to data gathering for assessing social impact.

> In addition to basic output metrics, we also include enterprise- or project-reported information, site visits by our staff, case studies and other reports by third parties, qualitative and quantitative surveys (including randomized controlled trials), data gathered using new approaches to mobile technology, and literature reviews. We fold in cost data to evaluate cost-effectiveness, and when possible, we gather data to provide a counterfactual to establish causality and attribution.
>
> (McCreless and Trelstad, 2012)

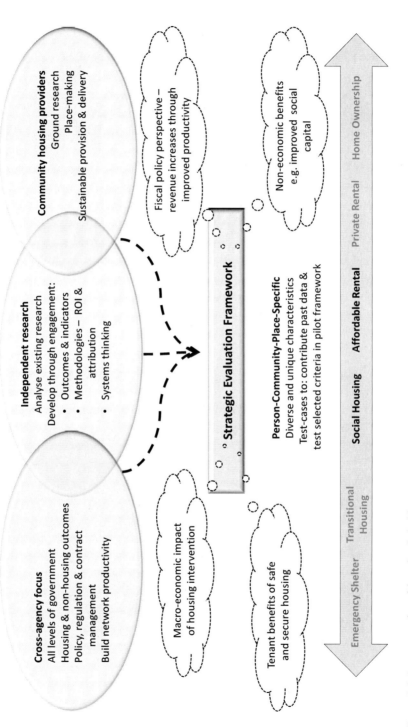

Figure 2.3 Integrating broad-based stakeholder perspectives.

Case study: a productivity-based framework for social housing delivery in Australia

Current Australian research is seeking to provide a new rationale for policy making, delivery and evaluation of social housing programmes. This includes learning lessons from international best practice. Country-specific financing and delivery mechanisms have, however, evolved out of particular cultural, political, economic, policy, legal and financial frameworks of different countries. This has led to limited options for exploring new directions in the delivery of social housing and limits the established, collective knowledge regarding general principles for best practice.

A central element of the approach being explored is productivity for the individual and for society more broadly. A broad range of relevant housing and non-housing outcomes can be potentially attributed to having safe and stable housing. Examples include higher resident well-being, better employment outcomes, stronger community ties and a sense of safety within a neighbourhood. The integration of information required to demonstrate this is complex, moving across stakeholders and agencies from tenant, to housing provider, to local government, to state government and to the federal government. It also requires significant data integration. The value of this consolidation will require clarification of causal links between outcomes and indicators and determination of the social return on investment.

The framework

The 'Strategic Evaluation Framework' for social housing delivery has been developed as part of Australia's Sustainable Built Environment National Research Centre (SBEnrc) project 'Rethinking Social Housing: Effective, Efficient, Equitable' (Kraatz et al., 2015). This framework aims to develop a platform that can be used by policy makers to determine the most effective and beneficial programme delivery options that address housing and non-housing outcomes, all explored through lenses of productivity.

The following case study discusses the development of this framework (Figure 2.4) that explores the benefits and costs of social housing through the four lenses of the tenant, the macro-economic costs and benefits, the fiscal perspective, and the non-economic focus (environmental and social capital).

The tenant benefits include enhanced capacity to engage in education and employment, to improve health and well-being, and to enable social engagement. The macro-economic impacts of housing intervention can include productivity improvements and growth through increased construction activity and greater workforce engagement. Fiscal benefits can include revenue increases if social housing has positive productivity impacts. It also includes potential impacts on government expenditure, such as better health leading to less sick leave and more people working longer, all of which adds to productivity and tax revenues, as well as to higher expenditure on education and training, and stronger engagement

24 Kraatz et al.

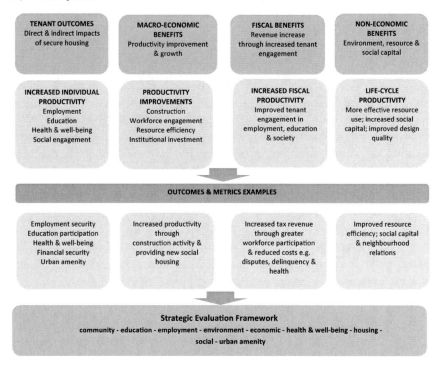

Figure 2.4 Strategic evaluation framework.

with work, stronger self-esteem and adherence to prevailing social mores, leading to lower expenditure on mental health and prisons. The non-economic benefits and returns can flow from improved resource efficiency, greater social capital and neighbourhood relations through a more stable social environment.

To address this broad context, the framework comprises four elements that integrate data, information and knowledge from diverse sources including:

- *outcomes and indicators matrix*: drawing on academic, government and industry data across the nine domains of community, economy, education, employment, environment, health and well-being, housing, social, and urban amenity;
- *causality and associations*: associations and/or causality; it will be necessary to demonstrate links between selected indicators in relation to social housing outcomes;
- *return on investment*: being able to demonstrate the broader social return on investment in social housing for each of the indicators and outcomes is a complex and essential element of the development of this framework;
- *data and datasets*: to provide the statistical basis from which to build the framework.

Significant gaps in knowledge are being identified in the course of this research in each of these areas. Highlighting this need and developing methodologies and avenues for addressing these gaps is an integral part of this research.

Outcomes and indicators

This research has established a provisional set of outcomes, drawn from previous research, and industry stakeholders. The 'outcomes and indicator matrix' is the result of an extensive literature review over a twelve-month period in which the researchers have drawn together previously used indicators from different disciplines that have links with social housing. The indicators have been sorted and placed into nine separate domains: employment, education, health and well-being, social, urban amenity, community, environment, economy, and housing objectives.

As part of researching these indicators and developing the matrix, this research has investigated the broad objectives of social housing provision. A key aim of this matrix is to provide both government agencies and community housing providers with the ability to measure outcomes and better articulate the broader community value of providing housing security to all.

The domains, outcomes and indicators have been compiled in a cascade utilising, at the upmost level, the Global Reporting Initiative.[2] This has been done to provide universality to the indicators, which intersect various policy and provision domains and to potentially enable them to be more readily aligned with existing organisational reporting. Burke and Hayward (2000) provide a useful example of cascading performance indicators (Table 2.2). They also highlight the need to have indicators for performance at the various levels of service provision in order to deliver indicators that have relevance and validity from a national to an agency or tenant level.

Table 2.2 Layers of performance indicators

Level	Purpose
National	How well is social housing meeting its objectives? How well are social housing agencies performing? How does the performance of social housing agencies compare with other sectors, for example, private rental sector?
State Housing Agency	How well is the agency or federation (of agencies) meetings its objectives? How does its performance compare with other like organisations?
State Housing Agency Business Unit	How well is a specific function or business performing, for example, housing finance, stock production, rental housing management?
SHA Regional offices	How well is a particular region performing, either overall or for a specific functional business?
SHA Work unit (teams)	How well is a work unit achieving its objectives? How does its performance compare with other similar work units?
SHA employee	Does the individual's work performance meet agreed targets?

Source: Burke and Hayward, 2000.

26   *Kraatz et al.*

A working set of outcomes and indicators across nine domains has been established in previous research undertaken by the team in 2014–15 (Kraatz et al., 2015). The intention is that this master set of indicators can be drawn upon by different agencies for different cohorts in different geographical locations across Australia.

Establishing causality and/or associations

A key challenge for this research is to be able to associate non-housing indicators and data, such as at a neighbourhood, household or individual level, to housing provision, such as types, styles, tenures, locations and conditions, by way of direct or indirect causal connections.

The relationship between housing and the various aspects of productivity being considered is complex, multidirectional and mediated by a host of intervening factors. A strong and logical method, grounded on previous research about the nature of connections between housing and non-housing domains, is required. To this end, the research team is developing a rigorous and defensible method drawing upon global literature.

Given the strong confirmed causal links between housing and health, the tradition of 'integrated environmental health impact assessment' will be further investigated as a concept to establish

> a means of assessing the extent, time trends or spatial distribution of health effects related to environmental exposures, and health-related impacts of policies that affect the environment, in ways that take account of the complexities, interdependencies and uncertainties of the real world.
>
> (Knol, 2010)

The 'Butterfly Model of Health' developed in the late 1990s builds on several previous models and reflects a 30-year trend. Its aim was to 'identify the direct relationships between human health and the so-called "determinants of health"' defined as 'factors, whether they be events, characteristics, or other definable entities, that brings about change in a health condition' (VanLeeuwen et al., 1999). This work explored the inter-relationships between the bio-physical and the socio-economic environments on people. This approach is one in a significant body of knowledge that acknowledges the links between social, environmental and health conditions. This tradition for building causal relationships includes the DPSIR framework developed for the World Health Organisation in the 1990s. This acronym stands for *driving forces* and the resulting environment *pressures* on the *state* of the environment and *impacts* resulting from changes in environment and the societal *response* to these changes in the environment to establish causal links.

In the UK, the Scottish Government has built upon this tradition to articulate associations between housing and health (The Scottish Government, 2008; 2011) through their modified DPSEEA[3] Model for the Good Places, Better

Health initiative commenced in 2008 (Scottish Government, 2015). In Canada, Buzzelli (2009) provides an annotated bibliography which explores causal links and/or associations between housing and a variety of non-housing outcomes such as health, education and corrections, at the household, neighbourhood and macro-economic levels. The work of the Housing Associations' Charitable Trust (HACT) in the UK, and the National Housing Conference in the US, will also inform the continuing SBEnrc research.

Return on investment

Identifying the return on investment (ROI) associated with social housing is driven by the need to better articulate the social and economic returns to the community of investment in social housing. To effectively do this, the research team is developing outcomes and indicators that go beyond the traditional specific housing indicators to embrace externalities not typically measured in relation to the investment in social housing itself.

This is important in the current context of social impact measurement being pursued by governments across Australia and internationally. Dunn (2014) defines social impact investing as investing in efforts that not only provide a return on investment, but also target specific social needs. Such measurement is also important in order to attract institutional investment to the delivery of social housing through establishing the expectation (supported by evidence) that investors will recover their money and potentially gain an income stream from the investment (Knowles in Dunn, 2014).

There are many methods through which indirect, non-market values have been quantified in the past. The different methodological approaches have fed into the nature of indicators used for measurement. Two methods for measuring outcomes and potentially determining return on investment have been identified for further research in the context of the strategic evaluation framework (Figure 2.5):

- social return on investment (SROI);
- well-being valuation analysis (WVA).

Both these methodologies have been used and developed in the UK. SROI was adopted to ensure that the potential (non-economic) value added is adequately assessed when determining the placement of funds and choice of service providers (Harlock, 2012). The Public Contracts (Social Value) Act was introduced in 2012. To implement this regulation, an analysis of social value is required when determining contract allocation. It provides a policy tool that levels the playing field between third-sector organisations and commercial operators. It does so by placing value on the less tangible, but important outcomes, that third-sector organisations can bring about. The WVA approach draws upon both the SROI method and traditional cost–benefit analysis (Fujiwara, 2014; Fujiwara and Campbell, 2011). Developed specifically for measuring the social value of housing

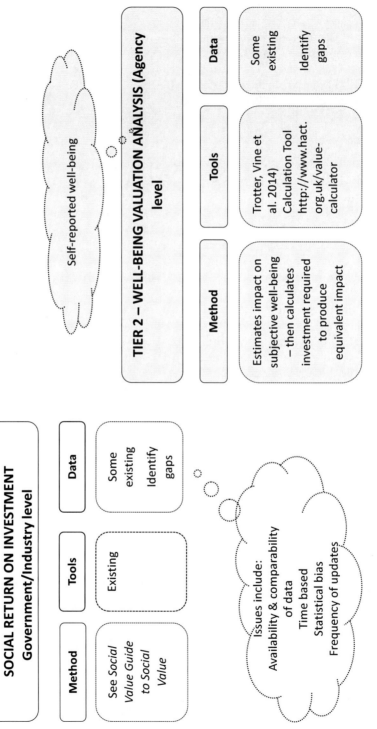

Figure 2.5 Two-tiered approach to return on investment.

associations in the UK, the method emerged in response to the perceived lack of appropriate tools for quantifying social value on a large, sector-wide scale.

In the Netherlands, social housing provision is monitored using the 'Aedex real estate index' (Bortel and Elsinga, 2007). The index

> Measures the profitability of housing associations and the difference between this figure and the profitability that could be achieved by pursuing a commercial strategy. This difference, also called a 'dividend to society' is assumed to be the profitability that housing associations do not realise because of their non-profit character.
>
> (Bortel and Elsinga, 2007)

This represents an innovative way of making housing providers financially accountable while still valuing their social impact. In Australia, several studies have adopted the SROI method. Two examples include *The Social Value of Community Housing in Australia* study (Ravi and Reinhardt, 2011) and that for the Victorian Women's Housing Association (Kliger et al., 2011; Social Ventures Australia, 2010). Kliger et al. adopted the SROI approach to discuss the premise 'that investment in affordable housing for low-income women provides both micro and macro-economic benefits for cities and communities' (Kliger et al., 2011). They examined 'the value produced by the volunteer and philanthropic group known as the Victorian Women's Housing Association (VWHA)'. This paper draws on the Social Ventures Australia study reported in 2010 (Social Ventures Australia, 2010).

Data and datasets

Data for this project is being obtained from several sources including both quantitative and qualitative methodologies (Randolph and Judd, 2001; Moore et al., 2002). Those based on quantitative methodology include official demographic data from Australia's key social security safety net provider, Centrelink; state-based housing agencies or census data (Australian Bureau of Statistics); systematic observations; sample surveys including structured interviews; experimental or quasi-experimental studies; and cost–benefit analyses. Those based on qualitative methodology can be drawn from participant observation; in-depth interviews and surveys; focus groups; action research; and critical or historical analyses.

These outcomes and indicators sit within the broader context of social housing delivery in Australia. Other statistical information (both national and state-based) is required to assist interpretation of the indicators. Data needs to be gathered from several sources including other indicator sets used for specific performance evaluation purposes. It also needs to be studied in conjunction with other statistical reports and national datasets, and over time (due to the longitudinal nature of data required). Some existing datasets that will be used are outlined below. Table 2.3 highlights several national and state-based datasets

30   *Kraatz et al.*

Table 2.3 Some relevant datasets

Source	Explanation
Household, Income and Labour Dynamics in Australia (HILDA) Survey	To support research questions falling within three broad areas of income dynamics, labour market dynamics and family dynamics.
National Social Housing Survey (NSHS)	Includes tenant satisfaction metrics
Australian Bureau of Statistics (ABS) data	Australian Census of Population and Housing; Survey of Income and Housing Costs; National Health Survey; Rental Investors Survey; Disability, Aging and Carers Survey; Mental Health and Wellbeing of Adults survey; Time Series profile (TSP) DatePack; Survey of Housing Occupancy and Costs 2009-10, ABS, 2011 Canberra
Australian Institute of Housing and Wellbeing (AIHW)	Housing assistance in Australia 2011, 2012 & 2014. This provides relevant demographic data. For example, almost 40% of social housing households have a person with a disability, age and sex distribution; National Social Housing Survey – A summary of national results 2012
Community Housing and Infrastructure Needs Survey (CHINS)	Relates to Indigenous Community Housing not "Community Housing" in the general
AURIN – The Australian Urban Research Infrastructure Network (AURIN) Portal	Browse for metadata for all the datasets available in the AURIN Portal – covers most aspects of urban environments in Australia, from health and well-being, to economic metrics and environmental indicators
Developmental Pathways Project – Telethon Institute for Child Health Research	Investigates pathways to health and wellbeing, education, disability, child abuse and neglect, and juvenile delinquency outcomes among Western Australian children and youth.

from which data will need to be drawn in order to consolidate cohort, geographical and longitudinal impacts of policy change with regards to the provision of social housing.

It is likely that data gaps will also be identified and international knowledge will be drawn on to help identify additional data that may be required to provide an effective picture. In the UK, the Well-being Valuation Analysis methodology draws upon several datasets including the British Household Panel Survey; Understanding Society; The Crime Survey for England and Wales; and the Taking Part survey.

> These datasets include people's responses to wellbeing questions, and questions on a large number of aspects and circumstances of their lives such as employment status, marital status, health status, whether they volunteer, whether they play sports, whether they live in a safe area, and so on, resulting in a wide range of values.
>
> (Trotter et al., 2014)

In Canada, again several sources of data exist including:

- Statistics Canada for data on population and demographics, dwelling counts, housing and shelter costs, and economic and migration estimates;
- Canada Mortgage and Housing Corporation for customised data on the cost of housing, the availability of housing, vacancy rates and housing Starts and Completions Survey;
- Human Resources and Social Development Canada for minimum wage database;
- The Homeless Individuals and Families Information System;
- Homeless and Social Housing Data.

All such datasets address a unique aspect of the social and economic benefits that flow from safe and secure housing. The intention is that integrated with other more dynamic organisational and tenant information and knowledge (McCreless and Trelstad, 2012), such data can help define the impacts and benefits of social housing in the context of the proposed framework.

Concluding remarks

This chapter has shown that social housing is a global challenge affecting millions of people and requiring the action of a multitude of stakeholders from the public, private and not-for-profit sectors. The need for an integrated approach to address issues across the housing continuum is also demonstrated. The authors have explored the need for the horizontal and vertical integration of data, information and knowledge within and across departments and organisations to achieve a more sustainable social housing system.

This has been explored in a case study from an Australian perspective. The next stage of this case study research is being undertaken through SBEnrc in two parallel parts in order to further develop key aspects of the presented framework. The first part is looking at data integration, being undertaken by Curtin University, with the Western Australian Housing Authority and Access Housing WA, to explore data sharing across government departments and organisations in order to develop a workable framework that can inform policy making and social housing delivery. The second part is investigating the integration of this framework with evolving government performance frameworks in times of fiscal constraint. This is being led by Griffith University, with partners including the New South Wales Land and Housing Corporation, the Queensland Department of Housing and Public Works and the National Affordable Housing Consortium.

A key challenge in the development of this type of research is managing the complex array of information; particularly when exploring impacts across the nine domains considered important to social housing outcomes in the framework presented in the case study. This complexity cannot be effectively addressed using only traditional approaches focused on the management of supply or demand that do not provide management strategies that meet the underlying

32 *Kraatz et al.*

needs and values of the affected populations. Further research is also needed into combining Geographic Information Systems (GIS), Bayesian network and system dynamics modelling approach (Sahin and Mohamed, 2013). This recognises that no single current methodology can efficiently and adequately integrate the demand, supply and asset management processes of housing while addressing the associated social, economic and environmental dimensions.

Notes

1 This is equivalent to USD32,182/year at the average exchange rate for 2015 (EUR1 = USD1.11).
2 GRI is an international independent organisation that helps businesses, governments and other organisations understand and communicate the impact of business on critical sustainability issues such as climate change, human rights, corruption and many other (GRI, 2016).
3 This acronym stands for driving forces, pressures, state, exposures, health effects, actions.

References

ABS, 2013. Housing occupancy and costs, 2011–12, ABS cat. no. 4130.0. Canberra: Australian Bureau of Statistics.

AEDES, 2016. *Dutch Housing in a Nutshell: Examples of Social Innovation for People and Communities*, The Hague: AEDES.

Australian Council of Social Services, 2014. *Letter to Senator Mark Bishop, Senate Economic References Committee: Enquiry into Affordable Housing*, Canberra: Senate Printing Unit, Parliament House.

Australian Institute of Health and Welfare, 2014. *Housing Assistance in Australia*, Canberra: Australian Institute of Health and Welfare.

Australian Institute of Health and Welfare, 2016. *National Housing Assistance Data Repository*. Australian Institute of Health and Welfare. Available at: http://meteor. aihw.gov.au/content/index.phtml/itemId/462772 [Accessed 16 November 2016].

Bortel, G. V. and Elsinga, M., 2007. A network perspective on the organization of social housing in the Netherlands: The case of urban renewal in The Hague, *Housing, Theory and Society*, 24(1), 32–48.

Bourne, R., 2016. The UK doesn't need more social housing – but we do need to build more homes. *The Telegraph*, 1:27PM GMT 21 January.

Burke, T. and Hayward, D., 2000. *Performance Indicators and Social Housing in Australia*, paper presented at the Housing Policy and Practice in the Asia Pacific: Convergence and Divergence Conference, Hong Kong, 13–15 July.

Buzzelli, M., 2009. *Is it Possible to Measure the Value of Social Housing?*, Canada: Canadian Policy Research Network.

Carlson, M., 2014. *Alternative Sources of Capital for the Social/Affordable Housing Sector in Canada*. Toronto, Canada: Housing Services Corporation.

Department of the Prime Minister and Cabinet, 2014. *Reform of the Federation White Paper: Roles and Responsibilities in Housing and Homelessness*, Canberra: Commonwealth of Australia.

Dunn, J., 2014. *Social Impact Investing: Good Returns from New Sector*. Available at: http://business.nab.com.au/afr-special-report-social-impact-investing-8140 [Accessed 20 April 2015].

European Investment Bank, 2016. *GBP 1 Billion EIB Backing for UK Social Housing*. Available at: www.eib.org/infocentre/press/releases/all/2016/2016-100-gbp-1-billion-european-investment-bank-backing-for-uk-social-housing.htm?lang=en [Accessed 1 May 2016].

Fujiwara, D., 2014. *Measuring the Social Impact of Community Investment: The Methodology Paper*, London: Housing Associations Charitable Trust (HACT).

Fujiwara, D. and Campbell, R., 2011. *Valuation Techniques for Social Cost-Benefit Analysis: Stated Preference, Revealed Preference and Subjective Well-Being Approaches: A discussion of the Current Issues*, London: HM Treasury.

Global Reporting Initiative, 2016. *About GRI*. Available at: www.globalreporting.org/information/about-gri/Pages/default.aspx [Accessed 1 August 2016].

Gopal, P., 2010. *Returns Up to 10% on Low-Income Housing Lure Google, Kroger*. Available at: www.bloomberg.com/news/articles/2010-10-06/returns-up-to-10-on-low-income-housing-lure-google-kroger [Accessed 24 May 2016].

Harlock, J., 2012. *Impact Measurement Practice in the UK Third Sector: A Review of Emerging Evidence*, Birmingham: Third Sector Research Centre.

Jacobs, K., Atkinson, R., Colic-Peisker, V., Berry, M. and Dalton, T., 2010. *What Future for Public Housing? A Critical Analysis*, Melbourne: AHURI.

Kliger, B., Large, J., Martin A. and Standish, J., 2011. *How an Innovative Housing Investment Scheme Can Increase Social and Economic Outcomes for the Disadvantaged*, paper presented at the State of Australian Cities Conference. Sydney, Australia, 29 November–2 December.

Knol, A. B., Briggs, D. J. and Lebret, E., 2010. Assessment of complex environmental health problems: Framing the structures and structuring the frameworks, *Science of the Total Environment*, 408, 2785–2794.

Kraatz, J., Mattan, A., Mitchell, J. and Newman, P., 2015. *Rethinking Social Housing: Effective, Efficient, Equitable*, Brisbane: Sustainable Built Environment National Research Centre. Available at: www.sbenrc.com.au/research-programs/1-31-rethinking-social-housing-effective-efficient-equitable-e3/ [Accessed 01 August 2016].

McCreless, M. and Trelstad, B., 2012. A GPS for social impact: Root capital and acumen fund propose a system for program evaluation that is akin to GPS, *Stanford Social Innovation Review*, Fall, 21–22.

Moore, G. T., Russell, C., Beed, T. and Phibbs, P., 2002. *Comparative Assessment of Housing Evaluation Methods: Evaluating the Economic, Health and Social Impacts of Housing*, Sydney: AHURI.

Moskalyk, A., 2008. *The Role of Public-Private Partnerships in Funding Social Housing in Canada*, Ontario: Canadian Policy Research Networks.

National Shelter, 2014. *Housing Australia Fact Sheet: A Quick Guide to Housing Facts and Figures*, Perth: National Shelter.

New South Wales Auditor-General's Office, 2013. Making the best use of public housing. In *New South Wales Auditor-General's Report: Performance Audit*. Sydney: New South Wales Auditor-General's Office.

Queensland Department of Housing and Public Works, 2014. *Housing 2020: Delivering a Flexible, Efficient and Responsive Housing-Assistance System for Queensland's Future*, Brisbane: Queensland Government.

34 *Kraatz et al.*

Randolph, B. and Judd, B., 2001. *A Framework for Evaluating Neighbourhood Renewal – Lessons Learnt from New South Wales and South Australia*, paper presented at the National Housing Conference, Brisbane, Australia, 24–26 October.

Ravi, A. and Reinhardt, C., 2011. *The Social Value of Community Housing in Australia, Community Housing Federation of Australia*, Australia: CHFA, PowerHousing Australia and Bankmecu.

Renewal SA, 2013a. *Affordable Housing*, Adelaide: Government of South Australia.

Renewal SA, 2013b. *South Australia: Housing Affordability: Demand and Supply by Local Government Area*, Adelaide: Government of South Australia.

Romans, E. C., 2014. *Forms of Housing Tenancy Regimes in Affordable Housing Public Policies in Barcelona, 2007–2014*, paper presented at the European Network of Housing Researchers 2014 International Conference, Edinburgh, Scotland, 1–4 July.

Rosenfeld, O., 2015. *Social Housing in the UNECE Region: Models, Trends And Challenges*, Geneva: United Nations Economic Commission for Europe.

Sahin, O. and Mohamed, S., 2013. A spatial temporal decision framework for adaptation to sea level rise, *Environmental Modelling & Software*, 46, 129–141.

Social Ventures Australia, 2010. *Victorian Women's Housing Association Investment in Affordable Housing for Women: The Social and Economic Returns*, Melbourne: Victorian Women's Housing Association.

The Scottish Government, 2008. *Good Places, Better Health: A New Approach to Environment and Health in Scotland: Implementation Plan*, Edinburgh, Scotland: The Scottish Government.

The Scottish Government, 2011. *Good Places Better Health for Scotland's Children: Childhood Mental Health and Wellbeing Evidence Assessment*, Scotland: NHS Scotland.

The Scottish Government, 2015. *Good Places, Better Health*. Available at: www.gov.scot/Topics/Health/Healthy-Living/Good-Places-Better-Health [Accessed 01 August 2016].

Trotter, L., Vine, J., Leach, M. and Fujiwara, D., 2014. *Measuring the Social Impact of Community Investment: A Guide to Using the Wellbeing Valuation Approach*, London: HACT Housing.

Tunstall, R., Lupton, R., Kneale, D. and Jenkins, A., 2011. *Teenage Housing Tenure and Neighbourhoods and the Links with Adult Outcomes: Evidence from the 1970 Cohort Study*, London: Centre for Longitudinal Studies, Institute of Education, University of London.

UK Department of Communities and Local Government., 2012. *Definitions of general housing terms*. Available at: https://www.gov.uk/guidance/definitions-of-general-housing-terms [Accessed 24 May 2016].

US Department of Housing and Urban Development, 2016. *HUD's Public Housing Program*. Available at: http://portal.hud.gov/hudportal/HUD?src=/topics/rental_assistance/phprog [Accessed 01 July 2016].

VanLeeuwen, J. A., Waltner-Toews, D., Abernathy, T. and Smit, B., 1999. Evolving models of human health toward and ecosystem context, *Ecosystem Health*, 5(6), 204–219.

WA Housing Authority, 2015. *Estimating Unmet Demand for Public and Affordable Housing*, paper presented at the National Housing Conference, Perth, Australia, 28–30 October.

Wong, T.-C. and Goldblum, C., 2016. Social housing in France: A permanent and multifaceted challenge for public policies, *Land Use Policy*, 54, 95–102.

Yates, J., 2013. Evaluating social and affordable housing reform in Australia: Lessons to be learned from history, *International Journal of Housing Policy*, 13(2), 111–133.

Young, L., 2015. *140,000 Canadian Families are Waiting for Housing. Here's What the Parties Plan to Do*. Available at: http://globalnews.ca/news/2268505/140000-canadian-families-are-waiting-for-housing-heres-what-the-parties-plan-to-do/ [Accessed 23 May 2016].

3 Resilient policies for wicked problems

Increasing resilience in a complex and uncertain world through information integration

Adriana X. Sanchez, Paul Osmond and Jeroen van der Heijden

Introduction

> Managing the future is a 'wicked' problem, meaning that it has no definitive formulation and no conclusively 'best' solutions and, furthermore, that the problem is constantly shifting.
>
> (Hjorth and Bagheri, 2006)

Urban settlements[1] represent some of the most complex and long-lasting artefacts created by humans and their relevance continues to grow. Nowadays, more than half of the world's population lives in them. In fact, the number of people living in urban settlements has been increasing globally at an average annual rate that is almost twice the growth rate of the total world population. The World Bank reported a global urban population growth rate of 2.1 per cent in 2015 versus a 1.2 per cent general population growth (The World Bank, 2016). Modest forecasts by the United Nations indicate that by 2050 the vast majority of people will live in cities; 66 per cent globally or 6.4 billion people (UN, 2014). This represents almost a doubling of the number of people living in cities when compared to 2008 statistics, mainly due to ongoing urbanisation and population growth.

Even though overall growth rates are declining, this tendency towards urbanisation means that cities will continue to grow in importance in terms of the impact their management and governance can have on national and global trends. Whether this effect is positive or negative will depend on the policies established to plan, develop and manage the cities of the world. Additionally, there is a growing awareness about the uncertain nature of long-term futures, fast pace of change and incompleteness of information. This awareness is coupled with rising fears about the effects of a changing climate, high ambitions to ensure resources security and the speed and scale of urbanisation. All of which is leading to the recent popularity that the term *'resilience'* is gaining across both public and private policies and initiatives. This chapter responds to this trend and explores

36   *Sanchez et al.*

how integrating information can contribute to urban resilience through the production of more meaningful and well-informed policy and action.

Urban resilience

While there are a number of resilience concepts available in the literature, this chapter proposes that resilience must go beyond dealing with individual threats and that a dynamic rather than static perspective is required. This responds to a global movement towards more integrated and dynamic views of resilience. Concepts based on socio-ecological resilience principles for example have grown in popularity across both academic and policy circles. This concept is underpinned by the idea that resilience is the outcome of the dynamic interaction between different components of a system (Adger et al., 2011).

This idea has also been observed in the global economic arena. The 2013 World Economic Forum focused on this shift to more dynamic views of the resilience of world economies (Madslien, 2013; Tapscott, 2013). Within an urban context, the Rockefeller Foundation's 100 Resilient Cities Program has more recently discussed city resilience in terms of being the product of a reflective, integrated and flexible approach that considers the individual systems that form a city as well as the whole (100 Resilient Cities, 2015). This program includes some of the most economically prominent cities across the globe. This view of resilience underpins the concept proposed by Sanchez et al. (2016b) under the title '*sustainable resilience*'; resilience that endures. From here on, the term resilience refers to

> Urban sustainable resilience is the capacity of socio-eco-technological complex dynamic urban systems to tolerate disturbances, which can be chronic or acute, and persist in a sustained manner through ongoing learning and adapting to changes to the environment and the needs of the system. It requires efficient, diverse, coordinated and cohesive strategies that proactively address short, medium and long-term challenges. Urban sustainable resilience is underpinned by sustainability and dynamic system resilience principles in order to define practical policy aspects that allow cities to tolerate disturbances, evolve with the changing environment (where environment refers to climate, social sentiment and technological context) and mitigate future sources of stress.
>
> (Sanchez et al., 2016b)

Sanchez et al. (2016b) propose that in order to strive towards more sustainably resilient cities, the policies that are used to plan, deliver and manage them also need to exhibit resilience qualities in order to achieve long-term objectives through sustained and adaptive action. This chapter will first discuss some of the challenges that lead city governments to develop resilience policies and initiatives, and seek support from other levels of government as well as the private sector and civil society. It will then briefly explore the role of information and the relevance of its integration across organisational boundaries and software domains to achieve resilient policies. Finally, it will provide two examples of how information integration can help achieve resilience and resilient policies, and provide some concluding remarks.

Wicked problems and policy responses

Cities exist against a backdrop of complex challenges such as multi-level and multi-actor governance systems, changing climates, rapid technological progress, finite infrastructure life span and complex network feedbacks, growing urban population and a number of other emerging problems. All of these drive different requirements in terms of information needed to act and levels of integration across organisational boundaries and software domains essential for effective and efficient urban governance.

Wicked problems are complex challenges, often characterised by multiple sources of stress and multiple temporal and spatial scales; and by responses being influenced by a number of complex interactions between components of the system that are difficult to predict (McEvoy et al., 2013). The concept of resilience used in this chapter supposes viewing urban settlements as complex, dynamic systems subject to a continuum of disturbances within a context of uncertainty. This means that cities are formed by social, ecological and technological networks that evolve and react to sources of stress (disturbances) as a result of interactions between and within them. It also means that preparing for single known trigger events might not be the best way to address urban resilience. For example, if resilience policies focus only on floods or bushfires as experienced in the past, the city may be unprepared for other emerging sources of stress never experienced before such as cyber-attacks, rising sea levels or cyclones and typhoons.

The following sub-sections will elaborate on the nature of these challenges and their implications for policy development and implementation.

Changing climate

The fifth assessment report of the Intergovernmental Panel on Climate Change (IPCC) asserts with very high confidence that regional climates are changing, often in ways that cannot be predicted. Over the past five decades the south-eastern part of the Australian continent for example has been experiencing decreasing rainfall annual averages. However, there is a high degree of uncertainty about future rainfall trends; predictions for Victoria (south-eastern state) vary between no changes in annual rainfall to a 40 per cent decline (Reisinger et al., 2014). Similar patterns of uncertainty are observed in other regions, with the number and intensity of extreme weather events also expected to change. While some of these conditions have been experienced previously, the changes in frequency and intensity will create new difficulties for cities. They will also compound with other climate-related challenges never experienced before, such as sea level rise for coastal cities, and non-climate-related challenges such as

38 *Sanchez et al.*

those further discussed below; all of this adding to the general context of uncertainty (Birkmann et al., 2010).

A survey carried out in Norway in 2007 showed that most adaptation policies at municipal level in that country arise from direct experiences of extreme weather events (Amundsen et al., 2010). However, 'the speed, severity, and complexity of known and unknown changes in climate' will be a constant challenge for urban settlements used to implementing solutions in a reactive manner based on previous experience (Adger et al., 2011). A changing climate requires a proactive approach. This level of unpredictability around weather events highlights the need to carry out long-term planning based on the potential for future damage rather than the degree of damage caused by past events (McIlwain et al., 2013). Climate change is characterised by multi-hazard phenomena, including concurrent and slowly creeping threats as well as potential tipping points. Additionally, while 'small-scale everyday hazards are often underestimated in urban areas', cities continue to be devastated by low-frequency events which are becoming more common as time passes (Birkmann et al., 2010). This multiplicity of potential future risks and the lack of understanding of the long-term effects are also problematic for policy development by raising the questions of adapting to what, how and within which spatial scale (Urwin and Jordan, 2008).

This rapidly changing and uncertain environment will require a more integrated approach to information management that allows decision-makers, the private sector and the general public to have access to up-to-date, relevant and easy-to-understand information that can be used to innovate, adapt and mitigate risks as new challenges and opportunities arise.

Complex governance system

Cities are governed by a complex web of public and private organisations that manage their service and infrastructure networks. Climate change adaptation and mitigation planning has started to be embedded in many planning processes across the world but, at least in countries such as Australia, this has mostly happened only at a conceptual level. Implementation is often solely focused on energy and water consumption initiatives, continues to be sparse and, especially in Australia although a global issue, is highly susceptible to political changes and legal conflicts (Reisinger et al., 2014).

Current urban governance systems pose challenges to increasing local adaptive capacity due to

- limited financial and human resources to develop, implement and assess effective policies;
- 'limited integration of different levels of governance';
- 'different attitudes toward the risks associated with climate change';
- 'different values placed on objects and places at risk'.

(Reisinger et al., 2014)

Resilient policies for wicked problems   39

In addition to these, there is a lack of local expertise and a clear role for local government when working with these kinds of policies and measures (Amundsen et al., 2010). The rapid growth of urban population often leads to a loss of governability and growth of informal governance networks, further adding to the challenges of developing comprehensive policies by already uncertain governance roles (Birkmann et al., 2010). This fragmented nature of governance systems often translates into policies by different actors not being compatible, resulting in local policies being undermined or contradicted by those at other governance levels or even within different fields of the same level (Anguelovski and Carmin, 2011). A series of resilience expert interviews carried out in Australia, the United States (US) and the United Kingdom (UK)[2] in 2016 suggested that this is often the result of inadequate access to information about policy processes at different levels of government and a lack of incentive mechanisms for individuals to take an integrated, holistic view of the issues at play.

Technological progress

Cities evolve in complex ways, often driven by technological and infrastructural innovations that increase consumption and production of resources and goods that create new difficulties and opportunities (Agudelo-Vera et al., 2011). The constant challenges arising and limited funds drive the adoption of inexpensive solutions before more complex and expensive ones are trialled (Tainter and Taylor, 2014). On many occasions, however, the adoption of uniform, generalised technical solutions may create new and more difficult problemss (Hassler and Kohler, 2014). Additionally, technology is starting to be developed at a pace that makes it difficult for decision-makers to have a clear understanding of the challenges and opportunities brought by their use, replacement and disposal.

Hurdles are not only brought by the pace of change but also by the volume and scattered nature of information about the implications of these technological advances for policy development and implementation. Due to the long cycles required for policy changes, this creates further difficulties for the system because new technologies are generated faster than policies can adapt. In addition to this, when new technologies and digital systems are adopted by one level of governance, these may not be compatible or interoperable with those used by other levels or fields, further hindering long-term holistic urban planning and management of metropolitan areas.

Infrastructure life span and complexity

The nineteenth century saw a culture of urban planning driven by goals of durability and continued use that generated a large percentage of today's urban critical infrastructure (e.g. railways, water and sewerage) (Hassler and Kohler, 2014). These are now legacy assets which are difficult to substitute due to their sunk cost and are expensive to maintain. City planning in the twentieth century was however largely driven by a culture of 'better new' delivering shorter infrastructure life spans and an

40 *Sanchez et al.*

expectation of replacement rather than maintenance. These two legacies come together with a lack of robustness and redundancies across infrastructure networks to increase the amount of irreparable construction (Hassler and Kohler, 2014). This is reflected in the US infrastructure scorecard produced by the American Society of Civil Engineers (ASCE). The estimated required investment needed to repair, maintain and, when irreparable, substitute existing infrastructure was USD3.6 trillion in 2013, with all but solid waste infrastructure being awarded a mediocre or poor rating. One of the key solutions proposed by ASCE was to promote sustainability, resilience and maintenance as a key part of infrastructure management strategies (Victor et al., 2013).

Urban environments are additionally inherently susceptible to cascading effects of changes occurring elsewhere in the urban network or system, and having a knock-on effect (Boyd and Juhola, 2015). This is often presented in the form of coupled natural-technical disasters (Birkmann et al., 2010). An example of this is the failure of linked critical and/or disaster-prevention infrastructure during and after an extreme weather event magnifying the effects of the natural disaster and making the response and recovery efforts more difficult. Some barriers to urban resilience policies are thus related to historical and infrastructural problems (Collier et al., 2013), as well as a lack of integration based on an understanding of the interactions between infrastructure networks. Hurricane Katrina in 2005 is perhaps one of the most well-known examples of this complexity compounding effect. It flooded 80 per cent of New Orleans, caused damages equivalent to 1.2 per cent of the US's gross domestic product (GDP) and displaced over 380,000 residents. The effects of this natural disaster were worsened by uncertainty about repairing levees, the complexity of the US governance system and an inadequate understanding of negative feedbacks between critical infrastructure networks, among other significant factors (Tainter and Taylor, 2014; Townsend et al., 2006).

In addition to the inherent complexity of these networks, they are commonly managed by independent entities who have little or no incentive to plan holistically and investigate feedbacks with other infrastructure types. This fragmentation often leads to each network being managed through specialised software platforms and internal standards which may not be compatible with those used by other networks. This means that even if incentives would exist for a holistic planning of a city's critical infrastructure, the legacy software and database may prove to be one more obstacle to overcome.

Emerging problems

In addition to the above, cities face other risks either in the form of direct physical harm, such as terrorist attacks, or in the form of, for instance, disruptions to public services through cyber-attacks, economic crisis, epidemics and conflicts which are perceived as imminent, emergent and catastrophic (Davoudi, 2016; Sharifi and Yamagata, 2014). The following section will explore how integrating information can help cope with emerging problems from a policy perspective.

Information integration for resilience policy

Information and its vertical and horizontal integration across organisational boundaries and software domains bring both challenges and opportunities for urban resilience and policy. Addressing differing information requirements for increasing adaptive capacity of local communities is one example of such challenges. The 'absence of a consistent information base and uncertainty about projected impacts and a lack of binding guidance on principles and priorities' is one of the limiting factors for effective climate change policy development and implementation highlighted by the IPCC's Australasian chapter (Reisinger et al., 2014). The work done by Amundsen et al. (2010) in Norway further supports these findings. This group identified 'unfamiliarity with and lack of data on climate change' as one of the main barriers to implementing effective climate change policies. They also highlight the need for access to knowledge about how policies can be implemented successfully.

The road towards resilient cities may require a process of transition which brings socio-technical networks[3] into the limelight. These have a significant impact on the resilience of the cities they are embedded in (Meerow et al., 2016). Considering these networks as part of the fabric that makes cities means reconceptualising 'cross-scale interactions as interdependencies between technical and social networks' (Ernstson et al., 2010). This emphasises that cities are formed by multiple networks (social, ecological and technical) whose functioning depends on the sharing of information, and it brings information integration and socio-technical systems and networks to the core of resilience thinking.

Conceiving cities as socio-eco-technical meta-systems has also consequences for hazard identification. The American Planning Association for example highlights that planners and policy-makers must consider a wide range of risks, from weather-related hazards to technological hazards. In the latter group, socio-technical interactions may cause pollution, terror, system failures, crime or economic risks that may affect the long-term resilience of the city (Coaffee and Bosher, 2008). In terms of opportunities and challenges, using information technologies to coordinate physical actions and infrastructure can be a key to developing smart and resilient network designs. In terms of planning, this implies allocating more funding to linking digital and physical infrastructure. Register (2014) exemplifies this point through an electronically coordinated electric power grid. Having a circular[4] and distributed model means that power can be sourced 'from several directions depending on availability in case some power sources go down or lines are cut, such as when wires are blown down with their poles by hurricanes'.

Melbourne's Chief Resilience Officer has recently emphasised socio-technical interactions to be at the centre of their Resilient Melbourne strategy. He mentions how these interactions can help make policy implementation more efficient and inclusive, for example by targeting traditionally marginalised residents such as the homeless community. Eighty per cent of Melbourne's

42 Sanchez et al.

homeless own a smartphone. This means that web-based applications that encourage a type of cohesive but decentralised self-organisation can serve as powerful tools to achieve more resilient cities and effective policies that encompass the most vulnerable of the urban population (Kent, 2016). An example of this is platforms that allow citizens to organise in a specific situation based on easily accessible information.

Adaptation is a continuous process and highly dependent on the sensitivity of the system to changes and feedbacks as well as the ability of actors to take action based on that information. High levels of sensitivity allow understanding when and where feedbacks occur, so adaptive action can be taken by those responsible in a timely manner (Adger et al., 2011). Urban governance that is more sensitive to system dynamics and proactive in adapting to uncertainties arising from socio-eco-technical interactions may help harness innovation within a politicised urban context to continually increase the resilience of the city (Ernstson et al., 2010).

The notion of sustainable resilience also points to the need for a cohesive and coordinated system and policy at the metropolitan level. In a city, the concept of cohesiveness is highly dependent on spatial and governance network scales. Without this cohesion, a component of the system such as a municipal or state agency may enact a policy program or initiative that may be effective for their scale but may also either reduce the capacity of other sectors to adapt or even increase the impact of disturbances on others (Adger et al., 2005). Information integration on the other hand enables self-organisation while maintaining a cohesive system.

Two main challenges for effective self-organisation that leads to enduring resilience are accessing information needed to prioritise interventions at different scales and being able to learn from policies as they are implemented. Here integration of information and actions across scales enables a more flexible environment that is conducive to adaptive policy. This means developing a capacity to solve local problems in a vertically and horizontally coordinated way, so individual component changes contribute to the overall resilience of the system (Bettencourt, 2015).

The following section presents two strategies related to information integration that could help enhance urban sustainable resilience.

Open data portals

Understanding what type of process may be progressing in, for example, asset management systems across different governance actors, such as municipal governments and state agencies, may help develop better-informed policies. These in turn may also promote a level of interoperability across networks that can give way to more integrated resilience strategies. Detailed information about the social, ecological and technological networks to be affected by urban interventions is greatly needed. However, accessing such information and datasets can be difficult and time-consuming (Bettencourt, 2015). Information availability therefore becomes central to effective and adaptive urban resilience policy.

Within this context, interoperability and accessibility are then key aspects of the long-term resilience of policies affecting a wide range of governance actors and infrastructure assets. Higher levels of accessibility to and the growth of large datasets (big data) can help strike a balance between short-term management processes and long-term needs, hence helping to develop policies that are more effective across all time horizons (Batty, 2013). Advances in information processing and dissemination may also facilitate collective learning and self-organisation (Godschalk, 2003).

Open data portals and ensuring interoperable relevant information is accessible to all pertinent stakeholder groups may therefore become part of resilience policies. New York City (NYC) has for example initiated their open data portal, where a wide variety of datasets from city government, health, education, industry, safety, environment, transport, social services, recreation and 'NYC BigApps' are available for download. This portal provides all city stakeholders with access to searchable information, such as street trees, in its basic format and through an interactive visualisation interface. This level of access has enabled a wide range of innovations in how the already available datasets are analysed, visualised and used for better public policy development. The New York Fire Department for example has used this data to develop a fire predictive model based on around 60 factors to carry out targeted inspections of buildings with the highest risk (NYC, 2016).

In Australia, the National Climate Change Adaptation Research Facility (NCCARF) recommended in 2013 creating an open-access national repository of data relevant to climate policy. This group suggested making available both raw data and 'data 'translation' functions' (Hussey et al., 2013). The Australian Government Data Portal contained, in 2016, over 9,000 datasets and was developed based on the Declaration of Open Government affirmed in 2010 (Australian Government, 2010; Australian Government, 2016). This includes most of the open datasets available at the state level. Individual state data portals also have additional functionalities and datasets. Victoria's open data portal for example includes a 'suggest a dataset' tool which enables the state government to approach the specific data owners and ask for the release of new datasets requested by the public. As of November 2016, this portal included 6,054 datasets (Victoria State Government, 2016).

The New South Wales (NSW) data portal has the 'apps4nsw' programme which is meant to 'encourage the use of NSW Government data to create innovative web and mobile applications' (NSW Government, 2016). The Western Australian Government's open data policy also includes the Shared Location Information Platform (SLIP) that can be used through Geographic Information Systems (GIS) (Government of Western Australia, 2016). This is similar to the South Australian Resource Information Geoserver (SARIG) (Government of South Australia, 2014). The Queensland Government additionally provides a listing of the open data strategies of each governing body. This lists datasets owned by each body and their release schedule as well as how each agency decides whether to release a dataset in response to changing data needs (Queensland Government, 2015).

44 *Sanchez et al.*

All of these efforts are steps forward in making available the information required for the development of better-informed, more sensitive, adaptive and cohesive resilience policies. However, there is still room for improvement. There are many datasets that are only available on request or released in long cycles, rather than all 'releasable' datasets being readily available and automatically published as they are generated. There are also no standards across datasets that ensure the same type of information is released in the same formats and with the same naming conventions. This means that people aiming to use different datasets to understand complex interactions may need to invest a significant amount of time and effort in unifying the naming conventions and classifications in order to carry out their analysis. Some datasets are also only available in pdf formats, further complicating the issue by adding the need to convert available data points into formats that can be used by data analysis software.

Standardised and comprehensive open data portals would not only allow governing bodies to develop more integrated, adaptive and better-informed policies, it would also allow citizens to become partners in this process. Effectively, it can increase the data processing and analysis capacity of policy-developing bodies in a more economically efficient and sustainable manner while also promoting self-organising processes in policy development that can contribute to its long-term resilience.

Data availability and transparency can further provide higher levels of sensitivity and help overcome adaptive policy challenges brought by a multi-level governance model. In Australia for example, one of the impediments to effective adaptation policy is the lack of timely availability of information, data and appropriate response strategies. This could be resolved by embedding mechanisms that allow reliable data to be made available as it is created so decision-makers and other relevant groups can access the data when needed and, importantly, in a form they can understand. The NCCARF has additionally recommended that Australia makes mandatory that all federally funded data is made available through a centralised repository, to which state and local organisations can also contribute. In this case, they also recommend legal indemnity for contributing parties. This, they argue, would allow public and private organisations to better align their incentives with policy strategies (Hussey et al., 2013).

Information and communication technologies can play an important role in ensuring processes to make data availability become standardised and facilitating the logistics of collecting and sharing this data (Bettencourt, 2015). This level of sensitivity, however, is only useful if actors have the ability to act based on this feedback. 'Without this ability there is no capacity for learning and for changing actions in the future' (Adger et al., 2011).

Urban information modelling

Cities are complex systems constituted of physical elements interrelated into elaborated spatial relations, with a complexity increasing as the shape and the structure are changing and evolving. To try to understand the dynamics

Resilient policies for wicked problems 45

and the processes shaping our cities, we must coherently make models according to the dynamic and complex nature of cities, but also models should remain understandable and simple enough to be operationally useful … Enriched 3D city models are the foundation bricks of future Smart Cities.

(Billen et al., 2015)

The urban 'lifestyle' and economic activities have been estimated to generate 70–80 per cent of the global greenhouse gas emissions which are driving climate change (Birkmann et al., 2010). Urban planning of sustainably resilient cities will therefore need to understand the interplays between climate change, human actions and future risk. This would allow urban planners to address future threats to urban settlement in a proactive manner to avoid urban communities living under constant threat with increasing vulnerabilities (Jabareen, 2013). Addressing wicked problems such as climate change requires implementing a mixture of measures that both attack the root-causes of future risks and adapt to residuals of the problems (Register, 2014). The same is true for challenges associated with infrastructure networks and technological progress. In order to develop such a comprehensive resilience policy, decision-makers would need to have access not only to information but also platforms that allow them to understand the interplays between urban networks and influencing factors.

Building Information Modelling (BIM) is a global phenomenon which is gaining significant momentum across the world … is a socio-technical system that extends to emerging technological and process changes within the architecture, engineering, construction and operations industry … it promises the ability to create models that combine data which was traditionally spread across multiple documents and databases along with the ability to share information between different models.

(Sanchez et al., 2016a)

The use of BIM for urban asset planning, design, construction and operations has been increasing over the last two decades, with many countries recently mandating its use (Sanchez et al., 2016a). Although so far most efforts have focused on single-structure modelling, this process can be applied to larger urban settings, linking currently disparate datasets through a single 3D model interface. Germany for example has been working towards standardised city models and the UK towards linking BIM models to the national survey map (Månsson and Lindahl, 2016).

There has been considerable research into using 3D city models based on GIS spatial query functionalities and whole-city BIM models; see for example Döllner and Hagedorn (2007), Borrmann (2010), Cheng et al. (2013) and Padsala and Coors (2015). In more recent times, the idea of urban information modelling has also emerged as the potential result of vertical and horizontal integration of information across governing bodies, private institutions and residents. This integration would create a complex urban knowledge network interfacing through

46 *Sanchez et al.*

a semantically enriched 3D model. 'This would also help to become more effective and efficient in the way of providing policies for city management and urban planning suited to the best interests of their citizen' (Billen et al., 2015).

Early attempts to develop such interfaces include the 3D viewer developed by Döllner and Hagedorn (2007) which integrated computer-aided design (CAD), GIS and BIM data. The idea of this tool was to serve as a unified virtual and graphic framework to access disaggregated and heterogeneous datasets in an ad hoc manner. Having a single 3D interface integrating all information modelling data related to a city, being CAD, BIM and GIS compatible, would provide decision-makers with a powerful tool to understand feedbacks between networks and overarching trends. They could then use this knowledge as a basis to develop more coordinated and cohesive policy strategies (Lapierre and Cote, 2007). The Newton et al. chapter in this book (Chapter 7) explores a more recent digital platform being developed in Australia by the Cooperative Research Centre for Low Carbon Living (CRC LCL) called Precinct Information Modelling (PIM).

These models could be used for integrating large sets of data from a wide range of sources and levels of detail. Such a platform would become an urban analytical tool through which individual governing bodies can draw information from different local or regional fields to improve their local planning (Billen et al., 2015). This would promote a type of self-organisation that maintains cohesiveness and allows for more effective coordination across governance and spatial scales.

For this vision to be successful, however, many changes will be required in order to achieve the level of interoperability and standardisation required to efficiently build an urban information model. These should be more than a simple geometric representation of the urban built environment. They would be 3D dynamic representations of physical and abstract objects, allowing to combine different information sources through a single interface. This will require integration platforms to be developed, defining open data standards and standardisation priorities that promote interoperability between models (e.g. CityGML and Industry Foundation Classes (IFC)), developing data production and integration strategies, developing methods to evaluate the model's usability, and developing and promoting data standard compliance systems, among others (Billen et al., 2015).

Conclusions: the road ahead

Cities are the centre of most nations' productivity and population growth. They face numerous complex challenges that will require more integrated strategies for city planning and management. This chapter has highlighted how information integration is beginning to move to the centre of resilience policy challenges and opportunities. However, in order to leverage existing processes and resources to deliver more effective, efficient and long-term policy action, several areas require further research and agreement.

Beyond base research into resilience indicators and tools, more general efforts are also required to promote information integration across the built environment that can help create more resilient cities and policies. In both cases presented in this chapter, open data portals and urban information modelling, standardisation and transparency need to exist in order to develop and use these technologies in an efficient manner. The UK Government has for example developed a strategy to achieve their vision of Digital Built Britain. This strategy includes providing funding to develop international open data standards that can be used to share data more easily across networks. Areas where they expect to pool resources and budget for cross-sector collaborations include 'standard Internet of Things protocols and resilience capabilities' and 'social media interfaces and integration' (HM Government, 2015).

These kinds of initiatives are important because without standards governing the interface and overall framework of data being created, released and used in a dynamic environment, the cost associated with integrating and administering new sources could quickly become astronomical (Roth et al., 2002). Additionally, research is also required into processes that may help achieve wider use of such open standards across all urban data producers as well as into governance and policy-making processes that are more sensitive and adaptive so actions can be taken based on the information available.

Another area of research that would help advance this vision is about how information and communications technologies (ICT) can help automate processes involved in the production, collection, validation and sharing of datasets that may be relevant to resilience policy development and implementation efforts.

On the policy side, efforts that promote the use of common standards and data protocols – such as naming conventions – across networks, and their release into a common repository would also help leverage greater levels of digitisation and integration already being observed. This could start with the creation of inventories of data and information availability, ownership and released format interoperability.

The creation of dynamic platforms where individuals can access the data and information available as required to understand urban trends and prioritise actions as well as interact with other actors is also needed.

Notes

1 Urban settlements refer to systems formed by a conglomeration of ecological, social and technical components. These form socio-technical, socio-ecological and eco-technological networks, where each component and their networks are dynamically changing and interacting with each other in often unpredictable ways.
2 This refers to ten interviews with resilience experts from government, industry and academia carried out between May and August 2016 by one of the authors of this chapter and to be published in 2017. The interviewed experts had at least ten years of experience in the fields of resilience and sustainability policy, and have current or recent roles advising resilience policy.

48 *Sanchez et al.*

3 These are networks where technologies and technological functions closely interact with social functions and social interests (Hodson and Marvin, 2010; Kling et al., 2003).
4 Circular refers to systems that are not structured as traditional radial grids where branches flow from a single central power source (Register, 2014).

References

100 Resilient Cities, 2015. *What is Urban Resilience?*. Available at: www.100resilientcities. org/resilience#/-_/ [Accessed 17 August 2015].

Adger, W. N., Arnell, N. W. and Tompkins, E. L., 2005. Successful adaptation to climate change across scales. *Global Environmental Change*, 15(2), 77–86.

Adger, W. N., Brown, K., Nelson, D. R., Berkes, F., Eakin, H., Folke, C., Galvin, K., Gunderson, L., Goulden, M., O'Brien, K., Ruitenbeek, J. and Tompkins, E. L., 2011. Resilience implications of policy responses to climate change. *WIREs Climate Change*, 2(September/October), 757–766.

Agudelo-Vera, C. M., Mels, A. R., Keesman, K. J. and Rijnaarts, H. H., 2011. Resource management as a key factor for sustainable urban planning. *Journal of Environmental Management*, 92(10), 2295–2303.

Amundsen, H., Berglund, F. and Westskog, H., 2010. Overcoming barriers to climate change adaptation: A question of multilevel governance? *Environment and Planning C: Government and Policy*, 28, pp. 276–289.

Anguelovski, I. and Carmin, J., 2011. Something borrowed, everything new: Innovation and institutionalization in urban climate governance. *Current Opinion in Environmental Sustainability*, 3(3), 169–175.

Australian Government, 2010. *Declaration of Open Government*. Available at: www. finance.gov.au/blog/2010/07/16/declaration-open-government/ [Accessed 8 March 2017].

Australian Government, 2016. *Datasets*. Available at: www.data.gov.au/dataset [Accessed 24 August 2016].

Batty, M., 2013. Big data, smart cities and city planning. *Dialogues in Human Geography*, 3(3), 274–279.

Bettencourt, L. A., 2015. Cities as complex systems. In: B. A. Furtado, P. A. Sakowski and M. H. Tóvolli (eds.) *Modeling Complex Systems for Public Policies*. Brasília: IPEA, pp. 217–236.

Billen, R., Cutting-Decelle, A.-F., Métral, C., Falquet, G., Zlatanova, S. and Marina, O., 2015. Challenges of semantic 3D city models: A contribution of the COST research action TU0801. *International Journal of 3-D Information Modeling*, 4(2), 68–76.

Birkmann, J., Garschagen, M., Kraas, F. and Quang, N., 2010. Adaptive urban governance: New challenges for the second generation of urban adaptation strategies to climate change. *Sustainability Science*, 5(2), 185–206.

Borrmann, A., 2010. *From GIS to BIM and Back Again: A Spatial Query Language for 3D Building Models and 3D City Models*, paper presented at 5th International 3D GeoInfo Conference, Berlin, Germany, 3–4 November.

Boyd, E. and Juhola, S., 2015. Adaptive climate change governance for urban resilience. *Urban Studies*, 52(7), 1234–1264.

Cheng, J., Deng, Y. and Du, Q., 2013. *Mapping between BIM Models and 3D GIS City Models of Different Levels of Detail*, Proceedings of the 13th International Conference on Construction Applications of Virtual Reality, London, UK, 30–31 October.

Coaffee, J. and Bosher, L., 2008. Integrating counter-terrorist resilience into sustainability. *Proceedings of the ICE: Urban Design and Planning*, 161(2), 75–83.

Collier, M. J., Nedović-Budic, Z., Aerts, J., Connop, S., Foley, K., Newport, D., McQuaid, S., Slaev, A. and Verburg, P., 2013. Transitioning to resilience and sustainability in urban communities. *Cities*, 32, S21–S28.

Davoudi, S., 2016. Resilience and governmentality of unknowns. In: M. Bevir (ed.) *Governmentality after Neoliberalism*. London: Routledge, in press.

Döllner, J. and Hagedorn, B., 2007. Integrating urban GIS, CAD, and BIM data by servicebased virtual 3D city models. In: M. Rumor, V. Coors, E. M. Fendel and S. Zlatanova (eds.) *Urban and Regional Data Management-Annual*. London: Taylor and Francis, pp. 157–160.

Ernstson, H., van der Leeuw, S. E., Redman, C. L., Meffert, D. J., Davis, G., Alfsen, C. and Elmqvist, T., 2010. Urban transitions: On urban resilience and human-dominated ecosystems. *Ambio*, 39(8), 531–545.

Godschalk, D. R., 2003. Urban hazard mitigation: Creating resilient cities. *Natural Hazards Review*, 4(3), 136–143.

Government of South Australia, 2014. *SARIG*. Available at: http://data.sa.gov.au/data/showcase/sarig [Accessed 29 February 2016].

Government of Western Australia, 2016. *SLIP*. Available at: http://data.wa.gov.au/slip2 [Accessed 29 February 2016].

Hassler, U. and Kohler, N., 2014. The ideal of resilient systems and questions of continuity. *Building Research & Information*, 42(2), 158–167.

Hjorth, P. and Bagheri, A., 2006. Navigating towards sustainable development: A system dynamics approach. *Futures*, 38(1), 74–92.

HM Government, 2015. *Digital Built Britain. Level 3 Building Information Modelling: Strategic Plan*, London: HM Government.

Hodson, M. and Marvin, S., 2010. Can cities shape socio-technical transitions and how would we know if they were? *Research Policy*, 39, 477–485.

Hussey, K., Price, R., Pittock, J., Livingstone, J., Dovers, S., Fisher, D. and Hatfield-Dodds, S., 2013. *Statutory Frameworks, Institutions and Policy Processes for Climate Adaptation: Do Australia's Existing Statutory Frameworks, Associated Institutions and Policy Processes Support or Impede National Adaptation Planning and Practice?* Gold Coast: National Climate Change Adaptation Research Facility.

Jabareen, Y., 2013. Planning the resilient city: Concepts and strategies for coping with climate change and environmental risk. *Cities*, 31, 220–229.

Kent, T., 2016. *Advancing our City with People at its Heart*. Available at: www.100resilientcities.org/blog/entry/advancing-our-city-with-people-at-its-heart#/-_Yz5jJmg%2FMSd1PWI%3D/ [Accessed 24 February 2016].

Kling, R., McKim, G. and King, A., 2003. A bit more to it: Scholarly communication forums as socio-technical interaction networks. *Journal of the American Society for Information Science and Technology*, 54(1), 47–67.

Lapierre, A. and Cote, P., 2007. Using Open Web Services for urban data management: A testbed resulting from an OGC initiative for offering standard CAD/GIS/BIM services. In: V. Coors, M. Rumor, E. Fendel and S. Zlatanova (eds) *Urban and Regional Data Management. Annual Symposium of the Urban Data Management Society*. London: Taylor & Francis, pp. 381–394.

McEvoy, D., Fünfgeld, H. and Bosomworth, K., 2013. Resilience and climate change adaptation: The importance of framing. *Planning, Practice & Research*, 28(3), 280–293.

50 Sanchez et al.

McIlwain, J. K., Azrack, J., Ricci, D. M., Angelides, P. A., Brandes, U. S., Brookman, M. D., Brown, J., Carey, K., Cox, T., Ford, K., Hager, C. M., Horowitz, S., Lam, D., Lashbrook, W. and Lowe, St., 2013. *After Sandy: Advancing strategies for long-term resilience and adaptability*, Washington, USA: ULI Foundation.

Madslien, J., 2013. *Davos 2013: 'Dynamic Resilience' in a Volatile World*. Available at: www. bbc.com/news/business-21086431 [Accessed 18 August 2016].

Månsson, D. and Lindahl, G., 2016. BIM performance and capability. In: A. X. Sanchez, K. D. Hampson and S. Vaux (eds) *Delivering Value with BIM: A Whole-of-life Approach*. London: Routledge, pp. 46–57.

Meerow, S., Newell, J. P. and Stults, M., 2016. Defining urban resilience: A review. *Landscape and Urban Planning*, 147, 38–49.

New South Wales (NSW) Government, 2016. *Data NSW*. Available at: http://data.nsw. gov.au/ [Accessed 29 February 2016].

NYC, 2016. *NYC Open Data*. Available at: https://nycopendata.socrata.com/ [Accessed 23 February 2016].

Padsala, R. and Coors, V., 2015. *Conceptualizing, Managing and Developing: A Web Based 3D City Information Model for Urban Energy Demand Simulation*, paper presented at Eurographics Workshop on Urban Data Modelling and Visualization, Delft, 23 November 2015.

Queensland Government, 2015. *Open Data Strategies*. Available at: https://data.qld.gov. au/article/department-strategies [Accessed 6 March 2017].

Register, R., 2014. Much better than climate change adaptation. In: S. Lehmann (ed.) *Low Carbon Cities: Transforming Urban Systems*. Abingdon, Oxford, UK: Routledge, pp. 75–84.

Reisinger, A., Kitching, R., Chiew, F., Hughes, L., Newton, P., Schuster, S., Tait, A. and Whetton, P., 2014. Chapter 25. Australasia. In: *Working Group II contribution to the IPCC Fifth Assessment Report Climate Change 2014: Impacts, Adaptation, and Vulnerability*. Stanford: Intergovernmental Panel on Climate Change.

Roth, M. A., Wolfson, D. C., Kleewein, J. C. and Nelin, C. J., 2002. Information integration: A new generation of information technology. *IBM Systems Journal*, 41(4), 563.

Sanchez, A. X., Hampson, K. D. and Vaux, S., 2016a. *Delivering Value with BIM: A Whole-of-life Approach*. London: Routledge.

Sanchez, A. X., van der Heijden, J., Osmond, P. and Prasad, D., 2016b. *Urban Sustainable Resilience Values: Driving Resilience Policy that Endures*, paper presented at CIB World Building Congress, Tampere, Finland, 30 May–2 June.

Sharifi, A. and Yamagata, Y., 2014. Resilient urban planning: Major principles and criteria. *Energy Procedia*, 61 (The Sixth International Conference on Applied Energy – ICAE2014), 1491–1495.

Tainter, J. A. and Taylor, T. G., 2014. Complexity, problem-solving, sustainability and resilience. *Building Research and Information*, 42(2), 168–181.

Tapscott, D., 2013. *World Economic Forum: Creating a Dynamic, Resilient World*. Available at: www.huffingtonpost.com/don-tapscott/world-economic-forum-crea_b_2535293. html [Accessed 2016 August 18].

The World Bank, 2016. *Population Growth (Annual %) and Urban Population Growth (Annual %)*. Available at: http://data.worldbank.org/indicator/SP.POP.GROW?end= 2015&start=1978 [Accessed 28 February 2017].

Townsend, F. et al., 2006. The Federal Response to Hurricane Katrina: Lessons Learned. In: *Chapter Five: Lessons Learned*. Washington: The White House.

UN, 2014. *World Urbanization Prospects – 2014 Revision*, s.l.: United Nations.

Urwin, K. and Jordan, A., 2008. Does public policy support or undermine climate change adaptation? Exploring policy interplay across different scales of governance. *Global Environmental Change*, 18(1), 180–191.

Victoria State Government, 2016. *Data.vic*. Available at: www.data.vic.gov.au/ [Accessed 23 February 2016].

Victor, R., Baskir, G., Bennett, J., Camp, J., Capka, R., Curtis, S., Davids, G., Frevert, L., Hatch, H., Herrmann, A., Hookham, C., Howe, F., Iarossi, B., Jacobson, D., Kito, S., Lehman, M., Lynch, O., Matin, S., May, J., McKeehan, B., Merfeld, P., Millar, R. and Taylor, P., 2013. *Report Card for American Infrastructure*, s.l.: American Society of Civil Engineers.

4 Internet of Things for urban sustainability

Fonbeyin H. Abanda and Joseph H. M. Tah

Introduction

The Internet of Things (IoT) represents the rapidly growing number of internet-enabled devices that can communicate with each other and have the ability to transmit and receive data over a network. The IoT is being applied in a diversity of areas to address significant societal challenges. For example, it has been widely acknowledged that cities are responsible for two-thirds of the world's energy and resource consumption, including 70 per cent of all greenhouse gas emissions. This is leading to a growing interest in the use of the IoT in improving the sustainability of cities. This has spawned applications in many areas typically associated with so-called intelligent buildings and smart cities. At the building level, it has been used in making buildings more intelligent to reduce energy consumption and carbon emissions. At the urban level, it has been used to improve the quality of life of city dwellers through measures that promote an eco-friendly or sustainable environment. Perhaps partly because of the nascent nature of the IoT, the exact functioning of these technologies is still very vague and there is only a limited understanding of the benefits as well as challenges associated with IoT.

This chapter explores the use of the IoT in improving the sustainability performance of cities. It also presents an overview of IoT-enabling technologies such as barcodes, radio frequency identification (RFID) and sensors typically used in the built environment. The chapter later examines the application of these technologies in selected areas such as energy management, environmental monitoring and waste management intended to improve the sustainability performance of cities. Selected case studies from two leading smart cities, Singapore and Barcelona, are examined in detail to establish the extent of the application of the technologies and identify the anticipated benefits. The chapter concludes with a discussion about IoT-related applications and the challenges as well as opportunities of implementing them more widely.

Sustainability and the smart built environment

The environmental impacts of the built environment have long been acknowledged and are no longer considered new. According to the recent

Intergovernmental Panel on Climate Change (IPCC), in 2010, the building sector accounted for 32 per cent of the world's energy use and around a third of 'black carbon emissions'; emissions from fossil fuels (Lucon et al., 2014). The IPCC projected that by 2050 energy demand in the building sector will double and carbon dioxide (CO_2) emissions will increase by somewhere between 50 and 150 per cent (Lucon et al., 2014).

The need to curb greenhouse gas emissions and minimise or eliminate material and/or energy waste from the construction industry has been top of the agenda of many governments for some time now. The United Kingdom (UK) government for example required in 2013 that the sector lower its emissions by 50 per cent by 2025 (HM Government, 2013). In 2015, a cohesive and legally binding agreement to hold global warming below 2ºC and strive towards keeping it below 1.5ºC was confirmed at the Conference of the Parties (COP) 21 in Paris (Graham, 2015).

The onus is now on various governments to act and implement strategies to achieve the set targets. With the technology, know-how and support platforms available and proven in the field, the built environment has one of the most cost-effective mitigation potential. Previous efforts in improving the environmental performance of this industry had focused on adopting innovative techniques such as off-site manufacturing, renewable energy, sustainable building materials and effective waste management strategies, among others. From an information management perspective, compared to technologies, the role of data management and intelligent data consumption for sustainability has however not received similar levels of research interest.

The built environment sector is a data-intensive domain. As argued by Egan (1998), an average car contains about 3,000 components compared to a house that has about 40,000 components. In addition, a house or building contains devices added by occupants to maintain a certain level of comfort. The situation is similar in infrastructure projects; they are made of components, including materials, contain devices that are characterised by several properties and generate large amounts of data. The challenges of dealing with these datasets are compounded by the way they are used, including the behaviour of users that are too often unpredictable.

Today, data is everywhere and generated by almost any one thing or process at all scales. The advent of information technologies such as the internet, closed-circuit television (CCTV), sensors and smart devices have made real-time collection of built environment data easier. The key challenge is how this voluminous amount of data can be used to make informed decisions to improve services in the built environment; 'Can built environment professionals have real-time access to data during the delivery and operation processes for making critical decisions?' (Sawnhey, 2015).

Existing decision-making systems have dealt with data as disparate entities according to the different domains of interest. The advent of the internet has revolutionised the sharing and exchange of information between computers over different geographical locations. Recent technologies, including social media, provide even greater opportunities to build on the strength of the internet.

54 *Abanda & Tah*

Connecting physical objects, humans and the internet takes data integration and decision-making to unimaginable heights. This will undoubtedly support the deployment of IoT. This chapter investigates the use of IoT in improving the sustainability performance of cities.

Internet of Things – what is it?

The IoT is a paradigm where objects, animals, people or anything are provided with unique identifiers and the ability to transfer data over a network without requiring human-to-human or human-to-computer interaction (Miranda et al., 2015). The term IoT might have been originally coined by Kevin Ashton (Ashton, 1999) to describe the vision of internet-connected sensors, devices and citizens. Since then, many other terms have been used. The term Internet of People (IoP) has been used to mean internet-enabled personal electronics (Miranda et al., 2015). In the United States (US), the Chicago smart city project is called Array of Things (AoT) (AoT, 2016). Although the backbone of AoT is the IoT, it should be considered as a project and not an equivalent or synonym of IoT. Cisco used 'Internet of Everything' (IoE) to mean connecting people, processes, data and things to respond to the need for real-time, context-specific information intelligence and analytics to address specific local imperatives for cities (Mitchell et al., 2013). It is not clear as to how IoT differs from IoE. Consequently, IoT will be used to mean both concepts.

The IoT relies on unique markers that have uniform resource identifier (URI) connected to the Web. As discussed in the previous section, the key to IoT is connecting objects, and hence their data, to the Web, in real-time for well-informed decision-making. Existing technologies such as barcodes, RFID and sensors can serve as the foundation on which to build the IoT. A barcode is an electronic data interchange medium that contains machine-readable dichromatic marks that encode information for object labelling using an arrangement of geometric symbols (Lu et al., 2013). A barcode uses a labelling system composed of a pattern of light and dark bars producing a graphical form of binary logic which can be scanned and interpreted by computers. RFID is a generic term for technologies that use radio waves to automatically identify a person, object or other information (Violino, 2005). An RFID is composed of four components (Violino, 2005):

- a transponder (more commonly just called a tag) that is programmed with information that uniquely identifies itself, thus the concept of 'automatic identification';
- a transceiver (more commonly called a reader) that handles radio communication through the antennas and pass tag information to the outside world;
- an antenna attached to the reader to communicate with transponders;
- a reader interface layer, or middleware, which compresses thousands of tag signals into a single identification and also acts as a conduit between the

RFID hardware elements and the client's application software systems, such as inventory, accounts receivable, shipping, and logistics.

RFID is widely used in the built environment industry. The RFID tags can be attached to bags of waste and key 'gates' be installed with RFID readers. The system can then collect the movement data of any waste container and a central server would store the data for use in the knowledge management system (Zhang et al., 2012). A sensor is a device or an object that detects events or changes in its environment or surroundings and then provides corresponding feedback.

Other technologies used for object identification that have been used for IoT are biometrics (Pau and Mihailescu, 2015), optical character recognition (OCR) (Quack et al., 2008) and smart cards (Jing et al., 2014). Existing smart phones have also been empowered with capabilities that facilitate the interaction of humans with the environment. These devices offer integrated cameras and a wide range of communication facilities such as Bluetooth, wireless local area network (WLAN) or access to the internet (Vinciarelli and Odobez, 2006). Nowadays, it is quite common that mobile phones constitute one of the devices people carry along anywhere they go. A recent study by the Pew Research Center revealed the global median ownership rate for smartphones stood at 43 per cent in 2015 (Poushter, 2016). Smartphone ownership in 2015 was 72 per cent in the US, 77 per cent in Australia and 88 per cent in South Korea. While the ownership rate for emerging and developing countries may be lagging behind that of developed countries, its recent surge has been extraordinary, rising from a median of 21 per cent in 2013 to 37 per cent in 2015 (Poushter, 2016). This trend will undoubtedly support the deployment of IoT; phones contain numbers tied to specific individuals. It is therefore natural to use a mobile phone as a personal input device for the IoT (Vinciarelli and Odobez, 2006).

Applications of IoT for sustainability in urban environments

Recently, there has been a surge in the use of IoT for improving the sustainability of cities. Sustainability is complex and covers many areas, and so does the application of IoT; they cover numerous and disparate organisations, communities and the society at large, cutting across all areas of human life. Stemming from the original definition of sustainable development in the *Our Common Future* report, also known as the Brundtland Report, Allen (2001) proposed five dimensions of urban sustainability. These are: economic, political, social, physical and ecological sustainability. The focus of this chapter is on ecological sustainability, defined by Allen (2001) as

> The impact of urban production and consumption on the integrity and health of the city region and global carrying capacity. This demands the long term consideration of the relation between the state and dynamics of environmental resources and services and the demands exerted over them.
>
> (Allen, 2001)

56 *Abanda & Tah*

To facilitate understanding, the different applications of IoT are examined under major themes including energy and renewable energy, waste and greenhouse gas emissions management and monitoring.

Energy management

Globally, lighting accounts for about 19 per cent of the total energy consumed in buildings (The Climate Change Group, 2012). Also, global emissions from buildings stand at about 6 per cent of total greenhouse gases (Castro et al., 2013). Furthermore, it has been argued that greenhouse gas emissions from lighting are equivalent to about 70 per cent of the emissions from the world's passenger vehicles (The Climate Change Group, 2012). Recent stringent polices especially in Western countries, such as efficiency standards, labels, subsidies and technological progress, have led to significant improvements in lighting technologies, thereby leading to significant energy savings and reduced emissions (Lucon et al., 2014). Improved technologies have led to typical saving potentials of the range of 40–50 per cent (Lucon et al., 2014). Coupling the improved lighting technologies with smart sensors will further improve their performance leading to significant energy savings. Linking and monitoring energy-consumption patterns and occupants' behaviour can be exploited for reducing greenhouse gases, operational energy and cost. For example, coupling light-emitting diode (LED) technologies with smart control systems can lead to up to 80 per cent energy savings (The Climate Change Group, 2012).

Renewable energy management

The distributed renewable energy resources are one of the major smart grid enablers that can be installed to supplement power during peak hours (Al-Ali and Aburukba, 2015). Each renewable technology can be assigned an Internet Protocol (IP) address and set up to upload its status and download control commands via the internet.

Greenhouse gas emission monitoring

Linking vehicles, commuter traffic and emissions to air quality can give traffic management the 'right' data to manage road guidance and parking. Costabile and Allegrini (2008) developed a system that links transport to air quality which automates the computation of transport-related air pollution. The system can recommend pollutant vehicles to be banned in certain areas should the pollutant level exceed a set threshold.

Waste management

Waste collection and disposal has been a problem for urban communities for generations. With regards to municipal waste, local authorities often adopt fixed

Internet of Things for sustainability   57

schedules for collecting waste from neighbourhoods. A major weakness with this approach is that bin lorries dispatched to neighbourhoods may find bins either overloaded or not full to capacity. Systems that capture the exact amount of waste in bin containers are therefore lacking. Sensors tagged to waste bins can allow real-time transmission of data to the local authorities to alert them that the bin is full and needs to be emptied. These types of bins are called *smart bins*. The use of intelligent waste containers or smart bins, which detect the level of load and allow for an optimisation of the collector trucks' route, can reduce the cost of waste collection and improve the quality of recycling (Zanella et al., 2014).

Rova, a Dutch waste management firm has developed IoT applications for gathering information from wheelie bins and then employing data analytics to predict which bins are likely to be full. The technology integrates a route optimisation software that guides the garbage trucks to the bins that most need emptying, avoiding those that are not yet full. The IoT system has led to about 20 per cent operational cost savings (Clark, 2015). Hong et al. (2014) developed a smart garbage bin that was operated as a pilot project in Gangnam district, Seoul, Republic of Korea, for a one-year period. The experiment showed that the average amount of food waste could be reduced by 33 per cent.

With regards to on-site construction, Zhang et al. (2012) developed a smart waste management system that does not only provide full logistical records for waste transportation but also provides waste collection arrangements and incident handling guidance to both management and operational staff. Some selected IoT applications for built environment sustainability have been summarised and presented in Table 4.1.

Table 4.1 Applications of IoT-based technologies for sustainability

Purpose	Description	Sources
Barcodes		
On-site construction waste management	Barcode-based system for tracking arrival time of construction M&E so as to deal with limitations of M&E storage on site.	Li et al. (2005)
	Barcode system to register the flow of materials so that performances of working groups in terms of material wastage can be easily measured.	Chen et al. (2002)
Waste recycling	A cloud-based barcode system developed to support the waste electrical and electronic equipment and used electrical and electronic equipment (UEEE).	Wang et al. (2014)
RFID		
Municipal waste management	An RFID system to support waste management service providers (e.g., municipalities, waste collectors) in tracking a waste identity (i.e., customer), weight of and missing/stolen bins quickly and accurately without human intervention.	Chowdhury and Chowdhury (2007)

58   *Abanda & Tah*

Table 4.1 continued

Purpose	Description	Sources
Hazardous waste management and tracking	RFID tags are embedded on hazards packages of hazardous waste for tracking to avoid transportation and disposal in wrong destination.	Namen et al. (2014)
Solid waste management	RFID antenna and reader on waste bins rather than the truck. As soon as a recyclable material is put into the bin it can be identified and the data relayed through a central system to a scrap dealer, and other related parties such as internet-based sales services. Thus, the recyclables could be picked out from the waste bin prior to general pick up.	Abdoli (2009)
	The smart waste management system provides full logistical records for waste transportation, waste collection arrangements and incident handling guidance to both management and operational staff.	Zhang et al. (2012)
Biomonitoring of green trees	RFID tags are embedded in or on trees in order to guarantee a real-time data communication. Data from pollution sensitive plants may be sent via wireless signalling concerning their management or environmental status. RFID tags help maintain diversity in the street tree population, assess the health of the urban forest, and communicate with property owners. Considering that inventories need to be updated regularly in order to help schedule tree maintenance work, determine planting sites, and manage invasive insects, RFID tags can be used as a safe system for tree identifications.	Luvisi and Lorenzini (2014)
Sensors		
Renewable energy harvesting	Constitute one of the components of an energy harvesting system from renewable sources.	Ferdous et al. (2016)
Energy efficiency of street lighting	A real-time adaptive lighting scheme detects the presence of vehicles and pedestrians and dynamically adjusts their brightness to the optimal level.	Lau et al. (2015)
Predicting the performance of building components	In order to predict the future performance of building components, it is useful to initially create a digital 3D model and connect it to sensors to see if the responses of the components are appropriate. Visual scripting tools are one method for bridging the hardware/software gap between sensors and 3D modelling software.	Kensek (2014)
Energy consumption	Sensors are embedded in meters or power plug meters for temporal tracking of energy consumption per item.	Martani et al. (2012)

Case studies

The use of IoT in energy and renewable energy management, greenhouse gas emission monitoring and waste management greatly relies on wireless devices such as RFID and wireless sensor networks to gather real-time data from 'things'. The following sections examine two in-depth case studies to illustrate how IoT has been applied at the city level. The case studies considered are Barcelona Smart City and Smart Nation Singapore, which are arguably among the best in the world.

Barcelona Smart City, Spain

The Barcelona Smart City programme[1] employs innovative solutions to manage services and resources for improving the life of its citizens. The key themes covered by the project are environmental, information and communications technology (ICT), mobility, water, energy, matter (waste), nature, public space, open government, information flows and services. Some of the projects that fit into the aforementioned themes are smart lighting, smart parking, smart water and waste management. The 'Internet of Everything' is the backbone of the project with over 500 km of underground fibre-optic networks installed progressively over the years. The main goal of the project was to improve efficiency of city services and to address sustainability and environmental concerns.

Sustainability

The project covers many areas of sustainability ranging from environmental pollution monitoring and management, to energy efficiency and waste management. However, based on the project's website and other publications, there is paucity of information as to which IoT technology has been used for different sustainability areas. The few IoT technologies that were uncovered were:

- energy saving (smart lighting; sensors): lighting system consisting of LEDs with sensors to regulate the intensity of lighting as required;
- waste management (smart bins; RFID): introduction of ultrasonic sensors and RFID devices in bins and containers to create a cheaper and more sustainable system of waste collection;
- environmental monitoring (sensors):
  - smart noise control (sensors): implementation of a solution that will optimise and centralise the collection, integration, processing and dissemination of information provided by the noise sensors of different suppliers; about 50 sensors distributed around the city to provide real-time information generated from sensors used for monitoring noise levels and other forms of environmental pollution;

- a system of sensors has been used for remote recording and monitoring of the environmental impact of the civil works in Avinguda de l'Estatut, one of the regions in Barcelona (Gea et al., 2013).
- water management (tele-management): sensor system that integrates temperature, salinity and humidity sensors that make it possible, through an irrigation controller, to send the necessary information to a management system.

Benefits

There are many qualitative and quantitative benefits reported by the projects related to the use of IoT. Some examples of quantitative benefits include the use of smart water technologies leading to a total saving of EUR42.5 million per year, 33 per cent or EUR36.5 million[2] in lighting and parking management (Boulos and Al-Shorbaji, 2014).

Smart Nation Singapore, Singapore

This project[3] brings together government and private actors with the goal of employing innovative technologies to improve the lives of citizens and create greater business opportunities. The different entities involved are the country's world-class universities and medical facilities.

Sustainability

The project addresses several aspects of sustainability including energy efficiency, waste management and environmental monitoring for pollution control through the use of the following technologies:

- energy saving (smart lighting; sensors): installation of sensors that learn human traffic flow and optimise the provision of lighting in common areas;
- water management (sensors): installation of sensors on the water supply network for collecting real-time data for the modelling and the analysis of water supply operations;
- environmental monitoring:
  - eight buoy-based monitoring stations with sensors for forecasting the onset of pollution;
  - enhancement of existing telemetric air quality monitoring network for monitoring air quality;
  - use of water level sensors and CCTV to monitor major drains and canals especially during intense rainstorms.

It is important to note that 1,000 sensors have been rolled out under Smart Nation Platform Phase 1.

Benefits

So far, quantitative benefits of the IoT applications for Singapore Smart Nation project are scarce. However, it is estimated that upon the completion of the project, the following benefits will be accrued:

- the new desalination testing technology for seawater will be 50 per cent more energy efficient than any current method (CIVIQ, 2013);
- seventy per cent of all journeys will be by public transport by 2030 (EDB, 2014);
- there will be a 35 per cent improvement of energy intensity level from 2005 levels by 2030 (EDB, 2014);
- eighty per cent of all buildings will be green or eco-friendly by 2030 (EDB, 2014);
- five per cent of peak electricity demand will be from renewable energy by 2020 (EDB, 2014).

Information integration

The existence of data in disparate entities poses interoperability and integration challenges. Thus, to mash the different data, the need for different data sources to interoperate and integrate is imperative. The Barcelona case study showcases how systems can integrate temperature, salinity and humidity sensors that make it possible, through an irrigation controller, to gather and use real-time information needed for more efficient service and resource management. Most studies however, including the Singapore case study and the selected studies in Table 4.1, seldom discuss how the various datasets have been integrated.

In cities, datasets are generated by many heterogeneous sources and stored in thousands of different information systems and databases. Connecting the different smart city services and devices brings both challenges and opportunities for service integration. A review of the literature revealed slightly different definitions of 'integration' with no uniform or generally acceptable definition. Similarly, there is no generally agreed definition for 'interoperability' (see for example Rezaei et al., 2014; Bahar et al., 2013; Charalabidis et al., 2008a,b; Gasser and Palfrey, 2007). The definition by Gasser and Palfrey (2007) is perhaps the most encompassing, including key aspects considered in most other definitions. It states that 'ICT interoperability' is the ability to transfer and render useful data and other information across systems (which may include organisations, applications or components) (Gasser and Palfrey, 2007). Data integration is a process of combining data from heterogeneous sources into meaningful and valuable information. Key activities involved in data integration include the cleaning, monitoring, transforming and delivering of data in a trusted and consistent format.

Cities contain enormous amounts of data already layered on top of each other. Although, it is a very daunting task to separate the data, new knowledge can be generated from mashed up data. In some circumstances, data, information, systems and technologies are purposely mashed up or superimposed for rich knowledge acquisition. For example, can postcode data be cross-referenced with monthly neighbourhood energy data to determine zones that consume more energy? While this question may seem simplistic and can easily be handled using traditional data integration techniques, Dong and Srivastava (2013) proposed a big data integration technique to deal with large volumes of data being generated from heterogeneous systems.

Big data thinking and analysis need to be considered in processing the large amount of data being generated from IoT applications. According to Dong and Srivastava (2013), big data integration differs from traditional data integration in at least five dimensions often called 'the 5 Vs of Big Data':

- volume of data tends to be too large because of the involvement of machines, such as sensors and CCTV, in the collection of data;
- velocity is about the dynamic nature of the data being continuously collected from sources;
- variety is about the heterogeneous nature of data sources;
- veracity is about having disparate qualities with significant differences in the coverage, accuracy and timeliness of data from the heterogeneous sources;
- value is about turning the data into added value without which it is useless.

Barriers to information integration

The applications of IoT in different areas were examined in previous sections. It emerged that the applications of IoT lead to qualitative as well as quantitative benefits in different sustainability areas. While IoT can be applied to most domains to improve urban sustainability, its application in the energy domain is crucial. The minimisation of energy loss in the built environment indirectly affects other sustainability parameters. Energy savings lead to emission reduction, less toxic or environmental harmful gases released into the atmosphere and lower use of material resources. In an urban setting, it is almost impossible to think of anything that is run without energy. Cars, buildings, motorbikes, computers, trains are some examples, just to a name a few. Mashing up or integrating information from these sources can lead to better decision-making about efficient resource management. The IoT is key to the integration of information from the aforementioned sources that often exist in silos. However, considerable challenges and barriers will need to be overcome in other to drive full deployment of IoT. These are:

- data storage: the scale of data collected from sensors is voluminous, vast and presents serious storage challenges (Barnaghi et al., 2012; Cooper and James, 2009). This is further exacerbated by the significant number of sensors required to collect data from objects, the heterogeneous nature of the data sources and the real-time dynamic nature of the data;

Internet of Things for sustainability 63

- querying IoT data: existing database management systems for interrogating databases rely on the structured format of the stored data. Most of such systems use structured query languages. However, without significant levels of standardisation across the industry, it is unrealistic to expect the vast amount of data being streamed from objects to be in a well-structured format ready for consumption. Extensible Markup Language (XML) which offers the possibility of representing structured as well as unstructured data has been suggested in the literature as a potential solution (Cooper and James, 2009). Query languages such as XQuery developed by the World Wide Web Consortium (W3C) can then be used to query the XML-based data (Walmsley, 2015);

- quality, trust and reliability of data: IoT data generated from heterogeneous sources are likely to be prone to errors. Also, the quality of data may diminish with an increasing number of sensors per object. Often, many sensors on an object can generate inconsistent or conflicting readings which can be difficult to filter for consumption. Such polluted or erroneous data can trigger unnecessary or even costly processes and actions (Chaves and Nochta, 2011);

- security and privacy: the vision of the IoT is to connect every object in the world to the internet (Cooper and James, 2009). Gartner, the world's leading information technology research and advisory company, estimated that in 2016 there were 6.4 billion connected 'things' in use worldwide, up 30 per cent from 2015, and this will reach 20.8 billion by 2020 (Gartner, 2015). People are already at the centre of this connected web of things. People want to control all 'things' and want to be informed about the state of 'things'. The role of people in the IoT is what has been termed 'humans in the loop'. Imagine returning home from a winter holiday to a reception in a warm comfortable house that has been heated! This can be achieved through remotely connecting to the home heating systems over the Web. While remotely accessing home devices to start up heaters to warm the home is great, there is a danger of 'bad guys' accessing the same home for 'evil' intentions. It is and will be difficult to protect personal data because of its sheer quantity and interconnectedness (Cooper and James, 2009; Cellary and Rykowski, 2015);

- heterogeneity and integration: as discussed in previous sections, the IoT data are generated from different sources, in different types and formats being streamed at different times from the same or different types of systems. The abundance of these data over the Web presents heterogeneity and interoperability challenges. Storing IoT data in two or more separate systems or software programs presents significant communication, exchange and integration challenges for informed-decision-making.

64 *Abanda & Tah*

Roadmap to progress

Recent studies about the direction of future research about IoT have focused on improving the technology. Hence, areas such as identification (e.g. development of new technologies that address global ID schemes, identity management and encoding/encryption), communication (e.g. design of energy efficient protocols by multi-frequency protocol) and network technologies (e.g. investigating smart chips) (Sundmaeker et al., 2010) have been earmarked as areas of further research. Hence it will be of no benefit to duplicate these earlier efforts. There are however several areas for potential future research within the applications sector. Based on the challenges associated with the applications of IoT already outlined, four main areas have been identified that provide opportunities for further research in improving the capability of the IoT. These are human behaviour modelling, big data, Building Information Modelling (BIM)/City Information Modelling (CIM) and the open/linked data.

Human behaviour is random, purpose-driven or comfort-related, and often very difficult to predict (Abanda and Cabeza, 2015). The limitations in modelling human behaviour within and out of buildings constitute significant setbacks for the integration of information used by people, which is at the heart of IoP. Human behaviour consists of activities or processes that are difficult to predict. There is therefore a need to investigate how process mining and machine learning techniques can be used to better model human behaviour for easy integration with the IoT or IoP.

The existing data storage systems and querying tools are a limitation for the management of the large volume of IoT data. Emerging big data technologies provide opportunities for managing such voluminous datasets. The main strength of big data is its ability to deal with the large volume, variety, veracity, value of data and velocity in data processing. Big data technologies such as Apache Hadoop can be used for distributed storage and process of very large datasets on different computer clusters.

The IoT will require integrating information generated from built environments and the people who dwell in them. Building data can best be captured using BIM while urban environment data can be captured using CIM. By embedding IoT technologies into integrated BIM–CIM systems, there is potential to generate new opportunities in making informed decisions about IoT data. There are recent studies about BIM for smart built environments (Zhang et al., 2015) already emerging. Interoperability is, however, often cited as a major problem with BIM and will present further challenges when considering integration with CIM systems.

Another area of interest is to explore how IoT can build on existing linked-data technologies. Many governments are putting data about services in different formats for citizens to explore and, in some cases, establish the legitimacy of actions taken by government officials or other agents (OGP, 2016). Open data or publicly available datasets about 'things' can be linked together or mashed up to form linked data for better decision-making. The domain of linked data should be investigated to establish how it can be employed in the IoT for improving urban sustainability. This is especially important in countries where governments are already making large sustainability-related datasets publicly available. This is for example the case in countries such as the UK (UK Open Data, 2016) and the US (US Open Data, 2016) who have open data portals.

Concluding remarks

This chapter discussed the challenges faced by the built environment with respect to making cities more sustainable in meeting the needs of citizens. A case for the need of emerging information technologies for managing urban sustainability data was developed based on these challenges. The emerging IoT was identified as one of the leading technologies that can be used in connecting anything and everything on the globe for better information management and decision-making.

Although the applications of IoT cut across many different domains, this chapter focused on its use for urban sustainability. To further understand how IoT has been used in improving urban sustainability, two exemplary case studies of smart cities that have adopted IoT were provided: the Barcelona and Singapore smart cities projects. It was, however, challenging to obtain detailed data about the underlying IoT-based technologies employed in the two case studies. Consequently, a further examination of the IoT-based technologies for urban sustainability was undertaken and some selected examples of studies presented.

Based on the challenges associated with the IoT, some key research directions were provided. Specifically, human behaviour modelling, data storage and querying, BIM–CIM integration, big data and linked/open data for IoT have been identified as areas requiring further research. The maximum exploitation of these emerging systems depends greatly on their degree of interoperability and the ease with which they can connect data and information from heterogeneous systems. The projected 20.8 billion connected 'things' (Gartner, 2015) and the fact that data will be produced 44 times faster (4,300 per cent increase) in 2020 than it was in 2009 (CSC, 2012), presents challenges as well as opportunities. Tapping into the strengths of emerging technologies, such as BIM, CIM and big data, to develop better-integrated and interoperable systems can overcome these hurdles and provide exciting opportunities in the IoT domain. Perhaps partly because of this, many governments are driving the adoption of IoT. In 2014, the UK Chief Scientific Adviser advised the UK government to foster and promote a clear aspiration and vision for the IoT (Government Office of Science, 2014). Recently, on 3 June 2016, the US Information Technology Industry Council (ITI) made calls for concerted government efforts in developing IoT technology (ITI, 2016). If this trend continues across the globe, IoT will only become a more encompassing and powerful tool for better connecting and managing built environment information.

Notes

1 http://smartcity.bcn.cat/en.
2 This is equivalent to USD47.2 million and USD40.5 million respectively, at the average exchange rate for 2015 (EUR1 = USD1.11).
3 www.smartnation-forbes.com/.

References

Abanda, F. H., and Cabeza, L. F., 2015. Investigating occupants' behaviour using emerging Building Information Modelling. In the *Proceedings of the ICSC15 – The CSCE International Construction Specialty Conference*, Vancouver, Canada, 7 – 10 June.

Abdoli, S., 2009. RFID Application in municipal solid waste management system. *International Journal of Environmental Research*, 3(3), pp. 447–454.

Al-Ali, A. R. and Aburukba, R., 2015. Role of the internet of things in the smart grid technology. *Journal of Computer and Communications*, 3, pp. 229–233.

Allen A., 2001. Urban sustainability under threat: The restructuring of the fishing industry in Mar del Plata, Argentina. *Development in Practice*, 11 (2&3), pp.152–173.

AoT, 2016. *Array of Things: What is the Array of Things?* Available at: https://arrayofthings.github.io/ [Accessed 18 July 2016].

Ashton, K., 1999. *That Internet of Things' Thing: In the Real World, Things Matter More Than Ideas*. RFID Journal. Available at: http://www.rfidjournal.com/articles/view?4986 [Accessed 18 July 2016].

Bahar, Y. N., Pere, C., Landrieu, J. and Nicolle, C., 2013. A thermal simulation tool for building and its interoperability through the Building Information Modeling (BIM). *Buildings*, 3, pp. 380–398.

Barnaghi, P., Wang, W., Henson, C. and Taylor, K., 2012. Semantics for the internet of things: Early progress and back to the future. *International Journal on Semantic Web and Information Systems (IJSWIS)*, 8 (1), pp. 1–21.

Boulos, M. N. K. and Al-Shorbaji, N. M., 2014. On the internet of things, smart cities and the WHO healthy cities. *International Journal of Health Geographics*, 13 (10), pp. 1–6.

Castro, M., Jara, A. and Skarmeta, A., 2013. *Smart Lighting Solutions for Smart Cities*, paper presented at the 27th IEEE International Conference on Advanced Information Networking and Applications, Barcelona, Spain, 25–28 March, pp. 1374–137.

Cellary, W. and Rykowski, J., 2015. Challenges of smart industries – Privacy and payment in visible versus unseen internet. *Government Information Quarterly*, in press.

Charalabidis, Y., Gionis, G., Hermann, K. M. and Martinez, C., 2008a. *Enterprise Interoperability Research Roadmap*, European Commission. Available at: http://cordis.europa.eu/pub/fp7/ict/docs/enet/ei-research-roadmap-v5-final_en.pdf [Accessed 18 July 2016].

Charalabidis, Y., Panetto, H., Loukis, E. and Mertins, K., 2008b. Interoperability approaches for enterprises and administrations worldwide. *The Electronic Journal for E-Commerce Tools and Applications*, 2 (3), pp. 1–10.

Chaves, L. W. and Nochta, Z., 2011. Breakthrough towards the internet of things. In D. C. Ranasinghe, Q. Z. Sheng and S. Zeadally (eds.) Unique Radio Innovation for the 21st Century. Berlin: Springer, pp. 25–38.

Chen, Z., Li, H. and Wong, C. T. C., 2002. An application of bar-code system for reducing construction wastes. *Automation in Construction*, 11, pp. 521–533.

Chowdhury, B. and Chowdhury, M. U., 2007. *RFID-based Real-time Smart Waste Management System*, paper presented at Australasian Telecommunication Networks and Applications Conference, Christchurch, New Zealand, 2–5 December.

Civiq, 2013. *Smart Cities: Technology Integrated Urban Spaces*. Available at: http:// civiqsmartscapes.com/img/whitepapers/smart_city_civiq2.pdf [Accessed 18 July 2016].

Clark, L., 2015. *Waste and Traffic Management Apps Come from Internet of Things*. Available at: http://www.computerweekly.com/feature/Waste-and-traffic-management-applications-come-from-Internet-of-Things [Accessed 18 July 2016].

Cooper, J. and James, A., 2009. Challenges for database management in the internet of things. *IETE Tech Rev*, 26, pp. 320–329.

Costabile, F. and Allegrini, I., 2008. A new approach to link transport emissions and air quality: An intelligent transport system based on the control of traffic air pollution. *Environmental Modelling & Software*, 23, pp. 258–267.

CSC, (2012). *The Growth of Global Data*. Available at: https://assets1.csc.com/insights/downloads/CSC_Infographic_Big_Data.pdf [18 July 2016].

Dong, X. L. and Srivastava, D., 2013. Big data integration. In *Proceedings of the 29th IEEE International Conference on Data Engineering*, Brisbane, Australia, 8–11 April, 2013. pp. 1245–1248.

EDB, 2014. *Smart-sustainable Cities as an Economic Opportunity*. Available at: http://www.i2r.a-star.edu.sg/horizons14/pdf/Smart-Sustainable%20Cities%20as%20an%20Economic%20Opportunity.pdf [Accessed 18 July 2016].

Egan, J., 1998. *Rethinking Construction: The Report of the Construction Taskforce*. Available at: http://constructingexcellence.org.uk/wp-content/uploads/2014/10/rethinking_construction_report.pdf [Accessed 18 July 2016].

HM Government, 2013. *HM Government Industrial Strategy: Construction 2025*. Available at: https://www.gov.uk/government/uploads/system/uploads/attachment_data/file/210099/bis-13-955-construction-2025-industrial-strategy.pdf [Accessed 18 July 2016].

Ferdous, R. M., Reza, A. W. and Siddiqui, M. F. 2016. Renewable energy harvesting for wireless sensors using passive RFID tag technology: A review. *Renewable and Sustainable Energy Reviews*, 58, pp. 1114–1128.

Gartner, 2015. *Gartner Says 6.4 Billion Connected 'Things' will be in Use in 2016, up 30 Percent from 2015*. Available at: http://www.gartner.com/newsroom/id/3165317 [Accessed 18 July 2016].

Gasser, U. and Palfrey, J., 2007. *Breaking Down Digital Barriers when and how ICT Interoperability Drives Innovation*. The Berkman Center for Internet & Society, Harvard University Research Center for Information Law, University of St. Gallen, USA. Available at: https://cyber.law.harvard.edu/interop/pdfs/interop-breaking-barriers.pdf [Accessed July 2016].

Gea, T., Paradells, J., Lamarca, M. and Roldán, D., 2013. *Smart Cities as an Application of Internet of Things: Experiences and Lessons Learnt in Barcelona*, paper presented at Seventh International Conference on Innovative Mobile and Internet Services in Ubiquitous Computing (IMIS), Taichung, Taiwan, 3–5 July.

Government Office of Science, (2014). *The Internet of Things: Making the Most of the Second Digital Revolution*. Available at: https://www.gov.uk/government/uploads/system/uploads/attachment_data/file/409774/14-1230-internet-of-things-review.pdf [18 July 2016].

Graham, P., 2015. *Newsletter 13 – COP21, A Global Deal*. Available at: http://www.gbpn.org/newsletter-13-cop21-global-deal?utm_source=newsletter&utm_medium=email&

68 *Abanda & Tah*

utm_campaign=Newsletter%2013%20-%20COP21,%20a%20global%20deal [Accessed 18 July 2016].

Hong, I., Park, S., Lee, B., Lee, J., Jeong, D. and Park, S., 2014. IoT-based smart garbage system for efficient food waste management. *The Scientific World Journal*, 2014, ID 646953, 13 pages.

ITI, (2016). *ITI Calls for Concerted Government Role in Developing Internet of Things Technology*. Available at: https://www.itic.org/news-events/news-releases/iti-calls-for-concerted-government-role-in-developing-internet-of-things-technology [18 July 2016].

Jing, Q., Vasilakos, A. V., Wan, J., Lu, J. and Qiu, D., 2014. Security of the internet of things: perspectives and challenges. *Wireless Networks*, 20 (8), pp. 2481–2501.

Kensek, K. M., 2014. Integration of environmental sensors with BIM: Case studies using arduino, dynamo, and the revit API. *Informes de la Construcción*, 66, 536, e044.

Lau, S. P., Merrett, G. V., Weddell, A. S. and White, N. M., 2015. A traffic-aware street lighting scheme for smart cities using autonomous networked sensors. *Computers and Electrical Engineering*, 45, pp. 192–207.

Li, H., Chen, Z., Yong, L. and Kong, S. C. W., 2005. Application of integrated GPS and GIS technology for reducing construction waste and improving construction efficiency. *Automation in Construction*, 14(3), pp. 323–331.

Lu, J.-W., Chang, N.-B. and Liao, L., 2013. Environmental informatics for solid and hazardous waste management: Advances, challenges, and perspectives. *Critical Reviews in Environmental Science and Technology*, 43 (15), pp. 1557–1656.

Lucon, O., Ürge-Vorsatz, D. Ahmed, A. Z., Akbari, H., Bertoldi, P., Cabeza, L. F., Eyre, N., Gadgil, A., Harvey, L. D. D., Jiang, Y., Liphoto, E., Mirasgedis, S., Murakami, S., Parikh, J., Pyke, C. and Vilariño, M. V., 2014. Buildings. In O. Edenhofer, R. Pichs-Madruga, Y. Sokona, E. Farahani, S. Kadner, K. Seyboth, A. Adler, I. Baum, S. Brunner, P. Eickemeier, B. Kriemann, J. Savolainen, S. Schlömer, C. von Stechow, T. Zwickel and J. C. Minx (eds.) *Climate Change 2014: Mitigation of Climate Change. Contribution of Working Group III to the Fifth Assessment Report of the Intergovernmental Panel on Climate Change*, Cambridge: Cambridge University Press.

Luvisi, A. and Lorenzini, G., 2014. RFID-plants in the smart city: Applications and outlook for urban green management. *Urban Forestry & Urban Greening*, 13, pp. 630–637.

Martani, C., Lee, D., Robinson, P., Britter, R. and Ratti, C., 2012. ENERNET: studying the dynamic relationship between building occupancy and energy consumption. *Energy Building*, 47 (0), pp. 584–591.

Miranda, J., Mäkitalo, N., Garcia-Alonso, J., Berrocal, J., Mikkonen, T., Canal, C. and Murillo, J. M., 2015. From the internet of things to the internet of people. *IEEE Internet Computing*, 19 (2), pp. 40–47.

Mitchell, S., Villa, N., Stewart-Weeks, M. and Lange, A., 2013. *The Internet of Everything for Cities*. Available at: http://www.cisco.com/c/dam/en_us/solutions/industries/docs/gov/everything-for-cities.pdf [Accessed July 2016].

Namen, A. A., Brasil, F. C., Abrunhosa, J. J. G, Abrunhosa, G. G. S., Tarré, R. M. and Marques, F. J. G., 2014. RFID technology for hazardous waste management and tracking. *Waste Management and Research*, 32 (9), pp. 59–66.

OGP, 2016. *What is Open Government Initiative?* Available at: http://www.opengovpartnership.org/ [Accessed 18 July 2016].

Pau, V. C. and Mihailescu, M., 2015. *Internet of Things and its Role in Biometrics Technologies and e-Learning Applications*, paper presented at Engineering of Modern Electric Systems

(EMES), 13th International Conference on Engineering of Modern Electric Systems (EMES), Oradea, Romania, 11–12 June.

Poushter, J., 2016. *Smartphone Ownership and Internet Usage Continues to Climb in Emerging Economies*, Pew Research Center. Available at: http://www.pewglobal.org/2016/02/22/smartphone-ownership-and-internet-usage-continues-to-climb-in-emerging-economies/ [Accessed 18 July 2016].

Quack, T., Bay, H. and Gool, L. V., 2008. *Object Recognition for the Internet of Things*, paper presented at IOT'08 1st International Conference on the Internet of Things, Zurick, Switzerland, 26–28 March.

Rezaei, R., Chiew, T. K., Lee, S. P. and Aliee, Z. S., 2014. A semantic interoperability framework for software as a service systems in cloud computing environments. *Expert Systems with Applications*, 41 (13), pp. 5751–5770.

Sawnhey, A., 2015. *Embracing Big Data in the Built Environment*. Available at: http://www.rics.org/en/about-rics/responsible-business/rics-futures/discussions/embracing-big-data-in-the-built-environment/?idkeep=True&DE_VM=8&DE_LNK=172098_400090 &DE_RND=1595229090&id=172098_400090 [Accessed 18 July 2016].

Sundmaeker, H., Guillemin, P., Friess, P. and Woelfflé, S. 2010. *Vision and Challenges for Realizing the Internet of Things*. European Commission, Information Society and Media. Available at: http://www.internet-of-things-research.eu/pdf/IoT_Clusterbook_March_2010.pdf [18 July 2016].

The Climate Change Group, 2012. *Lighting the Clean Revolution: The Rise of LEDs and What it Means for Cities*. Available at: http://www.newscenter.philips.com/pwc_nc/main/standard/resources/lighting/press/LED-report/TCG_LED_report%202012_final.pdf [Accessed 18 July 2016].

UK Open Data, 2016. *Welcome to Data.ac.uk*. Available at: http://www.data.ac.uk/ [Accessed 18 July 2016].

US Open Data, 2016. *The Home of the U.S. Government's Open Data*. Available at: https://www.data.gov/ [Accessed 18 July 2016].

Vinciarelli, A. and Odobez, J.-M., 2006. Application of information retrieval technologies to presentation slides. *IEEE Transactions on Multimedia*, 8 (5), pp. 981–995.

Violino, B., 2005. *What is RFID?* Available at: http://www.rfidjournal.com/articles/view?1339 [Accessed 18 July 2016].

Walmsley, P., 2015. *XQuery: Search Across a Variety of XML Data*. 2nd edition, Sebastopol: O'Reilly Media.

Wang, X. V., Lopez, B. N., Wang, L., Li, J. and Ijomah W., 2014. *A Smart Cloud-Based System for the WEEE Recovery/Recycling*, paper presented at ASME 2014 International Manufacturing Science and Engineering Conference MSEC2014, Detroit, US, 9–13 June.

Zanella, A., Bui, N., Castellani, A., Vangelista, L. and Zorzi, M., 2014. Internet of things for smart cities. *IEEE Internet of Things Journal*, 1(1), pp. 22–32.

Zhang, L., Atkins, A. S. and Yu, H., 2012. Knowledge management application of internet of things in construction waste logistics with RFID technology. *International Journal of Computing Science and Communication Technologies*, 5 (1), pp. 760–767.

Zhang, J., Seet, B.-C. and Lee, T. T. 2015. Building Information Modelling for smart built environments. *Buildings*, 5, pp. 100–115.

5 Digital technologies improving safety in the construction industry

Wen Yi, Peng Wu, Xiangyu Wang and Albert P. C. Chan

Introduction

In recent years, safety culture has emerged as one of the most important concepts in the construction industry (Biggs et al., 2005; Dingsdag et al., 2006; Fang et al., 2006). Fang et al. (2001) argued that decentralisation and mobility are two important factors leading to poor safety performance of the construction industry. Decentralisation refers here to the individual decisions that employees have to make when facing specific safety problems. Mobility implies that employees in the construction industry move among companies, sites and positions more frequently than those in other traditional industries (Fang et al., 2001). In the United States (US), the occupational fatality rate in construction ranks fourth after agriculture, mining and transportation; the construction industry has by far the largest number of fatal injuries (BLS, 2008). According to the Centre for Construction Research and Training, the construction safety and health research and training arm of North America's Building Trades Unions, there were at least 396 construction fatalities during 2005 as a result of falls. It is also reported that, in 2005, there were 36,360 non-fatal injuries resulting from falls, accounting for 23 per cent of the cases with days away from work in the construction industry (Centre for Construction Research and Training, 2007). In 2003, the fatal injury incidence rate in the United Kingdom (UK) was 4.0 per 100,000 construction workers. When collated with fatalities in all industries, construction accounted for 31 per cent of all work-related deaths in 2002–2003 (Haslam et al., 2005). Similar patterns are also found in Singapore. According to Ling et al. (2009), the fatality rate for the construction sector in Singapore was 'unacceptably high'. In 2006, this rate was 9.4 per 100,000 persons employed and represented 39 per cent of the total workplace fatalities across all industries (Ministry of Manpower, 2006).

There have been many attempts and studies to promote safety management and safety culture in the construction industry. While safety culture is a set of prevailing indicators, beliefs and values that the organisation holds in safety; safety climate is a summary concept describing the employees' beliefs about all safety issues (Biggs et al., 2008; Dingsdag et al., 2006; Dingsdag and Biggs, 2008; Guldenmund, 2000; Glendon and Stanton, 2000). Many studies have been conducted on these two aspects to improve the safety performance of the construction industry. For example, Zohar (1980) initiated the research on eight safety factors, including safety training, effects of required work pace, status of safety committee, status of safety officer, effects of safe conduct on promotion, level of risk, management attitudes and the effect of safe conduct on social status. Hinze et al. (2013) used the concept of leading indicators to measure safety performance. Traditional measurements, such as injury rate, days away from work and the experience modification rating, commonly known as lagging indicators, can provide additional data about incidents after the fact. However, questions remain regarding the value of these indicators in predicting future performance.

As such, leading indicators which aim to identify the potential for an accident before it occurs (Leveson, 2015) are advocated as more effective means for predicting future levels of safety performance. It should be noted that these studies provide insightful concepts to the construction industry at both theoretical and practical levels. However, as emerging information and communication technologies (ICT) have been implemented in the construction industry, there has been a shift in safety-related research towards the use of these emerging technologies in safety management and accident prevention. For example, Zhang et al. (2013) investigated the use of Building Information Modelling (BIM) in construction safety by conducting automatic safety checking of construction models and schedules. According to this research, the developed automated safety checking platform can inform construction engineers and managers about the reasons, high-risk locations and time periods as well as about required safety measures.

Cheng and Teizer (2013) used real-time data collection and visualisation technology for construction safety and activity monitoring. This research found that important construction information related to both safety and activity in field operations can be automatically monitored and visualised in real-time, thus offering benefits such as increased situational awareness to workers, equipment operators or decision-makers from a remote location. Ding and Zhou (2013) designed a web-based system for early warning of safety-related risks from a hybrid data fusion model based on multi-source information, including monitoring measurements, calculated predictions and visual inspections. Practitioners can use such technologies in construction safety to improve the accuracy of safety warning, improving safety management productivity and reducing the cost of safety management activities. This chapter provides a brief overview of advanced technologies in construction safety, including BIM and ICT, as well as two examples of their use for safer scaffolding and early warning of construction safety.

Technology application in construction safety

Many advanced technologies have been implemented in the construction industry in recent years to improve safety performance. Within this topic, there is a significant body of literature related to the use of BIM in improving safety performance. These studies mainly focus on hazard identification, training for

72 Yi et al.

hazard identification, impact of design alternatives on safety, evaluation of alternative construction plans and simulation of human behaviour using alterative construction planning scenarios (see for example Zhang et al., 2015; Kim et al., 2015). BIM is a 3D-based framework designed to integrate and digitise complete building information so as to express all building components and their relationships (Zhang and Hu, 2011). Hadikusumo and Rowlinson (2002) used BIM to construct 3D components of the potential hazards and identify prevention strategies.

BIM can also be used as a simulation technique to test various aspects of construction safety performance. Slaughter (2005) used it as a dynamic process simulation model to systematically evaluate construction safety aspects of design and technology alternatives. Notable benefits of using BIM in construction safety through applications such as hazard identification and visualisation include: (1) increased cognitive ability of spatial information; and (2) better reliance on past experience and memory (Park and Kim, 2013). The number and scope of studies investigating the use of BIM in construction safety have continued to increase significantly over the years, highlighting the benefits that BIM can bring to construction safety.

Other advanced technologies, such as radio frequency identification (RFID) and Geographic Information System (GIS), have also been tested and implemented in the construction industry for safety management; sometimes in conjunction with BIM (see for example Rueppel and Stuebbe, 2008; Sattineni and Azhar, 2010; Guo et al., 2014; Sanchez et al., 2015). Lee et al. (2012), for example, presented an RFID-based construction management safety which can monitor workers and other objects, such as equipment, to alert workers when they are in danger. According to Teizer et al. (2010), traditional safety practices do not allow alerting workers when they are too close to heavy equipment, and there is a need to provide blind spot support to equipment operators. As such, real-time warning and alert technology using RFID has great potential in improving construction safety performance.

Bansal (2011) used GIS-based navigable 3D animations in safety planning processes that facilitated understanding of construction sequence and predicted places and activities which had higher potential for accidents. Irizarry and Karan (2012) integrated GIS and BIM, enabling managers to visualise the 3D model of tower cranes in their optimal locations. The visualisation allows managers to spot areas of high risks to employees who work on site. Other visualisation technologies have also been proven to be effective in improving safety performance through safety training. Li et al. (2012) designed a new safety assessment method, 4D interactive safety assessment, which aims to provide individual construction workers with 4D virtual risky scenarios concerning the project and a range of possible actions for selection. The method uses game technology for simulating high-risk activities. These examples show that the construction industry can benefit greatly from the adoption of ICT in construction safety. Automation tools and approaches have great potential to be integrated into the current safety management practices and create a smart working environment. In such work places, context-aware and self-adapting sensor networks and decision-support systems monitor and complement workers' activities (Riaz et al., 2006).

BIM for safety scaffolding

BIM is an object-oriented, data-rich, intelligent technology, which can be used to simulate the construction project in a virtual environment. Such information can be utilised for cost estimation, project planning and scheduling, and control, operation management and maintenance of the constructed asset. The financial and productivity benefits of BIM to the built environment industry are increasingly understood and widely acknowledged. However, there are still significant opportunities to improve construction safety via BIM.

Planning for safety sets the context at the design stage to minimise occupational injuries and accidents in the construction industry. Traditional safety planning in construction projects largely depends on manual input or provision of knowledge-based implementation, and is time-consuming, labour-intensive and inefficient. The increasing application of BIM across the built environment industry is revolutionising the way safety management is approached. BIM can provide a powerful new platform for developing and implementing 'prevention through design' concepts that can facilitate both engineering and administrative safety planning, and control tasks at the design and construction stage of a project (Chi et al., 2012; Fleming et al., 2007). Information modelling-enabled virtual safety controls can be used to detect potential safety hazards (Malekitabar et al., 2016) and 3D visualisation and 4D simulations increase the ease and understanding of construction processes. These features are inherently embedded in BIM and thus can enable more effective safety planning before and during construction.

Scaffold-related incidents

It is well-known that falling from height is one of the leading causes of accidents in the construction industry in many countries (National Institute for Occupational Safety and Health, 2000). Scaffold-related falls, in particular, constitute a large percentage of injuries and deaths for construction workers worldwide (Cutlip et al., 2000). In the US, nearly half of the fatal work-related accidents in the construction sector are caused by falls from height, and at least 80 fatal accidents and 9,500 non-fatal injuries occur annually due to scaffolding mishaps (BLS, 2008). In a survey conducted over half of the total scaffolding workforce in the UK, 2,389 accidents were identified as related to falls reported between 2001 and 2015 (National Access and Scaffolding Confederation 2016). According to Safe Work Australia (2010), 445 serious scaffolding-related workers' compensation claims were issued in the Australian construction industry in 2008. Collapsing or falling from scaffolding is also a challenging issue in China where approximately 78 fatal accidents were caused by the fall of workers during construction in 2010. Reports published in Korea in 2009 indicate that falling from elevation constitutes the largest proportion of

74 Yi et al.

work-related fatalities (181 deaths) and 59 fatalities resulted from temporary structures assembled to work at height (Korea Occupational Safety and Health Agency, 2009). This is, however, not a new problem. An older estimate from Spain showed falls from height accounted for about 40 per cent of serious accidents and approximately 30 per cent of these involved scaffolding construction between 1988 and 1995 (Boix et al., 1997). Thus, effective prevention strategies to improve the safety condition of height-related temporary work structures merits greater attention.

Scaffolding work requires erecting, altering or dismantling a temporary structure to support a platform from which a person or object could fall. Scaffolding systems are essential on construction sites to assist workers with transport and placement of bulk materials and equipment. Traditional approaches to conducting scaffolding design are heavily based on documentation and are less effective and reactive. Limited levels of design automation can lead to compliance issues of code-of-practice, design regulations, and workplace health and safety. Recent research has revealed BIM to have benefits for safety practices, planning and management, through the application of clash detection to construction sequencing and risk analysis to site planning (Sanchez and Joske, 2016). Links to structural analysis also serve both as a design aid and to support on-site compliance checking. Despite such advances, more work is still needed to develop tools and knowledge that help reduce the likelihood of accidents arising from non-compliant scaffolding structures. Such research can facilitate the design and construction of smarter and safer scaffolding through the use of rule-based modelling systems that link with BIM technology.

Case 1: BIM for safer scaffolding

Case 1 is based on the Australian Sustainable Built Environment National Research Centre (SBEnrc) Project 3.27 *Using Building Information Modelling (BIM) for Smarter and Safer Scaffolding Construction* which commenced in 2014. This project proposed a BIM-automated approach to reduce or eliminate risks in scaffolding engineering. It developed a digital modelling tool and integrated construction and safety constraints that can be applied directly to the design, analysis, assembly, inspection and disassembly of temporary structures. The research demonstrated a BIM system that dynamically generates scaffolding design by taking into account the design and safety rules/regulations, and reacts to the modification of project features for safety purpose (SBEnrc, 2015).

BIM for 3D parametric-driving scaffold design

Parametric design is a dynamic design method. Designers define the geometric figures of the parts by applying dimension parameters and constraints. The geometric figures and models are renewed as the dimension parameters and constraints change. Parametric model construction, constraint relation extraction, as well as methods for solving constraints are crucial steps in parametric design.

In a BIM model, objects are defined by built-in and user-specified parameters, and external data such as physical, aesthetic and functional data are extracted from databases. Parametric modelling enables parameters to be processed through the use of mathematical formulas and computational algorithms before being passed among objects. Discussions with contractors in Australia have revealed that there are a number of decisions which play a role in how the rules derived from manufacturers' specifications and safety standards affect scaffolding composition. These decisions reflect the influence of specific project constraints, including the component types and sizes readily obtainable, and practices that individual contractors have observed to increase on-site productivity, generally by reducing assembly times. In order to accurately reflect project requirements and contractor preferences, it is essential to provide users with control over input parameters relating to the chosen scaffolding system, applicable design codes and construction standards for the given location and conditions. At the same time, however, certain restrictions must be put in place that limit allowable inputs to options within the range of compliance defined by scaffolding specifications and standards. This ensures that minimum safety requirements are always met and designs can be 'deemed-to-satisfy'.

The parametric models include modular scaffold and mobile scaffold (Figure 5.1). Different preferences are accommodated within the prototype through user-controlled variables that act as modifiers on fixed project and scaffolding parameters, altering the inputs to design rules and thus the scaffolding composition generated. Combined with individual project constraints, these preferences manage the rule algorithms that control face conditions and corner junctures. The prototype is based on specifications for a type of scaffolding known as modular scaffolding.

Figure 5.1 Parametric-driving modular and mobile scaffold designs.

76 Yi et al.

BIM-based falling prevention planning and opening detection

BIM was used for detailed falling-prevention planning, including temporary safety railings and floor-opening coverings (Zhang et al., 2015). The different contexts are determined by acquiring the corresponding spatial and geometric information of each object following these steps:

1   The internal gap between the inner edge of the length of the platform and the face of the building or structure immediately beside the platform are detected to define where edge protection is needed.
2   Holes in scaffold platform are detected to prevent fall through openings.
3   Openings in edge protection at points of access to stairways or ladders are detected to determine where additional opening protection is required.

After object identification, safety rules for different types of openings are input into the scaffold model, and different conditions are categorised according to specific geometric attributes. Then, corresponding rules are executed and visualised for supporting decision-making. After applying and visualising an automated version of rule checking, human input is optional to assist in the final decision-making process. Finally, the checking results and visualisation are updated in the BIM model. Each hazard is detected and the proper protection method is chosen. The geometry of the created safety equipment is based on identifying the unprotected leading edges, holes and openings in scaffolding, among other factors.

A test model was created including different types of openings that could be a potential fall hazard. The identified openings have different sizes and geometric shapes (polygonal, rectangular and circular). The holes are located on platforms and open edges. The rule-checking steps are listed as follows:

1   Customise the spatial (opening sizes: length × width) and geometric (object's location) information in BIM model (see Figure 5.2).
2   Automatically check the model and detect holes and opening in scaffold (see Figure 5.3).
3   Install guardrail system at edge and cover access opening (see Figure 5.4).

BIM for layout of the workplace

The planning of site layouts is a routine task that should be conducted at the initial phases of construction projects. Site layout planning concentrates on optimal utilisation of limited spaces. A BIM-based 3D site layout plan was selected to study possibilities of safety-related automated checking. The geometry information is sufficient for automatic analysis such as collision detection. Elements of a BIM-based site layout plan are:

•   the construction site area, adjoining streets and other immediate surroundings that the construction site may impact;

Improving safety 77

- temporary site facilities, structures and equipment;
- temporary site arrangements, such as area reservations for material storage;
- visualisation of safety-related issues, such as illustration of risk distance between the scaffold and vehicles or cranes.

The visualisation was based on the site engineer's site layout plan. Firstly, a BIM-based visualisation was carried out relating to the erection and demolition plans of scaffolding based on the installation in practice. After this, BIM-based modelling and visualisation of site layout and traffic route plans near scaffolding were carried out based on textual specification describing the planned site layout (see Figure 5.5). Potential clashes between scaffolding, the building and a crane are shown in Figure 5.6, with each potential conflicting building component shown in colour

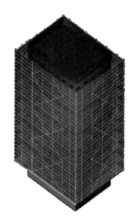

Figure 5.2 Building scaffolds.

Figure 5.3 Rule-based hole and edge detection.

78 Yi et al.

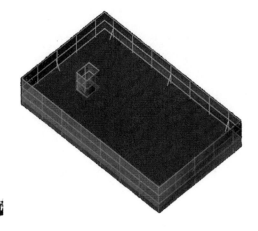

Figure 5.4 Guardrail systems at edge and cover access opening.

Figure 5.5 Modular scaffold installation in BIM.

code. Optimal mobile crane location can be decided based on visualisation and mathematical algorithms. Figure 5.7 shows the traffic simulation and verification. A warning system to alert the user of an unsafe distance between vehicle and scaffolding work is displayed in Figure 5.8. Corresponding manual work is required to rearrange the traffic route and moving track. The geometry of the warning system is based on identifying the clearance distance of traffic and scaffold work.

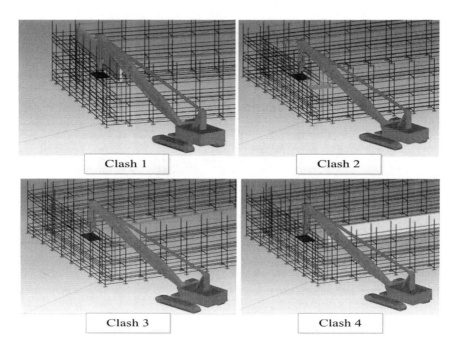

Figure 5.6 Clash detection in BIM between scaffold, building and mobile devices.

Figure 5.7 Site traffic simulation and verification.

Figure 5.8 Unsafe distance warning in site traffic planning.

This case study has demonstrated a BIM-based approach that can facilitate smarter and safer scaffolding design to improve robust scaffolding design and prevent severe safety incidents. The prototype captures practical knowledge related to scaffolding construction and safety guidelines by encoding design configurations from different construction settings. It provides decision-support by presenting choices as explicitly-defined input parameters. This allows the design intent and assumptions behind a particular scaffolding solution to be communicated among different project stakeholders, establishing a shared understanding of project-specific requirements and constraints.

ICT for early-warning system

Construction is commonly recognised as an information-dependent industry where the development of information technologies continuously provides innovative solutions to the industry by enabling automation, improving productivity and promoting effective decision-making mechanism (Erdogan et al., 2010). ICT offers significant potential for improving the construction industry. Efficient adoption of ICT solutions could lead to more effective management of project implementation and to a more productive industry. ICT has been widely adopted in addressing managerial and operational issues, such as global virtual working environments, inter-organisational collaborative design, corporate-level data-driven decisions and safety control and monitoring (Lu et al., 2014). Both managerial and technological improvement of ICT applications in construction projects and organisations are prominent.

A variety of ICT tools such as the internet, wireless technology, BIM, electronic data management systems and virtual/augmented reality, have been employed and proven to be beneficial for safety identification, inspection, training and education. ICT-based data acquisition tools/equipment are central to achieving a continually increasing capacity to collect, transmit, store, analyse and visualise data and information during the construction process. Accurate and timely control over workers, materials, equipment, work environment and construction methods is important for the prevention of risk and emergency response.

Heat stress in the construction industry

Heat stress is a well-known occupational hazard in many industries in hot countries. This trend is likely to continue to gain prominence as climate change brings more frequent heat waves and increasing risks of more severe and widespread heat-related hazards. The effects of working in hot weather can potentially have a severe influence on the welfare of workers if the risks have not been properly managed or considered. Common heat-related illness (HRI) include heat cramps, heat fainting, heat exhaustion and heat stroke. Heat stroke is a systemic inflammatory response to core temperatures exceeding 40 °C and has associated clinical manifestations of mental status changes such as anxiety, confusion, bizarre behaviour and loss of coordination (Bouchama and Knochel, 2002; McDermott et al., 2008). It is most often accompanied by acute symptoms such as extreme fatigue, hot skin or heavy perspiration, nausea, vomiting and diarrhoea, and reddened face is frequently an early indicator (Yeo, 2004). Potential chronic complications related to heat stroke are acute renal failure, liver failure, brain injury, respiratory failure, ischaemic bowel injury and pancreatitis (Jones-Laskowski, 2000).

Heat stress can lead to reduction of work enthusiasm, decreased productivity and increased non-fatal injuries and fatal accidents. Occupational heat stress results from a combination of factors, including climatic conditions (i.e. ambient temperature, humidity, solar radiation and air movement), work demands and clothing requirements. Construction workers normally have to undertake outdoor physical work and wear impermeable protective safety clothing. They are vulnerable to heat stress in summer, leading to heat stroke causing deaths and injuries. In the US, the HRI claim incidence between 1995 and 2005 by industry sector was highest in construction (Bonauto et al., 2007). A recent survey in Hong Kong reported at least 5 per cent of construction workers suffered from heat stroke and 23 per cent experienced related symptoms (Hong Kong Confederation of Trade Unions, 2012). Published reports in Japan show the construction industry accounted for the overwhelming majority (84 per cent) of all heat stroke cases between 1989 and 2000 (Japan International Centre for Occupational Safety and Health, 2001).

Across the world, governments and industry organisations have expressed concern about this issue and promulgated a series of fundamental practices for working in hot weather. General guidance notes on the prevention and management of heat stress risk in the construction industry suggest workers work only during cooler parts of the day and avoid heavy manual labour in a hot environment (Labour Department of Hong Kong, 2010; US Department of

82 Yi et al.

Labor, 2011). However, this may not always be possible due to tight working schedules and confined work conditions. Other suggestions mention that risk of heat stress can be reduced by engineering control (i.e. shelters, blowing fans, ventilation, air cooling and insulation) (Construction Industry Council of Hong Kong, 2013; HSE UK, 2011). There are, however, few reports that explore the effectiveness of such measures. Some measures such as rest breaks and sufficient water intake have already been applied in construction sites and have been shown to be effective (Morioka et al., 2006; Chan et al., 2012). Although the literature abounds with general guidance notes on heat stress hazards in construction, there is a lack of scientific research specific to the construction industry on which to base safety measures for combating heat stress.

The issue of managing workplace heat stress has, however, been a concern for some academics too. Many attempts have been made to establish safety evaluation models for assessing the risk of heat stress. For instance, Ren et al. (2009) proposed a framework for evaluating the hazards of heat stress in an underground mine. Zheng et al. (2012) evaluated work safety in hot environments through an expert scoring of work intensity, environment condition and worker status. Nevertheless, no in-depth and systematic studies have been initiated into early-warning and safety evaluation of heat stress recognition, assessment, control and management of individual behaviour. The following case study aimed to fill this gap by using ICT to detect impending attacks of heat stress and by recommending pertinent health and safety measures.

Case 2: early-warning system

Case 2 is based on the Research Grants Council of the Hong Kong Project PolyU510409 *Experimental Research on Health and Safety Measures for Working in Hot Weather* and PolyU510513 *Developing a Personal Cooling System (PCS) for Combating Heat Stress in the Construction Industry*, which commenced in 2015. This case shows an automated early-warning system supported by ICT tools to protect construction workers from heat stress. The early-warning system involves the following processes:

1 Collecting timely information and undertaking risk assessments of heat stress.
2 Generating an accurate and timely warning to trigger prompt health and safety intervention.
3 Disseminating heat strain assessment and symptoms of heat illness to site supervisor/foreman.

Figure 5.9 shows the prototype of the early-warning system, which consists of a smart wristband and a smart phone application.

Figure 5.10 illustrates the process map of the early-warning system. The worker can input their personal information and job nature into the application software. Once starting the daily work, the smart phone measures the temperature, relative humidity, work duration and tracks the location. The smart wristband worn by

Improving safety 83

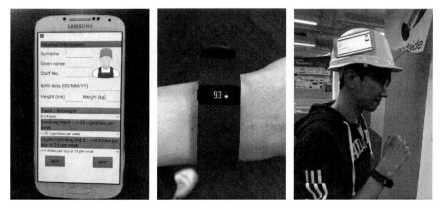

Figure 5.9 Prototype of the early-warning system.

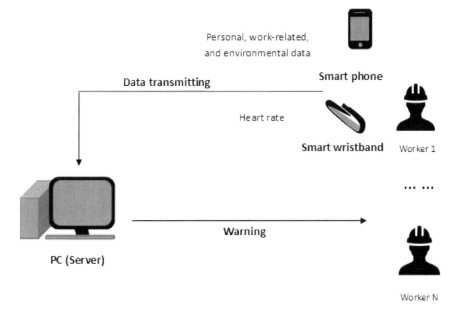

Figure 5.10 Process map of the early-warning system.

the workers monitors their heart rate in real-time. The wireless communication technology transmits environmental, physiological and work-related data to the server/Web for heat-stress assessment. Warning alerts could be sent back to the worker if certain thresholds are exceeded. Supervising staff and safety professionals may use this system as a surveillance method to monitor the health and safety of frontline workers when working in hot and humid conditions. The early-warning system is composed of four layers: (1) data collection layer; (2) data transmission layer; (3) data processing layer; and (4) warning and control layer.

84 *Yi et al.*

Data collection layer

Heat strain describes the overall psychological and physiological response resulting from heat stress. Physiological parameters including core temperature, heart rate, skin temperature and sweating rate have been identified as heat strain indicators (ISO, 2004). As the earliest response of physiological strain and general indicator of body stress, heart rate is adopted as a safe index for early warning. The heart rate of workers could be measured and monitored by a smart wristband every 5–10 seconds.

Fatigue is defined as a reduction in physical and mental work capacity. Previous studies have established algorithms for predicting the fatigue of construction workers by 'work-related factors' such as type of exercise and time, 'environmental factors' such as ambient temperature and relative humidity, as well as 'personal factors', such as age, physique, percentage of body fat, drinking/smoking habits and clothing (Chan et al., 2012). The duration of the operation and the level of activity of the construction workers are tracked by a stopwatch embedded in the system. Environmental variables such as ambient temperature and relative humidity could be measured every minute by sensors built into the phone. The developed application software could record the personal and work-related information.

Data transmission layer

The data transmission layer plays an important role in linking the data collection layer and data processing layer. The real-time heart rate measured by the smart wristband could be conveyed to the smart phone by Bluetooth. The Bluetooth-enabled smart phone and smart wristband could communicate with each other automatically as both wearable devices are within range of each other. A set of environmental and physiological data recorded in the smart phone, including air temperature, relative humidity, heart rate, age, body fat and smoking and alcohol drinking habits, could additionally be transmitted to the server via Global System for Mobile Communications (GSM). GSM is the most popular and successful mobile cellular communication system worldwide. Due to its widespread coverage, information transmission among GSM smart phones and server is easily achieved.

Data processing layer

The data processing layer is the core of the early-warning system that analyses the collected data and evaluates the heat strain of workers. Heat strain is assessed by the heart rate and fatigue of workers from both physiological and psychological aspects. On the one hand, the American Conference of Governmental Industrial Hygienists (ACGIH) recommends threshold limit values (TLVs) for heat stress and suggests a threshold heart rate value of 180 beats per minute (bpm) minus the age of an individual over a period of no more than three minutes (ACGIH, 2014). On the other hand, an earlier study by Yi and Chan (2015) established an artificial neural network model which could provide a reliable and scientific prediction of fatigue of construction workers in hot environments. As a practical and cost-effective approach to quantifying the subjective feeling, the rating of perceived exertion (RPE) was adopted to evaluate the fatigue and physical effort of the workers. A ten-point single-item RPE scale with the ratings between 0 (nothing at all) and 10 (extremely hard) was used. The statistical analysis of the data processing layer provides an effective means to monitor the heat strain level of construction workers.

Alert and control layer

Warning alerts would be sent to both the worker and supervisory staff immediately and directly once psychological or physiological thresholds are exceeded. For example, an alert warning will be issued when the heart rate of a 20-year-old worker exceeds 160 bpm for a period of three minutes, or their predicted RPE exceeds 7, which corresponds to 'very hard'. Thus, the worker could take measures, such as taking breaks in the shade to cool down and drink water, towards preventing heat stress, and safety professionals and site supervisors can pay more attention to those workers at risk.

Concluding remarks

Construction safety is a serious problem in the built environment industry worldwide. This industry has been recognised, for many years, as being among those having a high likelihood of accidents and, unfortunately, deaths. Advanced technologies have been considered to be an effective way to enhance safety performance of construction companies. This chapter reviewed the use of digital technologies, including BIM, RFID, GIS and other technologies, in enhancing construction safety performance. It was found that these technologies have significant potential benefits in hazard identification and visualisation, such as improved working memory ability, increased cognitive ability of spatial information and better reliance on past experience and memory. The use of these technologies can create a smart working place, where context-aware and self-adapting sensor networks and decision-support systems monitor and complement workers' activities.

Two case studies on the application of BIM for safer scaffolding and ICT for an early heat stress warning system were used to exemplify the use of more integrated information management tools to create safer work sites. The BIM for safer scaffolding case study showed how BIM can be used to facilitate the design and construction of smarter and safer scaffolding through the use of rule-based modelling systems. The second case study showed an early-warning system automated by ICT to protect construction workers from heat stress. This application collects timely information and undertakes risk assessments of heat stress as well as generates an accurate and timely warning. It is expected that these two case studies can help practitioners understand the importance and benefits of using ICT to achieve higher levels of information integration that support an improving construction safety performance.

Digital technologies have been employed in a variety of applications to address problems related to construction safety. GPS, smart sensors, laser scanning and RFID have been utilised to collect safety-related data including location, environment and identity information. Ultra wideband (UWB), 3G and wireless networks have also been utilised to swiftly convey safety data/information to the corresponding persons and places. Analytical technologies, such as GIS and data mining, and visualisation technologies, such as virtual reality and BIM, have been integrated to analyse the safety data/information and visualise the evaluation of construction safety. Attempts have also been made to evaluate the risk of safety hazards, monitor the equipment, materials, environment and the performance of workers, and replace human operations by robots. Most studies have concentrated on the safety problems and less attention has been paid to occupational health issues. Nevertheless, occupational diseases and injuries in construction workers lead to low productivity, physical disability and even death. Further research on using innovative technologies to monitor unsafe behaviour of workers, such as construction ergonomics and risk assessment, would provide the industry with much-needed insight and tools to make their workforce safer and more productive.

Construction workers undertaking physically demanding tasks with manual handling of materials and awkward postures are susceptible to suffering musculoskeletal disorders. This has led to the construction sector being recognised as one of the most dangerous industries for musculoskeletal disorders (Chen et al., 2005). Several VR-based safety training programmes aiming to identify work-related risks have been developed. Further research on the development of safety-training systems that integrate the rectification of harmful postures/movements is expected.

A safety-related issue which was not discussed in this chapter but requires further research is linked to noise. Construction is a noisy business. Activities in construction are characterised by the ubiquity of power tools and the intensive use of heavy equipment, leading to a large potential for elevated noise exposure. Exposure to continuous or intermittent loud noise results in noise-induced hearing loss (NIHL). NIHL is often permanent and is one of the most serious occupational diseases and injuries in the construction industry (Leensen et al., 2011). An accurate assessment of noise levels on construction sites and their impacts on construction workers is urgently needed. Future research is therefore required to develop an automated system utilising digital technologies, which detects noise levels, monitors noise-exposure duration and provides appropriate alerts to both workers and supervisors.

References

American Conference of Governmental Industrial Hygienists (ACGIH), 2014. *TLVs® and BEIs®: Threshold Limit Values for Chemical Substances and Physical Agents and Biological Exposure Indices*, Cincinnati, OH: American Conference of Governmental Industrial Hygienists.

Australian Sustainable Built Environment National Research Centre (SBEnrc), 2015. *Using Building Information Modelling (BIM) for Smarter and Safer Scaffolding Construction*. www.sbenrc.com.au/research-programs/3-27-using-building-information-modelling-bim-for-smarter-and-safer-scaffolding-construction/ [Accessed 1 March 2017].

Bansal, V. K., 2011. Application of Geographic Information Systems in construction safety planning. *International Journal of Project Management*, 29(1), 66–77.

Biggs, H. C., Dingsdag, D. P., Sheahan, V. L., Cipolla, D. and Sokolich, L., 2005. *Utilising a Safety Culture Management Approach in the Australian Construction Industry*, paper presented at Queensland University of Technology's Research Week 2005, Brisbane, Australia, 3–7 July.

Biggs, H. C., Dingsdag, D. P. and Roos, C. R., 2008. Development of a practical guide to safety leadership: Industry-based applications. In K. A. Brown, K. D. Hampson, P. S. Brandon and J. Pillay (eds), *Clients Driving Construction Innovation: Benefiting from Innovation*, Brisbane: CRC for Construction Innovation, pp. 154–158.

BLS, 2008. *National Census of Fatal Occupational Injuries in 2007*. August 20, 2008 edition. Washington, DC: Bureau of Labor Statistics, United States Department of Labor. Available at: www.bls.gov/iif/oshcfoi1.htm [Accessed 6 March 2017].

Boix, P., Orts, E., López, M. J. and Rodrigo, F., 1997. Trabajo temporal y siniestralidad laboral en España en el periodo 1988–1995. *Cuadernos de Relaciones Laborales*, 11, 275–320.

Bonauto, D., Anderson, R., Rauser, E. and Burke, B., 2007. Occupational heat illness in Washington State, 1995–2005. *American Journal of Industrial Medicine*, 50(12), 940–950.

Bouchama, A. and Knochel, J., 2002. Heat stroke, *The New England Journal of Medicine*, 346, 1978–1988.

Center for Construction Research and Training, 2007. *The Construction Chart Book: The US Construction Industry and its Workers*, 4th edn, Silver Spring, MD: Center for Construction Research and Training (CPWR).

Chan, A. P. C., Yam, M. C. H., Chung, J. W. Y. and Yi, W., 2012. Developing a heat stress model for construction workers. *Journal of Facilities Management*, 10(1), 59–74.

Chen, Y., Turner, S., Hussey, L. and Agius, R., 2005. A study of work-related musculoskeletal case reports to The Health and Occupation Reporting network (THOR) from 2002 to 2003, *Occupational Medicine*, 55(4), 268–274.

Cheng, T. and Teizer, J., 2013. Real-time resource location data collection and visualization technology for construction safety and activity monitoring applications. *Automation in Construction*, 34, 3–15.

Chi, S., Hampson, K. D. and Biggs, H. C., 2012. *Using BIM for Smarter and Safer Scaffolding and Formwork Construction: A Preliminary Methodology*, paper presented at CIB W099 International Conference on Modelling and Building Health and Safety, Singapore, 10–11 September.

Construction Industry Council of Hong Kong, 2013. *Guidelines on Site Safety Measures for Working in Hot Weather, Version 2*. Available at: http://www.hkcic.org/eng/info/publication.aspx [Accessed 3 March 2016].

Cutlip, R., Hsiao, H., Garcia, R., Becker, E. and Mayeux, A., 2000. A comparison of different postures for scaffold end-frame disassembly. *Applied Ergonomics*, 31, 507–513.

Ding, L. Y. and Zhou, C., 2013. Development of web-based system for safety risk early warning in urban metro construction. *Automation in Construction*, 34, 45–55.

Dingsdag, D. P. and Biggs, H. C., 2008. The use of lead indicators in safety culture research: measuring construction industry safety performance. In K. A. Brown, K. D. Hampson, P. Brandon and J. Pillay (eds) *Clients Driving Construction Innovation: Benefiting from Innovation*, Brisbane: CRC for Construction Innovation, pp. 146–153.

Dingsdag, D. P., Biggs, H. C., Sheahan, V. L. and Cipolla, D. J., 2006. *A Construction Safety Competency Framework: Improving OH&S Performance by Creating and Maintaining a Safety Culture*, Brisbane: CRC for Construction Innovation.

Erdogan, B., Abbott, C. and Aouad, G., 2010. Construction in year 2030: Developing an information technology vision. *Philosophical Transactions of the Royal Society A*, 368, 3551–3565.

Fang, D., Chen, Y. and Wong, L., 2006. Safety climate in construction industry: A case study in Hong Kong. *Journal of Construction Engineering and Management*, 132(6), 573–584.

Fang, D. P., Huang, X. Y. and Hinze, J., 2001. *Safety Management in Construction*, Beijing: Waterpower Press.

Fleming, T., Lingard, H. and Wakefield, R., 2007. *Guide to Best Practice for Safer Construction: Principles*, Brisbane: CRC for Construction Innovation.

Glendon, A. I. and Stanton, N. A., 2000. Perspectives on safety culture. *Safety Science*, 34, 193–214.

Guldenmund, F. W., 2000. The nature of safety culture: A review of theory and research. *Journal of Constructional Steel Research*, 34, 215–257.

Guo, H., Liu, W., Zhang, W. and Skitmore, M., 2014. *A BIM-RFID Unsafe On-Site Behaviour Warning System*, paper presented at ICCREM 2014: Smart Construction and Management in the Context of New Technology, Kunming, China, 27–28 September.

Hadikusumo, B. and Rowlinson, S., 2002. Integration of virtually real construction model and design-for-safety-process database, *Automation in Construction*, 11, 501–509.

Haslam, R. A., Hide, S. A., Gibb, A. G. F., Gyi, D. E., Pavitt, T., Atkinson, S. and Duff, A. R., 2005. Contributing factors in construction accidents. *Applied Ergonomics*, 36, 401–415.

Health and Safety Executive (HSE) UK, 2011. *Keep your Top on: Health Risks from Working in the Sun*. Available at: http://www.hse.gov.uk/pubns/indg147.pdf [Accessed 3 May 2015].

Hinze, J., Thurman, S. and Wehle, A., 2013. Leading indicators of construction safety performance. *Safety Science*, 51(1), 23–28.

Hong Kong Confederation of Trade Unions, 2012. 三成建築工人酷熱工作遇險 工會爭取一系列措施保平安 (*Almost 30% of Construction Workers in Hong Kong Have Suffered Heat Related Illness*). Available at: http://www.hkctu.org.hk/cms/article.jsp?article_id=803&cat_id=21 [Accessed 24 April 2017].

Irizarry, J. and Karan, E. P., 2012. Optimizing location of tower cranes on construction sites through GIS and BIM integration. *Journal of Information Technology in Construction*, 17, 251–366.

ISO, 2004. ISO 9886. Ergonomics. *Evaluation of Thermal Strain by Physiological Measurements*, Geneva: International Organisation for Standardisation Organisation.

Japan International Center of Occupational Safety and Health, 2001. *Japan Construction Safety and Health Association Visual Statistics of Industrial Accidents in Construction Industry.* www.jniosh.go.jp/icpro/jicosh-old/english/statistics/jcsha/index.html [Accessed 1 March 2017].

Jones-Laskowski, L., 2000. Responding to summer emergencies. *Nursing*, 30(5), 34–39.

Kim, H., Lee, H. -S., Park, M., Chung, B., Hwang, S. 2015. Information retrieval framework for hazard identification in construction, *Journal of Computing in Civil Engineering*, 29(3), 04014052.

Korea Occupational Safety and Health Agency, 2009. *Analysis of Industrial Disaster.* Available at: http://www.kosha.or.kr/board [Accessed 3 July 2013].

Labour Department of Hong Kong, 2010. *Prevention of Heat Stroke at Work in a Hot Environment.* Available at: http://www.labour.gov.hk/eng/news/PreventionOfHeat StrokeAtWork.htm [Accessed 17 August 2014].

Lee, H., Lee, K., Park, M., Baek, Y. and Lee, S., 2012. RFID-Based Real-Time Locating System for Construction Safety Management. *Journal of Computing in Civil Engineering*, 26(3), 366–377.

Leensen, M. C. J., van Duivenbooden, J. C. and Dreschler, W. A., 2011. A retrospective analysis of noise-induced hearing loss in the Dutch construction industry. *International Archives of Occupational and Environmental Health*, 84, 577–590.

Leveson, N., 2015. A system approach to risk management through leading safety indicators. *Reliability Engineering & System Safety*, 136, 17–34.

Li, H., Chan, G., and Skitmore, M., 2012. Visualizing safety assessment by integrating the use of game technology. *Automation in Construction*, 22, 498–505

Ling, F. Y. Y., Liu, M., Woo, Y. C., 2009. Construction fatalities in Singapore. International Journal of Project Management, 27(7), 717–726.

Lu, Y., Li, Y., Skibniewski, M., Wu, Z., Wang, R. and Le, Y., 2014. Information and communication technology applications in architecture, engineering, and construction organizations: A 15-year review. *Journal of Management in Engineering*, 31, A4014010.

McDermott, B. P., Lopez, R. M. and Casa, D. J., 2008. Exertional heat stroke basics: What strength and conditioning coaches need to know? *The Journal of Strength and Conditioning Research Corner*, 30(3), 29–32.

Malekitabar, H., Ardeshir, A., Sebt, M. H. and Stouffs, R., 2016. Construction safety risk drivers: A BIM approach. *Safety Science*, 82, 445–455.

Ministry of Manpower, 2006. *Workplace Safety and Health Act*, Singapore: Ministry of Manpower.

Morioka, I., Miyai, N. and Miyashita, K., 2006. Hot environment and health problems of outdoor workers at a construction site. Industrial health, 44(3), pp. 474–480

National Access and Scaffolding Confederation, 2016. *Safety Report 2016*, London: Maintaining High Standards in Scaffolding.

National Institute for Occupational Safety and Health, 2000. *Worker Deaths by Falls*, Washington, DC: DHHS (NIOSH) publication 2000-116.

Park, C. S. and Kim, H. J., 2013. A framework for construction safety management and visualization system. *Automation in Construction*, 33, 95–103.

Riaz, Z., Edwards, D. J. and Thorpe, A., 2006. SightSafety: A hybrid information and communication technology system for reducing vehicle/pedestrian collisions. *Automation in Construction*, 15, 719–728.

Ren, Z. G., Liu, G. Z., He, C. and Wang, S. W., 2009. Study on the indicators for ranking of thermal hazard in heat stress workplace of underground mine. *Journal of Safety Science and Technology*, 5(3), 41–46.

Rueppel, U. and Stuebbe, K. M., 2008. BIM-based indoor-emergency-navigation-system for complex buildings. *Tsinghua Science and Technology*, 13, 362–637.

Safe Work Australia, 2010. *Australian Workers' Compensation Statistics*. www.safeworkaustralia.gov.au/sites/SWA/about/Publications/Documents/874/Australian-Workers-Compensation-Statistics-2011-12.pdf [Accessed 1 March 2017].

Sanchez, A. X., Hampson, K. D. and Mohamed, S., 2015. *Perth Children's Hospital Case Study Report*, Perth: Sustainable Built Environment National Research Centre (SBEnrc).

Sanchez, A. X. and Joske, W., 2016. Benefits Dictionary, in A. X. Sanchez, K. D. Hampson and S. Vaux (eds) *Delivering Value with BIM: A Whole-of-life Approach*, London: Routledge.

Sattineni, A. and Azhar, S., 2010. *Techniques for Tracking RFID in a BIM Model*, paper presented at 27th International Symposium on Automation and Robotics in Construction, Bratislava, Slovakia. 25–27 June.

SBEnrc, 2015. *Using Building Information Modelling (BIM) for Smarter and Safer Scaffolding Construction*. http://www.sbenrc.com.au/research-programs/3-27-using-building-information-modelling-bim-for-smarter-and-safer-scaffolding-construction/ [accessed 21 April 2017].

Slaughter, S., 2005. *The link between design and process: Dynamic process simulation models of construction activities, 4D CAD and visualization in construction developments and applications*, Gainesville, FL: Swets and Zeitlinger, pp. 145–164.

Teizer, J., Allread, B. S., Fullerton, C. E. and Hinze, J., 2010. Autonomous pro-active real-time construction workers and equipment operator proximity safety alter system. *Automation in Construction*, 19(5), 630–640.

US Department of Labor, 2011. *Protecting Workers from Heat Illness*. Available at: www.osha.gov/Publications/osha-niosh-heat-illness-infosheet.pdf [Accessed 24 June 2014].

Yeo, T. P., 2004. Heat stroke: A comprehensive review. *AACN Clinical Issues: Advanced Practice in Acute and Critical Care*, 15, 280–293.

Yi, W. and Chan, A. P. C., 2015. *An Artificial Neural Network Model for Predicting Fatigue of Construction Workers in Hot and Humid Environment*, paper presented at ISEC-8 Implementing Innovative Ideas in Structural Engineering and Project Management, Sydney, Australia, 23–28 November.

Zhang, J. P. and Hu, Z. Z., 2011. BIM-and 4D-based integrated solution of analysis and management for conflicts and structural safety problems during construction: 1. Principles and methodologies. *Automation in Construction*, 20(2), 155–166.

Zhang, S., Teizer, J., Lee, J., Eastman, C. M. and Venugopal, M., 2013. Building Information Modelling (BIM) and safety: Automatic safety checking of construction models and schedules. *Automation in Construction*, 29, 183–195.

Zhang, S., Sulankivi, K., Kiviniemi, M., Romo, I., Eastman, C. M. and Teizer, J., 2015. BIM-based fall hazard identification and prevention in construction safety planning. *Safety Science*, 72, 31–45.

Zheng, G., Zhu, N., Tian, Z., Chen, Y. and Sun, B., 2012. Application of a trapezoidal fuzzy AHP method for work safety evaluation and early warning rating of hot and humid environments. *Safety Science*, 50(2), 228–239.

Zohar, D., 1980. Safety climate in industrial organizations: Theoretical and applied implications. *Journal of Applied Psychology*, 65(1), 96–102.

6 Information integration and interoperability for BIM-based life-cycle assessment

Ruben Santos and António Aguiar Costa

Introduction

Construction is a critical sector of the world economy. On-site activity accounts for 9 per cent of gross domestic product (GDP) in the European Union (EU), 6.5 per cent in the United Kingdom (UK), 5.5 per cent in France, 6 per cent in India, 6.8 per cent in Australia and 3.4 per cent in United States of America (US) (Rhodes, 2015; European Commission, 2016; Hampson et al. 2014). The construction sector is also responsible for about 40 per cent of global energy consumption and CO_2 emissions (ICLEI, 2008; Kulahcioglu, Dang, and Toklu, 2012). With that in mind, the EU has been devising and releasing directives to achieve higher levels of sustainability in the building sector (European Commission, 2010; European Commission, 2012). The architecture, engineering, and construction (AEC) sector in particular has been striving to develop methodologies that are more collaborative, integrated and information-based, to increase life-cycle awareness and improve construction processes and outputs.

Building Information Modelling (BIM) is a methodology that can support such vision. BIM is defined as a 'shared digital representation of physical and functional characteristics of any built object (including buildings, bridges, roads, etc.) which forms a reliable basis for decisions' in ISO 29481-1:2010 (ISO, 2010). It enables greater cooperation between different parties during a project's life-cycle, while saving time, money and resources (Grilo and Jardim-Goncalves, 2010; Fan, 2014). As a result, BIM can provide, through powerful simulations, key information regarding the energy consumption of a project, its life-cycle costs, interior comfort and even environmental impact due to construction and operation of an asset. Aware of its potential, several countries are encouraging BIM implementation in the construction industry and mandating its use in the public sector. Examples include Norway, Finland and the UK; the latter mandating BIM level 2[1] across public capital works since 2016 (Cabinet Office, 2011). The British movement for digitalisation spurred a widespread change at European level, especially in France and Germany. The French case is particularly relevant given the recent investment of EUR20 million towards BIM implementation and the digitalisation of the construction and infrastructure

92 *Santos & Aguiar Costa*

industry (Bâtiment Numérique, 2015). Germany developed the Planen-Bauen 4.0, also focused on this new paradigm (German BIM Steering Group, 2015).

The European Commission has also promoted the integration of BIM's data rich models and processes with sustainable methodologies that represent life-cycle thinking (European Commission, 2016), such as life-cycle assessment (LCA). This move may represent a key element in the reduction of wasted resources, energy savings and improved sustainable constructions. LCA is a 'compilation and evaluation of the inputs, outputs and potential environmental impacts of a product system throughout its life-cycle' according to ISO 14040:2006 (ISO, 2006). This methodology would enable BIM tools to effectively assess the environmental impact of projects. The AEC sector is increasingly incorporating this method in decision-making processes through initiatives such as the Building Life-cycle Assessment guide developed by American Institute of Architects (Bayer et al., 2010). This allows projects to choose the most suitable products while striving to optimise the sustainability of construction processes (Asdrubali et al., 2013; Cabeza et al., 2014, Bayer et al., 2010).

The following sections will discuss the BIM–LCA integration process and explore integration and interoperability challenges, the actors that are involved in this process and how they can contribute to the success of BIM-LCA integration.

Life-cycle assessment

As mentioned above, LCA is a methodology for assessing the environmental impact of the entire life-cycle of a product. It evaluates the environmental impact of a building (output) based on the resources and energy it consumes throughout its life-cycle (input) based on different environmental categories such as climate change, stratospheric ozone depletion, acidification, nitrification, human toxicity and eco-toxicity. Over the past few years, the LCA methodology has been more widely acknowledged, with the International Organization for Standardisation (ISO) publishing the standard ISO 14040:2006. This ISO promotes the adoption of LCA and presents a framework for the process-based LCA as seen in Figure 6.1 (ISO, 2006).

The framework proposed by ISO 14040 includes the following phases:

1 Scope definition: outlines the subject and the intended use of the study.
2 Life-cycle inventory analysis phase (LCI phase): inventory of input/output and collection of the necessary data to meet the goals of the study.
3 Life-cycle impact assessment phase (LCIA phase): provides additional information to assess the product system's LCI.
4 Life-cycle interpretation: the results of an LCI and/or an LCIA are analysed and complied in a report.

Hence, when conducting an LCA study, the LCA specialist must first define the goal and scope of the study, which will have an impact on the inputs (boundary

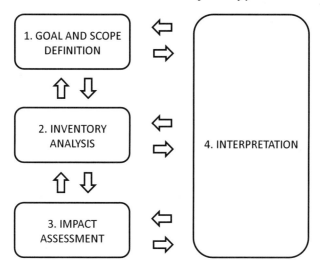

Figure 6.1 LCA methodology phases (based on ISO, 2006).

definition). Then, as shown in Figure 6.2, all flows within the defined boundary must be considered (LCI phase), from upstream to downstream, and its environmental impact (LCIA phase) must be evaluated at all stages (Kulahcioglu et al., 2012; Zabalza et al. 2009; Ramesh, Prakash and Shukla, 2010). In the end, the LCA specialist must interpret the results of the LCA study (interpretation phase).

Based on the ISO 14040:2006 framework, the American Institute of Architects (AIA) developed the 'Building Life-cycle Assessment in Practice' guide, which adapted the process-based LCA method to the building sector (Bayer et al., 2010). It proposes the following steps:

- step 1: definition of the project's sustainability targets;
- step 2: conduct LCA or not, based on project scope, time, and resources. If not, resort to alternative methods;
- step 3: definition of goals and scope of LCA study;
- step 4: selection of an appropriate LCA tool;
- step 5: life-cycle inventory (LCI) analysis;
- step 6: life-cycle impact assessment (LCIA);
- step 7: results and interpretations.

While the ISO 14040:2006 standard only provides a framework for single products, the AIA guide provides a clear application of LCA methodology to the building sector. It also proposes guidelines to integrate LCA in building design.

The AIA guide's steps must always be carried out if an LCA specialist intends to conduct an LCA study. However, there are different approaches to its

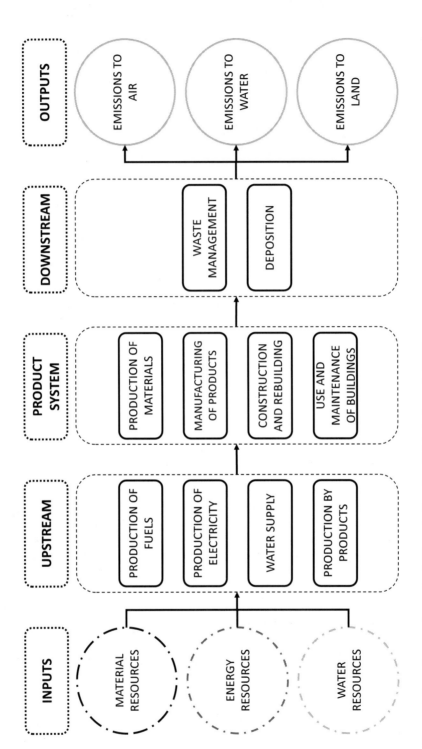

Figure 6.2 Input/output of a product's life-cycle (based on Kulahcioglu et al., 2012).

implementation which vary depending on the goal of the process-based LCA study (Bayer et al., 2010). Examples of such approaches include:

- cradle-to-grave: full LCA, from extraction to disposal phase;
- cradle-to-gate: assessment of a partial product lifetime, from extraction to a specific phase. In AEC sector this represents the manufacturing and construction phase;
- cradle-to-cradle: special kind of cradle-to-grave method, but the end-of-life (EOL) disposal is replaced by the product recycling process;
- gate-to-gate: partial LCA that evaluates only one phase of the whole process (e.g. construction phase).

Life-cycle cost analysis and life-cycle energy analysis

Besides LCA, there are other methodologies that assess the impacts of construction projects, such as life-cycle cost analysis (LCCA) and life-cycle energy analysis (LCEA) (Cabeza et al., 2014). The main difference between LCA, LCCA and LCEA is their focus and how each methodology quantifies the impacts of the construction phase. LCA focuses more on environmental data, LCCA on cost data and LCEA on energy data. LCCA considers the total cost of facility ownership, including owning, operating, maintaining and disposing of a building (Fuller and Petersen, 1996). This method is often used to support decision-making in construction and rehabilitation, as it provides a basis for initial investments accounting for future costs. LCEA is an approach that takes into consideration both direct and indirect energy use as input for the project, including manufacture, use and demolition phases. This analysis contains:

- embodied energy: sum of all the energy needed to create a product (indirect energy);
- operating energy: sum of all the energy needed to maintain a product (direct energy);
- demolition energy: sum of all the energy needed to end the product service and its disposal.

The operation phase accounts for 80–90 per cent of total energy consumption and environmental impact of traditional buildings (Asdrubali et al., 2013; Buyle et al., 2013). This means that it is imperative for designers to make use of as many tools and methodologies as possible that improve energy efficiency as well as employ 'green thinking' from the very beginning of projects. Thus, methodologies that promote life-cycle thinking, such as LCA, LCCA and LCEA should be paired with other AEC methodologies to increase the overall efficiency and sustainability of the construction process (environmental, economic and energy aspect). Governments should endorse such methodologies and develop policies that promote life-cycle thinking, aiming for a more energy efficient society that does not hinder future generations' development.

These methodologies become more relevant when we examine the investment needed to increase building performance, primarily going towards energy efficiency. According to the International Energy Agency (IEA) Special Report (International Energy Agency, 2014), by 2035 USD130 billion will be required worldwide to improve energy efficiency and meet future energy demand. The same report also states that the investment required each year to supply the world's energy needs will rise gradually towards USD2 trillion, accumulating to USD48 trillion by 2035. Of this, only 17.5 per cent will go towards low-carbon technologies such as renewables and nuclear. This investment, however, will help to lower energy consumption by almost 15 per cent by 2035 with respect to the main scenario foreseen in the IEA report. This report also forecasts an increase in investment in low-carbon energy supply of almost USD900 billion by 2035 (IEA, 2014). This increase of energy consumption will also increase greenhouse gas emissions, leading to severe worldwide consequences such as global warming.

More recently, in December 2015, Paris held the United Nations Climate Change Conference (COP21). World leaders agreed to limiting global warming to less than 2°C when compared to pre-industrial levels while striving to limit it to 1.5°C (UNFCCC. Conference of the Parties (COP), 2015). This means that the idea of a global mean-temperature increase is already accepted and considered unavoidable, demonstrating that the only action left is to mitigate the damage. The conference also proposed an annual fund of USD100 billion for mitigation and adaptation to be established by 2020 to limit global warming. To achieve COP21 goals, greenhouse gas emissions have to be measured and reduced, either by tackling CO_2 sources or by creating CO_2 sinks such as forests. To effectively reduce CO_2 emissions, each country must present a national inventory report of anthropogenic emissions by sources and removals by sinks of greenhouse gases. Internationally accepted methodologies, such as LCA, should be used to assess such information and development of that reporting process. Nonetheless, even though COP21 resulted in a worldwide commitment to limit the increase of global mean temperature, as of yet, there have been no tangible measures taken towards controlling the greenhouse gases emission rate.

To achieve such ambitious goals, the AEC sector must resort to a methodology that not only promotes a cooperative environment and eases information exchange, but also has the capacity to integrate environmental impact assessment methodologies such as LCA and LCEA. The BIM methodology can be the answer to such needs. BIM is well documented to enable collaborative environments between different specialties and the creation of different simulations within the same model that help improve communication. Although several studies have been developed in which synergy between BIM software and that of energy simulations and building performance was analysed, very few have analysed the potential to integrate LCA with BIM. If BIM models could also automatically perform an LCA study to quantify the environmental impact of future buildings and infrastructure, it would provide the necessary information to measure the

CO$_2$ emissions generated by the built environment. However, to make this possible, challenges such as the incorporation of the required information in BIM models must be overcome.

BIM–LCA integration

The quality of LCA data is of major importance, and ensuring that the process of data collection is transparent, representative and precise, presents significant challenges. In order to guarantee its quality, it is important to understand how it is structure and organised. LCA data is organised into three different types: generic, average or specific (Silvestre et al., 2015). Generic data usually represents different manufacturers and is developed by using only a portion of the information used for specific data, resulting in a less reliable source of information; examples include the European Life-Cycle Database (ELCD), ESUCO and Ecoinvent databases. Average data result from merging data provided by different manufacturers for the same declared unit such as average Environmental Product Declarations (EPDs). Specific data correspond to data collected at the manufacturer's plant of a single product or brand. However, they do not characterise the impacts of similar products from other producers who may provide the same or a similar product with different impacts and EPDs (Silvestre et al., 2015).

The EPD is part of the specific LCA data. They contain information that is legislated and harmonised according to the EN 15942:2011 standard, which has the purpose of facilitating the communication of a product environmental performance for business-to-business (B2B) (British Standards, 2011). According to Annex A of EN 15942:2011 (British Standards, 2011), the EPDs include:

1   General information of the product.
2   Parameters describing the environmental impacts of the product.
3   Parameters describing the resource use and primary energy use of the product.
4   Parameters describing the resource use, secondary materials and fuels, and use of water of the product.
5   Information regarding waste categories of the product.
6   Output flows of the product.
7   Additional technical information.
8   Additional information on release of dangerous substances to indoor air, soil and water during the use phase.

BIM tools have the ability to incorporate all LCA information in BIM-based objects, allowing for an automatic or semi-automatic LCA, saving time and money and increasing the designers' awareness of the full impacts of their choices. While some tools already exist in the market, such as Tally (Tally, 2014), they require the user to manually identify the materials considered for the project. They act as the intermediary between BIM-based projects and LCA databases,

98   Santos & Aguiar Costa

which is helpful for conducting an LCA study but entails data duplication. For example, Tally uses GaBi's LCA database, which has over 8,000 products used worldwide. During the design phase a lot of data are already included in the BIM model, but when existing LCA software is used, a significant part of this information, such as the construction elements' materials, must be redefined. LCADesign, developed through the Australian Cooperative Research Centre (CRC) for Construction Innovation, is another tool that performs an LCA study based on an automatic BIM quantity take-off, integrating an Australian life-cycle inventory database (Tucker et al., 2003). Nevertheless, unlike Tally, it would be necessary to update its database with their products and energy mix in order to use LCADesign in other countries.

Existing tools such as Tally and LCADesign are able to perform an LCA study using BIM models, but there is no single commercial tool that performs an automatic LCA analysis within a BIM software. The following sections will explore the work done at the University of Lisbon to lay the foundations for an automated LCA study, in which BIM objects possess all the necessary information. As such, the interoperability of BIM–LCA integration is of major importance, as the success of such a tool would depend on information transparency and the user's ability to manipulate the original information to fit their goals.

Exchange information requirements

As an initial step, one must understand how the information required to perform an LCA study can be incorporated in a BIM model. As stated above, parametric objects can offer an adequate answer to such an issue. Parametric modelling makes it possible to incorporate information regarding different specialities in a single object, as well as define respective parametric relationships and constraints (Lee et al., 2006). That information can be either geometric or semantic and is based on the designer's needs. For instance, if a designer intends to run an energy simulation, the BIM objects within the model must contain energy-related information, namely their energy consumption and efficiency. It is also possible to apply the same method in the case of environmental assessment.

Then, the information required to perform an LCA study in an automatic way using the BIM model must be identified. Ideally, it would be pertinent to include all the information contained in the EPDs in the BIM objects; however, such volume of information would definitely increase the size of the BIM files. Therefore, a compromise must be achieved between accuracy and file size/ computational time. The work done by the University of Lisbon proposes a three-level approach based on the three different types of LCA data (generic, average and specific). According to the goal of the LCA study or the level of development of the project, designers would choose which type of LCA data would be more suitable. For example, if we are in the early stages of a project, it would be more suitable to perform an LCA study with generic LCA data, as specific materials to be used are yet unknown. However, if the choice is being made about which products/materials are to be used in the project, it would be more appropriate to use specific LCA data.

In order to be LCA-data-compatible, BIM objects should contain fields for the necessary inputs of each level; this would for example include environmental impact categories described in ISO 14040:2006. Ideally, manufacturers should then be responsible for developing BIM-based objects of their products, providing the specific LCA data, which can be found in the products' EPD. Based on these specific databases it would then be possible to compile information and get average values, which can be used to develop generic and average LCA BIM objects. This chapter will not discuss the information that should be considered at each of the different levels of detail.

In addition to a product's specific information there are other data that would enrich the LCA study. This includes site-specific information that should be added by designers such as:

- transportation type – such as truck, train or ship;
- transportation distance between construction site and the manufacturers;
- energy used in each process; in a column it is necessary to consider the energy used for concrete mixing, steelwork, formwork, concreting and dismantling of formwork;
- water used in each process;
- site location; as each country has its own specific energy mix, which will affect the outputs generated (e.g. CO_2 emissions).

Information integration plays a crucial role in the BIM–LCA approach and an automatic LCA tool must either be able to connect to existing LCA databases or have that information already loaded. Also, it would be preferable that the tool could be capable of generating generic and average LCA data.

In short, to successfully develop an LCA tool that is integrated in a BIM software, manufacturers must contribute by developing their BIM-based products, designers must provide the site-specific information and the tool itself must be connected to LCA databases or already have some information, such as generic and energy mix data (Figure 6.3). If these steps are taken, BIM–LCA integration will allow designers to understand the environmental impacts of their projects in an expeditious manner and therefore contribute to a more sustainable society. However, BIM–LCA tool developers must take into consideration interoperability issues by, for example, promoting the use of open formats such as Industry Foundation Classes (IFC). This would make it possible to use such a tool in a wide range of BIM-based environments without concerns about losing information throughout the project development when different specialities use different software packages.

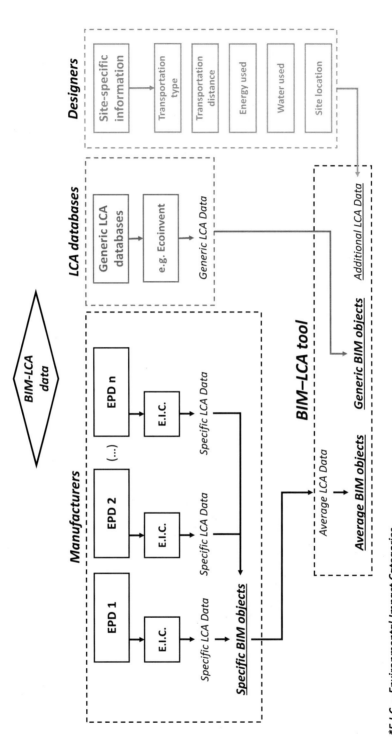

E.I.C. – Environmental Impact Categories

Figure 6.3 BIM–LCA data.

Information integration and interoperability

BIM can support a vision of sustainable growth based on higher resource efficiency through data sharing in the built environment industry. Data sharing is a complex process in which effective rules and controls need to be defined to ensure secure and reliable transactions. This process is generically termed interoperability. Interoperability can be achieved without universally accepted standards, but it compels the project to agree to a set of rules. If no standard is used or available, this process requires a high level of expertise and resource investment, and the ability to re-use the embedded information throughout the life-cycle of the built asset may be limited.

Efforts to address this problem have been undertaken, mainly related to the development of standard formats[2] such as IFC (buildingSMART, 2016; Sanchez et al., 2016; Hampson et al., 2002). IFC is a non-proprietary object-oriented interoperable format for representing information related with a building model. In order to promote the interoperability of BIM-LCA integration, manufacturers should use the IFC schema to create BIM objects of their products.

If a more generalised used of IFC solves the interoperability issue, the next challenge is to understand how this information can be incorporated in IFC-based objects. There are already international recommendations that provide some clues regarding the information that an IFC file may include and how we should proceed if we want to add new information. A good example is the Australian NATSPEC[3] Guide, which includes the NATSPEC National BIM Guide and the NATSPEC BIM object/element matrix, providing suggestions for digital building information exchange (NATSPEC, 2016). The NATSPEC BIM object/element matrix, in particular, is a spreadsheet that can be used for identifying and tracking BIM information during various stages of development of the project (Figure 6.4).

NATSPEC's BIM object/element matrix is organised in three main features: level of development (left), BIM object or element (middle), and general information use (right). The example shown in Figure 6.4 is an adaptation of the 'B2010 Wall-Exterior' sheet of NATSPEC, comprising the most relevant information that readers should retain.

The first column, *level of development (LOD)*, represents the 'progression of an element from conceptual to specified' (BIMForum, 2015). While an LOD 100 object is a generic geometrical representation containing conceptual information, an LOD 300 object will have the exact geometry and contain specific information. An LOD 500 object will be similar to LOD 300 but with additional information such as mounting specification and on-site verification. It is important to highlight that EPDs can be only assigned to LOD 400 and LOD 500 objects, as only these LODs contain information regarding the brand of chosen materials. In lower LODs, only generic and average LCA studies can be performed, obliging the use of generic/average LCA data.

The middle column concerns the information category and item of the BIM object. *Information category* specifies the type of information contained in the

Wall – Exterior	BIM Object or Element		General Information Use
	Item Category – Wall Exterior		Building System
	Description: A 2D and 3D element. A veritcal surface element often attributed to the building envelope. An exterior wall shall prevent the intrusion of the elements. An exterior wall may be a structural or non-structural element.		Item System Category – Uniformat

Level of Development A/A Document E202 – 2008 Developed by Graphisoft 2001	Information Category for Information Item (See Master Information Tab)	Information Item (information about the specific object or element)	IFC Support
LOD 100 – Conceptual			
Overall Building Massing Indicative of Area, Height, Volume, Location, and Orientation	Building Program & Project Meta Data	Facility ID	IfcWall->IfcBuilding.Name
	Building Program & Project Meta Data	Facility Name	IfcWall->IfcBuilding.LongName
	Building Program & Project Meta Data	Facility Description	IfcWall->IfcBuilding.Description
	Physical Properties of BIM Objects and Elements	Facility Length	IfcWall->IfcQuantityLength.Name="Length"
	Physical Properties of BIM Objects and Elements	Facility Weight	IfcWall->IfcQuantityLength.Name="Width"
	Physical Properties of BIM Objects and Elements	Facility Height	IfcWall->IfcQuantityLength.Name="Height"
	Physical Properties of BIM Objects and Elements	Facility Area	IfcWall->IfcQuantityArea.Name="GrossSideArea"
	Physical Properties of BIM Objects and Elements	Facility Volume	IfcWall->IfcQuantityArea.Name="GrossVolume"
	GeoSpatial and Spatial Location of Objects and	Position Type	IfcWall.ObjectPlacement
	GeoSpatial and Spatial Location of Objects and	Location Constraint	(clarify) – Ifc Constraint
	GeoSpatial and Spatial Location of Objects and	Code Constraint	(clarify) – Ifc Constraint
	Costing Requirements	Conceptual Cost	IfcWall->IfcCostValue.CostType="Conceptual"
	Costing Requirements	Conceptual Unit Cost	IfcWall->IfcCostValue.CostType="Conceptual" + UnitBasis
	Costing Requirements	Future Cost Assumptions	IfcWall->IfcCostValue.CostType="Whole life"
	Energy Analysis Requirements	Energy Performance Basis	
	Sustainable Material LEED or Other Requirements	Green Assumptions	IfcWall->IfcEnvironmetalImpactValue of ifcPropertySet with local LEED agreement
	Sustainable Material LEED or Other Requirements	Green Strategies	IfcWall->IfcEnvironmetalImpactValue of ifcPropertySet with local LEED agreement
	Sustainable Material LEED or Other Requirements	LEED Initiatives Bronze, Silver, Gold	IfcWall->IfcEnvironmetalImpactValue of ifcPropertySet with local LEED agreement
	Phases Time Sequencing and Schedule Requirements	Phasing (OmniClass Table –32)	IfcProject->IfcTask.Name (stages) + IfcClassificationReference to OmniClass
	Phases Time Sequencing and Schedule Requirements	Overall Duration	IfcProject->IfcTask->IfcScheduleTimeControl.ScheduleDuration

Figure 6.4 NATSPEC BIM object/element matrix – exterior wall (adapted from NATSPEC, 2016).

Interoperability for BIM-based LCA 103

object, based on OmniClass (functional and performance characteristics), while *Information item* identifies the different variables in each category (e.g. in physical properties category, the items are length, width, height, area and volume).

The last column, on the right, concerns the *general information use* that defines in which IFC classes the information must be inserted, using ISO 16739:2013. Figure 6.4 shows these fields. While some fields already exist in an IFC-based object, others do not, such as the example of 'IfcWall->IfcEnvironmentalImpactValue or ifcPropertySet with local LEED[4] agreement', which corresponds to the 'Sustainable material LEED or other' category. Manufacturers will have to create new classes (*ifcPropertySet*) in the BIM object in order to incorporate the necessary information to conduct an automatic LCA study.

Another initiative that might help to structure BIM objects' information is the *NBS BIM Object Standard*[5] developed by NBS National BIM Library in the UK, which aligns the information contained in BIM objects with COBie[5] (Figure 6.5). In the NBS guide, there is a distinction between generic objects (when designers are not sure of what they intend to use in their projects) and manufacturer objects (when a manufacturer intends to represent a product in a BIM object). Like NATSPEC, the UK NBS object guide also uses ISO 16739:2013 to define objects' IFC 2x3 property sets (e.g. IfcDoorStyle). The NBS object guide also proposes the consideration of new classes based on COBie in BIM objects, related

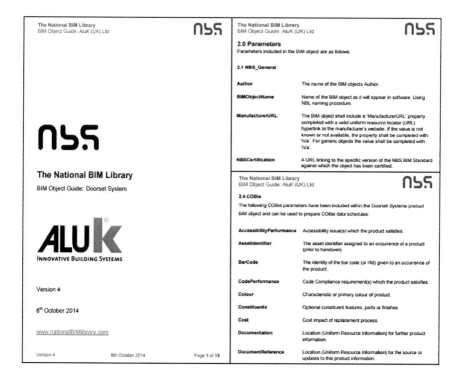

Figure 6.5 Example of NBS BIM object guide.

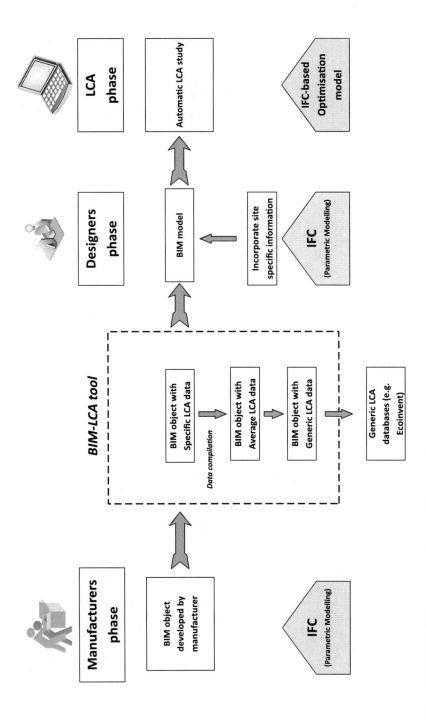

Figure 6.6 BIM–LCA integration framework.

to properties such as AssetType, Manufacturer, ModelNumber, and WarrantyDurationUnit. Most of these properties will be left blank in the case of generic objects. There are cases where products' properties are already defined in other standards, such as BS 8541-4, which define the level of information that objects must have according to specific uses. Manufacturers would benefit from using such standards, to incorporate economic or environmental impact information in their BIM object in a structured manner.

Summing up, for BIM–LCA integration to be successful, three phases must be fulfilled (Figure 6.6). First is the *Manufacturer's phase* which concerns the development of BIM objects by manufacturers with specific LCA data. In order to do so, there are currently some software available to edit IFC-based objects and insert new classes; see for example Vectorworks and ArchiCAD IFC Manager. Then, an automatic BIM–LCA tool should generate generic and average BIM–LCA objects based on specific LCA data and generic LCA databases. In the second phase, the *Designer's phase*, designers would provide additional information through their BIM model in order to conduct a more accurate LCA study. The last phase would be the *LCA phase*, in which an LCA study of the project would be performed, after manufacturers and designers have provided their share of information. In this phase, based on designers' solutions, the LCA tool would connect with LCA databases to extract the required information. As such, the LCA database's format will have to be interoperable with BIM-based software.

Conclusions and future research trends

The present chapter revealed a developing trend in the AEC sector around the integration of LCA and BIM processes and described inherent challenges of this integration such as interoperability of the data exchanged between different actors. The chapter first discussed the potential use of LCA methodology in the built environment and then analysed the literature in the field of BIM–LCA integration, from which it became evident that little effort has been made thus far. As a contribution to the research and development (R&D) community and professionals in the field of sustainable construction, the chapter also identified the required information to perform an LCA study and how that information could be incorporated in a BIM model.

In order for this BIM–LCA integration to be successful, three actors were identified: manufacturers (who develop IFC objects based on their products), designers (who add site-specific information) and LCA developers (who not only create LCA databases but also ensure that those databases are constantly updated). With regard to interoperability issues, the above-mentioned actors should cooperate and use the already existing BIM object guides, so that there is no information loss and the right information is added. This cooperation would ultimately lead to the development of an automatic BIM–LCA tool.

The authors believe that for this tool to be successful, three phases must be fulfilled: *Manufacturer's phase, Designer's phase* and *LCA phase*. An important aspect that must be assured, for an automatic LCA tool to work properly, concerns the format of the information present in the LCA databases. That information must be in a format that is readable by most software, the XML schema being a plausible solution. The developers of LCA databases should be aware of this and start working on an interoperable format that could automatically feed other tools. Governments should endorse the development of a BIM–LCA framework that could assist manufacturers, designers and developers in the automatic integration of BIM with LCA, decreasing interoperability issues. The specialists responsible for developing such a framework would have to consider already existing standards, such as ISO 14040:2006 and ISO 16739:2013, in order to find common ground that would suit all three actors. The R&D sector should also be involved in the development of the BIM–LCA framework and promote the use of open standards such as IFC. The R&D sector ought to raise the awareness of manufacturers and LCA database developers regarding the potential of BIM–LCA integration, as it represents new market opportunities in the AEC sector.

As to future trends in the field of digital integration and more sustainable facilities, the authors believe that optimisation based on environmental, economic and energy analysis will certainly improve the efficiency and performance of buildings, with designers being able to visualise the impacts of their choices. The growing trends towards Big Data is another field of research that may prove to be a source of important development in this area. It is expected that researchers will start studying the potential of data mining for predictive analysis. This may prove crucial to achieving more effective facilities management. Other fields that will likely be further developed are machine learning and time series analysis for the prediction of products' price fluctuation. This may assist the improvement of asset management decision-making processes.

Notes

1 BIM level 2 – 'Federated file-based electronic information with some automated connectivity' – http://bim-level2.org/en/.
2 Standard formats relate to the specifications used to exchange information between different software.
3 NATSPEC is an Australian 'not-for-profit organisation that is owned by the design, build, construct and property industry through professional associations and government property groups' (NATSPEC, 2016).
4 LEED – Leadership in Energy and Environmental Design.
5 NBS – National BIM Library, a BIM library of the United Kingdom, with a collection of BIM objects ranging from building fabric systems to mechanical and electrical objects – www.nationalbimlibrary.com/.

References

Asdrubali, F., Baldassarri, C. and Fthenakis, V., 2013. Life cycle analysis in the construction sector: Guiding the optimization of conventional Italian buildings. *Energy and Buildings*, 64, 73–89.

Bâtiment Numérique, 2015. *Plan Transition Numérique dans le Bâtiment (Operational Roadmap)*. Paris: Bâtiment Numérique.

Bayer, C., Gamble, M., Gentry, R. and Joshi, S., 2010. *Guide to Building Life Cycle Assessment in Practice*. Washington, DC: The American Institute of Architects.

BIMForum, 2015. *Level of Development Specification*. Available at: http://bimforum.org/wp-content/uploads/2015/11/Files-1.zip [Accessed 28 October 2016].

British Standards, 2011. *BS EN 15942:2011 – Sustainability of construction works – Environmental Product Declarations – Communication format business-to-business*. London: BSI.

buildingSMART, 2016. *buildingSMART Home Page*. Available at: www.buildingsmart.org/ [Accessed 28 October 2016].

Buyle, M., Braet, J. and Audenaert, A., 2013. Life cycle assessment in the construction sector: A review. *Renewable and Sustainable Energy Reviews*, 26, 379–388.

Cabeza, L. F., Rincón, L., Vilariño, V., Pérez, G. and Castell, A., 2014. Life cycle assessment (LCA) and life cycle energy analysis (LCEA) of buildings and the building sector: A review. *Renewable and Sustainable Energy Reviews*, 29, 394–416.

Cabinet Office, 2011. *Government Construction Strategy*. London: HM Cabinet Office.

European Commission, 2010. Directive 2010/31/EU of the European Parliament and of the Council of 19 May 2010 on the energy performance of buildings (EPBD). *Official Journal of the European Union*, 153, 13–35.

European Commission, 2012. Directive 2012/27/EU – Energy Efficiency Directive. *Official Journal of the European Union*, 315, 1–56.

European Commission, 2016. *The European Construction Sector: A Global Partner*. Brussels: European Commission.

Fan, S. -L., 2014. Intellectual property rights in Building Information Modeling application in Taiwan. *Journal of Construction Engineering and Management*, 140(3), 04013058.

Fuller, S. K. and Petersen, S. R., 1996. Life-cycle costing manual for the Federal Energy Management Program. In *NIST Handbook*; Gaithersburg: U.S. Dept. of Commerce, Technology Administration, National Institute of Standards and Technology.

German BIM Steering Group, 2015. *Planen-Bauen 4.0*. Available at: http://planen-bauen40.de/ [Accessed 28 October 2016].

Grilo, A. and Jardim-Goncalves, R., 2010. Value proposition on interoperability of BIM and collaborative working environments. *Automation in Construction*, 19(5), 522–530.

Hampson, K. D., Kraatz, J. A. and Sanchez, A. X. 2014. The Global Construction Industry and R&D. In K. D. Hampson, J. A. Kraatz and A. X. Sanchez (eds) *R&D Investment and Impact in the Global Construction Industry*, London: Routledge, pp. 4–23.

Hampson, K. D., Drogemuller, R.M. and Manley, K. J., 2002. *Building an ICT Infrastructure for a National Research Centre*, paper presented at *eSMART Conference*, Manchester, 18–21 November.

ICLEI, 2008. *Green Public Procurement Training Toolkit*. Brussels: DG Environment-G2 European Commission.

International Energy Agency, 2014. *Special Report: World Energy Investment Outlook*, Paris: International Energy Agency.

ISO, 2006. *ISO 14040:2006 Environmental Management – Life Cycle Assessment – Principles and Framework*, Paris: International Organization for Standardization.

ISO, 2010. *ISO 29481-1:2010 – Building Information Modelling – Information Delivery Manual – Part 1: Methodology and Format*, Geneva: International Organization for Standardization.

Kulahcioglu, T., Dang, J. and Toklu, C., 2012. A 3D analyzer for BIM-enabled life cycle assessment of the whole process of construction. *HVAC&R Research*, 18(1/2), 283–293.

Lee, G., Sacks, R. and Eastman, C. M., 2006. Specifying parametric building object behavior (BOB) for a Building Information Modeling system. *Automation in Construction*, 15(6), 758–776.

NATSPEC, 2016. *NATSPEC Home Page*. Available at: http://bim.natspec.org/ [Accessed 01 February 2016].

Ramesh, T., Prakash, R. and Shukla, K. K., 2010. Life cycle energy analysis of buildings: An overview. *Energy and Buildings*, 42 (10), 1592–1600.

Rhodes, C., 2015. *The Construction Industry: Statistics and Policy, Standard Note SN/EP/1432*, London: House of Commons Library.

Sanchez, A., Hampson, K. D. and Vaux, S., 2016. *Delivering Value With BIM: A Whole-of-life Approach*, London: Routledge.

Silvestre, J. D., Lasvaux, S., Hodková, J., de Brito, J. and Duarte Pinheiro, M., 2015. NativeLCA – a systematic approach for the selection of environmental datasets as generic data: application to construction products in a national context. *The International Journal of Life Cycle Assessment*, 20 (6), 731–750.

Tally, 2014. *Tally: Know your Impact*. Available at: http://choosetally.com/ [Accessed 1 February 2016].

Tucker, S. N., Ambrose, M .D., Johnston, D. R., Seo, S., Newton, P. W. and Jones, D. G., 2003. *Integrating Eco-Efficiency Assessment of Commercial Buildings into the Design Process: LCADesign*, paper presented at CIB 2003 International Conference on Smart and Sustainable Built Environment, Brisbane, 19–21 November.

United Nations Framework Convention on Climate Change (UNFCCC). Conference of the Parties (COP), 2015. *Adoption of the Paris Agreement*. Paris: United Nations Office.

Zabalza, B., Aranda Usón, I. A. and Scarpellini, S., 2009. Life cycle assessment in buildings: State-of-the-art and simplified LCA methodology as a complement for building certification. *Building and Environment*, 44 (12), 2510–2520.

Part 2

Processes

7 Precinct Information Modelling

A new digital platform for integrated design, assessment and management of the built environment

Peter Newton, Jim Plume, David Marchant, John Mitchell and Tuan Ngo

Introduction

Cities are among the most complex systems on earth and for most nations their largest national assets and economic engines. In the twenty-first century, they are expected to be home to approximately two-thirds of a forecast nine billion people. Planning and managing the built environments of large, rapidly growing cities in a manner that delivers sustainable, liveable, resilient and equitable urban development constitutes one of the grand challenges facing contemporary societies. There are many critical transitions required for the creation of sustainable built environments (Newton, 2008; Newton et al., 2009; Newton, 2012). *Digital transformation* is one such transition; including the manner in which it can deliver the innovation required across all sectors of industry, including the built environment sector.

Much has been written about the capacity of information and communications technologies (ICT) to radically improve the performance and productivity of built environments, practitioners and their outputs. Over 20 years ago, it centred on capitalising on the emergence of high-speed digital broadband networks to enable real-time design collaboration and closing the information loop across the supply chain (Newton et al., 1993; Newton, 1995); and more recently creating a new platform for information modelling at building scale through Building Information Modelling (BIM) (Newton et al., 2009), with the Cooperative Research Centre (CRC) for Construction Innovation playing a catalytic role in Australia. Focus is now shifting to Precinct Information Modelling (PIM) and in a more general sense to urban information modelling. As focus for a 2013 scoping study by the Australian CRC for Low Carbon Living (Newton et al., 2013), PIM was conceived as a digitally enabled information platform comprising a set of standards and protocols that could harmonise and direct the fragmented activity involving urban modelling of big spatial datasets at precinct scale. Conceptualisations by the International Organization for Standardization (ISO)

112 Newton et al.

in relation to urban sustainability (Figure 7.1) and smart infrastructure frameworks (Figure 7.2) highlight the complexity and dimensionality of the challenge.

The spatial scales for built environment modelling range from product scale objects to building objects to neighbourhood representations and upwards to cities and regions. As scales change so do opportunities for different types of information modelling application: BIM → PIM → GIS, with the need for data interoperability across these key analytical lenses (Figure 7.3). Geographic Information Systems (GIS) emerged in the 1970s, followed more recently by BIM

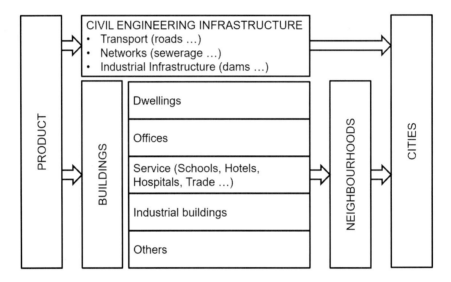

Figure 7.1 The urban sustainability framework: products, buildings, infrastructure, neighbourhoods and cities.

Source: Newton et al. (2009), used with permission.

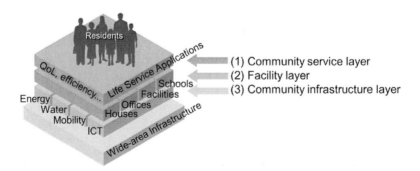

Figure 7.2 ISO model of multiple information layers required for precinct representation and performance assessment of smart sustainable built environment.

Source: Ichikawa (2013), used with permission.

Precinct Information Modelling 113

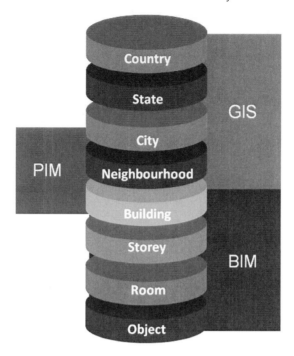

Figure 7.3 Spatial information platforms for the built environment.

to cater for the needs of planners and architects respectively. Urban design spans planning and architecture as well as the GIS-BIM information and modelling environments, operating primarily at neighbourhood/precinct scale. PIM as proposed here is central to this scale of urban innovation.

Precincts constitute the critical operational scale at which a city is assembled (greenfields), re-built (brownfields, greyfields) and operated; and where residents spend large proportions of their day either in domestic or workplace settings (Figure 7.2). They are the 'building blocks' of our cities (Sharifi and Murayama, 2013) and represent the scale at which urban design makes its contribution to city performance. Precincts constitute the origins and destinations for homes, schools, workplaces and recreation, and the trip generators associated with connecting each. In aggregate, they are a microcosm of urban life. It has been argued, however, that the unsustainable nature of today's cities is due in part to poor planning at the neighbourhood level (Codoban and Kennedy, 2008). For example, the high levels of car usage and traffic congestion are a reflection of an absence of mixed-use development, variety in housing types, medium density, and walkability and public transit access having been designed into urban neighbourhoods in recent decades. Purely in CO_2 terms, variability in the housing and transport attributes of different suburbs means that neighbourhood-scale carbon emissions can vary by as much as 50 per cent across low-density, car-dependent cities (Newton et al., 2012; Crawford and Fuller, 2011). Precincts constitute a critical focus for the achievement of any carbon neutrality target for cities since it is the scale at which an optimal combination of urban design innovation, urban technology innovation and behaviour change can jointly occur.

At precinct scale, the volume of data and information required to effectively model the built environment expands significantly beyond that required for an individual building: more than 7,000 individual objects for a city building, based on an examination by the authors of a range of typical commercial BIM models produced by architects, and not including detailed services information. It is at this scale where a convergence of digital technologies is required to support built environment planning and management. PIM is emerging as a critical platform for more effective planning, design and management of relevant spatial data at that scale. The operational structure of PIM is explained in a later section of this chapter. Since it is held in an accessible open standard format, any precinct can be modelled to accommodate the disparate needs of the range of analysis and operational activities that support more sustainable performance of precincts throughout their life-cycle (Figure 7.4). This is a fundamental, but still almost universally absent, component of urban development at either building or precinct scale.

A review of precinct scale assessment and rating systems, undertaken by the authors, reveals an increasing demand for tools that respond to the broad 'goals for built environment performance' established by national, state, metropolitan and local governments: sustainability, resilience, liveability, productivity and equity (Newton et al., 2013; Figure 7.5). These goals stimulate the development of performance indicators across core built environment systems such as energy, water, transport and waste, and more broadly into other areas such as human health and urban microclimate (Department of Infrastructure and Regional Development, 2011). 'Assessment tools', alternatively termed design decision support, model the performance of the precinct across core built environment systems such as energy, water, transport and waste, as well as the interactions and interdependencies that are in play, such as water–energy nexus and carbon mitigation–adaptation nexus. These tools rely on varying levels of computational sophistication as well as data and are continually evolving. 'Rating tools' (such as Green Star Communities, EnviroDevelopment, IS Rating, Building Research

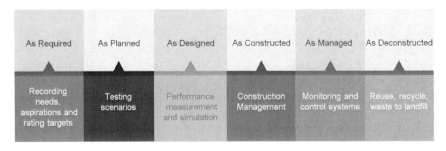

Figure 7.4 Temporal information platform for the built environment: precinct life-cycle.

Establishment Environmental Assessment Methodology (BREEAM) and Leadership in Energy and Environmental Design (LEED)) take the outputs from the assessment modelling as a basis for assigning weights ('importance') and ratings for use in industry 'labelling' or certification. Currently, most precinct assessments and ratings are evaluated against sets of benchmarks established by industry groups and/or governments. A PIM digital platform is designed to effectively support all the operations depicted in the precinct performance assessment framework (Figure 7.5). To date, only a small number of tools, such as MUtopia (Ngo et al., 2014) and ESP (Trubka et al., 2016) have been developed to undertake precinct assessment by taking advantage of PIM, object-oriented modelling and 3D visualisation. PIM-enabled data assembly and software application feature in the case studies reported at the end of this chapter. Before then, the following sections will address the geospatial context that has enabled PIM to emerge and the operational structure of PIM circa 2016.

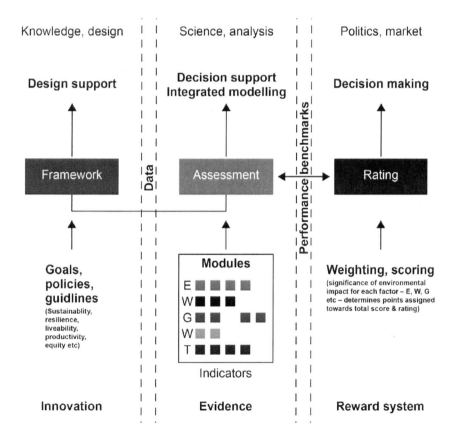

Figure 7.5 Precinct performance assessment framework.

Source: Adapted from Newton et al. 2013, used with permission.

116 *Newton et al.*

Towards a digital built environment

It is important to recognise that PIM is only a part, albeit a very significant part, of a much bigger conception of the emerging role of digital technologies throughout the world. To appreciate that context, the idea of a digital built environment (DBE) encompasses all the digital technologies that are in widespread use today, and indeed still evolving, to help us understand and manage the physical built environment. As argued elsewhere:

> We envision an inevitable shift towards a world in which our interaction with the physical world is increasingly facilitated through digital technologies that rely on data and information, either to inform the decisions that we take, or, where appropriate, form the basis for the autonomous response of entities acting for our benefit in the physical world.
>
> (Plume, 2015)

This is a bold vision. It recognises that the vast amount of information we now create, collect and hold in diverse databases and models, using a plethora of technology, tools and innovations, can be linked to the physical world that we inhabit in order to activate our engagement and interaction with our environment. Imagine a scenario in which you are able to sense, through your own smartphone, not only all aspects of the current operational state of the physical world around you, but also its wider demographic, socioeconomic, environmental, regulatory and institutional context, and perhaps also the intent that lies behind the planning and design of that built environment. With all that information available, new apps will be developed that make use of that data to shape our experience and enjoyment of the world in ways that we cannot yet imagine. Similarly, the physical built environment will detect our presence and use whatever personal information we choose to reveal about ourselves to formulate responses to our needs and aspirations at that time and location.

This scenario, of course, paints a picture from our perspective as end-users and citizens, but there is an even more significant role for those who act as custodians of the built environment, particularly the diverse responsibilities of built environment professionals who range across a very broad spectrum from planners, architects and designers, engineers of many hues, constructors and trade specialists, through to asset, facility and operational managers. All these disciplines seek access to fully integrated information about the built environment to enable the planning, design, construction, management and operation of the constructed world. Although much of that information already exists, the vision of a fully integrated digital built environment would facilitate its access.

The challenge is to develop robust ways of integrating all that information across the entire life-cycle of built facilities and infrastructure, ranging across all scales of development from urban land, buildings, service utilities (e.g. water, energy), transport network infrastructure, civil and landscape engineering projects, all the way up to an urban or regional scale.

At the heart of this vision are two traditional modelling or digital prototyping technologies, falling broadly into either the spatial domain or the design/construction domain (Plume et al., 2015). The modelling technologies for the spatial domain are often characterised as GIS. This generalisation, however, underrates the scope of technologies now associated with spatial modelling. These include satellite imaging, photogrammetry, laser scanning and Light Detection and Ranging (LiDAR), city modelling, smart city technologies (the Internet of Things), global navigation satellite systems (GNSS) and emerging initiatives like the Digital Earth. The modelling technologies associated with the design and construction domain are similarly characterised as BIM. This again fails to recognise the ever-widening scope of its application to all forms of civil, transport, mining and utility infrastructure and the expansion of its scope to encompass software technologies that accommodate both policy and processes associated with managing the built environment.

Most of these tools and approaches are developed as proprietary technologies and are often unable to share information across those software platforms in a reliable fashion. Although open information exchange formats exist in both domains, their development relies on limited funding and struggles to keep pace with the rapid innovation that is possible in the commercial sphere. At a project level, where industry is under constant pressure to innovate and deliver, there is a tendency to rely on a proprietary suite of software tools supplied by a single vendor. Robust open standards can break this reliance on exclusive proprietary software systems and serve to encourage wider commercial software innovation based on vendor-neutral data formats.

buildingSMART is a worldwide not-for-profit industry organisation that has developed open standards for exchanging BIM data, as well as standardised processes and technologies needed to support collaborative design (buildingSMART, 2016). There has been a shift, since around 2013, to apply these standards to transport infrastructure projects (roads and railways) and civil structures such as bridges and tunnels. The Open Geospatial Consortium (OGC) is the complementary global standards organisation that serves the spatial sector by developing standards for the delivery and management of spatial data across the internet (OGC, 2016). It is well known for its CityGML and Indoor GML standards.[1] Currently, OGC is actively working towards a new standard known as InfraGML, designed specifically to address the modelling of infrastructure elements of the built environment.

These modelling approaches have been traditionally seen as complementary, each addressing exclusively their own disciplinary needs, but there is a growing recognition that in order to address the pressing needs of global urbanisation, climate change, carbon accounting and management and urban resilience, it is critical to find ways of integrating information across these two domains. There are a number of initiatives in different parts of the world to bring these two domains together (JBIM, 2010; Galbraith, 2015; Gomez Zamora and Swarts, 2014; Hobson, 2013; Mommers, 2014).

118 *Newton et al.*

In order to realise the full vision for an effective, digitally enabled built environment, there needs to be a way of accessing that information when required. That leads to the set of facilitating technologies that are generally associated with the spatial industry: the internet for transporting the information; the semantic web to enable smart ways to find and retrieve unstructured contextual information; geolocation technologies to enable searching based on geographic context; and radio-frequency identification (RFID) with sensors to facilitate the Internet of Things to create a 'sensate environment'.

An operational structure for a PIM platform is proposed in the next section, followed by two case studies that illustrate its implementation.

Operational structure for a PIM platform

Precincts typically are composed of multiple cadastral entities (legal ownerships), containing built facilities and infrastructure (roads, railways, bridges, service utilities) and/or natural features. They are also occupied and used by people, and governed by organisations. At present, an interested party may be able to collect data about particular aspects of a precinct. The elements of that dataset are, however, often disaggregated within various databases, modelled using differing data definitions, held under different ownerships and, consequently, have potential access impediments. A shared PIM, based on an open standard, can be created and maintained to address these issues.

Figure 7.6 illustrates the precinct modelling strategy that is being developed as part of a collaborative research project (detailed in two use-case projects). The following sub-sections explain each part of this figure.

Data schema

All data models require a formal definition of the structure that will be used to store the data. This is referred to as the 'data schema'. The proposed PIM schema extends the current version of the international standard for building information (IFC4, 2013) to include new infrastructure and cadastral entities. In order to do this, the PIM research team proposes a new 'facility' concept that serves to collect together all built 'things': linear entities such as roads, railways and waterways, bridges and tunnels, as well as individual buildings or complexes of buildings. A further generalisation is proposed for subdivisions of these entities. While storeys are vertical subdivisions of a building, many linear infrastructure entities such as roads and railways can be defined in terms of horizontal spatial segments. For example, a road is composed of segments (length of road between intersections) and the intersections themselves. It is perhaps useful to go one level deeper in the spatial hierarchy and consider traffic lanes within road segments in the same way we think of spaces within a building storey. Similarly, an important bridging concept between the urban and building-level scales is the idea of a planning zone. The Industry Foundation Classes (IFC) standard already has a spatial zone entity, which can be used for schematic precinct planning. The PIM team

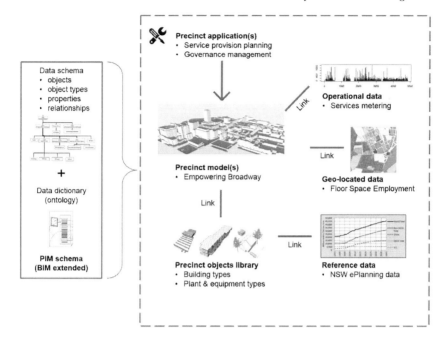

Figure 7.6 Precinct Information Model for an urban retrofit project.

proposes the need for a new specialisation of this spatial zone entity to hold the legal and spatial definition of property. In this way, cadastral entities at the building scale are the lots on which built facilities exist, and at the urban scale they are the fundamental spatial units of local environment and regional plans.

Data dictionary

The challenge when extending a data schema to include new concepts, such as those described for infrastructure and cadastral entities, is to limit the complexity of the schema as a whole so that it remains sufficiently expressive, without becoming too cumbersome to implement in software. The second component of the PIM schema, shown in the box at the left of Figure 7.6, is a data dictionary (ontology) that provides a solution to this issue. Within a PIM data model, instances of a generically defined PIM entity such as the spatial zone can be categorised by means of a reference to the relevant concept in the data dictionary. An example of such a dictionary is that developed by buildingSMART (bsDD, 2015). This is a repository of concepts, their definitions in multiple languages and the relationships between concepts. Each concept in the data dictionary is uniquely identified with its own globally unique identifier, so when used against an entity instance in the PIM model, the definition of that entity is unambiguous. For PIM purposes, all the land use and development types as published by the New South Wales state government (NSW Government, 2015) have been added

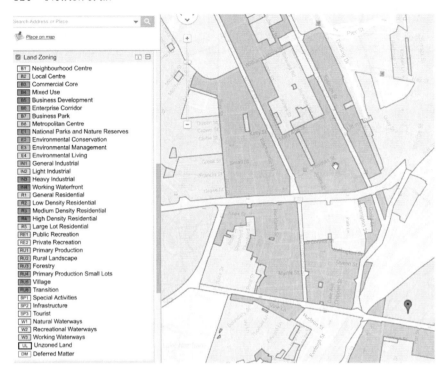

Figure 7.7 Land use map of the 'Empowering Broadway' precinct.

Source: NSW Department of Planning, public web site.

to the dictionary as a preliminary trial of this methodology. Figure 7.7 shows the use of these concepts to depict local environment planning zones for a precinct in the central Sydney metropolitan area.

The authors envisage that as governments adopt open data formats, each authority will maintain their own database and make this data publicly accessible in a standardised, useable form. The definitions of land use and development types are then expected to be uniform across all authorities referenced from an open data dictionary, bsDD or otherwise, and only the choices of development types that are permitted for a given land use will vary between authorities.

Levels of development

Precinct information models may be broadly characterised at three levels of development (Figure 7.8).

Figure 7.8 (a) Zonal functional typology – schematic master plan; (b) Built facility type – development proposal; (c) Elemental object typology – detailed asset model.

122 *Newton et al.*

Zonal functional typology

A precinct model at initial concept stage will designate geographic zones or simple volumes and spaces to represent the broad-level activities in the precinct. These zones have intrinsic geometric properties such as dimensions, volume and area; specified functional usage such as residential or commercial; possible performance properties including energy, carbon, water and waste; or cost properties such as construction and operational costs.

Built facility typology

At a more detailed level, the functional uses of space are modelled in 3D form based on approximations of the scale of development required and with more attention to the relationships between the spaces. Traditionally, building types are classified in this way for costing purposes. In Australia for example, the National Construction Code (NCC) Building Class Table (ABCB, 2016) and the Rawlinsons (2011) extended *Building* Types Classification both provide this type of classification. PIM however requires more comprehensive built asset type definitions, in particular for infrastructure entities or civic spaces.

Elemental object typology

At the most detailed level, all infrastructure and buildings are increasingly authored in BIM tools; with all spatial and physical elements described precisely including detailed properties of the specified products.

Interaction with the precinct information model data

The precinct model itself is stored in an object-oriented database that can be accessed remotely via secure login. Users can be assigned to groups with access rights to all or part of the database. The data can also be accessed via web requests; this functionality allows software developers to create either web-based or standalone applications that interact with relevant data entities in the model for particular purposes. This is done for example to allow precinct stakeholders to review and add their requirements and performance objectives against the existing precinct entities in the model. Figure 7.9 shows the PIMViewer software that has been written as an example of such an application. It can open a model across an internet connection to the server, edit that model and save the amended model either to a model file on the user's computer or merge it back into the source model on the server.

Precinct object library

In BIM, a type is a standard object that can be accessed from a library of objects. The type defines the default properties, including geometry where relevant, that

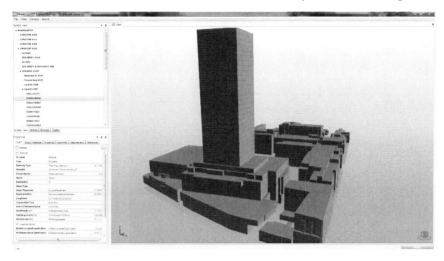

Figure 7.9 PIMViewer application.

will be used when instances of that type are added to the project model. Properties of an individual instance of a type may be set to vary from the default properties as required. For the PIM research, the IFC BIM data schema is extended to add several new types to correspond to the additional spatial and infrastructure entities discussed earlier in this chapter. The use of types in association with data dictionary concepts should allow for more flexibility to accommodate many use cases across geographic jurisdictions and language groupings in a much more flexible, consistent and robust manner.

References to associated data held in other data sources

Three sources of reference data are shown on the right-hand side of Figure 7.6. Operational and reference data is linked to associated entity instances in the PIM project model using references. There are a number of optional properties of a reference: a description of the referenced data source, a uniform resource locator (URL) that defines where to look on the web, and what parameters are required to access a particular set of data from the target site. A contemporary example of this approach is given in van Nederveen et al. (2015). The URL mechanism will be subject to review for PIM if and when an alternative standardised construct becomes widely adopted by open data providers, especially government authorities. Geo-located data is data that includes longitude and latitude coordinates, and standard orientation. The site and cadastral entities in the PIM data schema both include these properties. On the basis of this commonality between the PIM project model instances and the referenced data, selections from the reference source can be viewed in association with, or merged into the model and correctly located in space.

124   *Newton et al.*

Required functionality of a precinct information model

In summary, to be effective and useful for the many stakeholders who may interact with precinct information, the data model should include the following features:

- standardised identification, definition and management of precinct entities including units of measurement;
- stakeholder concepts; ownership, operational responsibility and user status;
- a standardised method to enable a given precinct model to be easily co-located within a common spatial context to other precincts or larger scale models;
- support for multiple performance objectives related to the precinct entities;
- method(s) to link entity instances that are within the current precinct information set to appropriate reference data that exists elsewhere;
- a filtering capability to enable ad hoc and/or defined queries (model view definitions) to provide relevant data to industry software tools;
- support for analytics at various levels of granularity (by land use zone, built facility type or component product element);
- metadata associated with data entities to indicate data provenance and reliability / level of accuracy.

Application of PIM to built environment precinct design projects

One of the grand challenges for sustainable urban development this century is the regeneration of the established and poor-performing built environments in the inner and middle suburbs of large, fast-growing cities. Precincts represent the scale at which this renewal needs to occur. The key arenas where such precincts exist are:

- central business districts (CBDs); CBDs and major activity centres of cities;
- brownfields; abandoned industrial and commercial property, usually in central and increasingly sought after locations;
- greyfields; the extensive, ageing but occupied residential areas of the inner and middle ring suburbs where the land is the asset, and the properties are prime for more intensified forms of redevelopment.

(Newton, 2016)

The case studies briefly outlined below involve PIM-based data definition, assembly and modelling of precinct-scale regeneration being explored for the Broadway precinct in Sydney's CBD and Fishermans Bend, a 240-hectar brownfield redevelopment adjacent to Melbourne's CBD.

Case Study 1: precinct infrastructure retrofit

The Broadway precinct, located on the edge of the Sydney CBD, is a unique opportunity to implement a precinct model as the project is retrofitting a dense urban area undergoing significant change. It is the focus of the Empowering Broadway Project (CRC LCL, 2015) that is examining ways to develop the governance protocols associated with multiple owners intending to share common energy and water services (distributed energy generation and integrated water technologies utilising stormwater harvesting, wastewater treatment and re-use) involving long-term commercial contracts.

The precinct spans several existing development types:

- the redevelopment of the old Carlton United Brewery site, which created Central Park, a brownfield redevelopment providing a 'clean sheet' for new facilities, but requiring significant site remediation and utilities investment. Buildings include premium residential, shopping and retail, and commercial offices with significant green infrastructure and excess capacity associated with a state-of-the-art water treatment and trigen energy plant;[2]
- an education campus offering trade-based courses, dating to the mid 1800s, with a large proportion of heritage and/or old buildings, poor environmental performance of the building stock in general and increasing energy consumption;
- an ambitious and expanding university, proactively upgrading its facilities to support the new learning needs of the digital age with a large existing and dense building stock, having to identify complex renewal/growth options through replacement of existing structures, and having increasing energy needs.

The stakeholders are initially focusing on issues related to developing a governance framework with respect to the mechanics of sharing local energy and water resources. The presumption is that the operating profiles of the three participants will lead to aggregate lower levels of energy consumption, and significantly reduced operational costs and embodied carbon profiles. Thus, the complementary goals of the project are to investigate the *commercial* feasibility of shared precinct infrastructure water + energy, and to develop *a precinct model that hosts the data* for those assessments and scenario modelling.

In order to support this work, the precinct model must consist of four categories of information, serving to define the required scope of the PIM:

- a representation of the physical entities that make up the precinct containing appropriate general property data;
- information relating to the planned performance characteristics of those precinct entities;
- the ability to hold actual performance data, often collected in real time and perhaps aggregated in an appropriate manner;
- information about the people associated with the precinct, their roles, ownership and responsibility with respect to those entities.

This information is required across all built asset types for any PIM. In all of these categories, acquiring definitive, up-to-date and consistent data is a major challenge, not only with the direct participants of the organisations/owners engaged in the project, but external government authorities and service providers.

The Broadway project has benefited from a City of Sydney dataset developed to track changes in development patterns over its jurisdiction. The floor space employment survey that is undertaken every five years records a site by its cadastre, all buildings on the site, all storeys in the building and then all spaces (rooms) on the storey. Importantly it uses two classifications to identify precisely the '*type of business enterprise*' and the specific '*activities*' hosted in the rooms.

The dataset for the Broadway precinct comprises some 379 sites, 440 buildings, 1,404 storeys, 830 businesses and 16,000 rooms. The value of this dataset is its comprehensive coverage, the consistency of representation and the classification of the business types and the activities undertaken in their facilities. For PIM purposes, this has served as a valuable base dataset, but lacks the basic building fabric and services data that is needed for carbon and other calculations. The conversion of this data from its GIS implementation into open BIM format (IFC) results in models as those visualised in Figure 7.10.

To date, a major part of work undertaken has been in the physical modelling and the primary classification systems used by state and local governments. Scenario modelling of current and future demand based on facility populations and consumption rates, as well as utility supply data, for existing and new services is scheduled to occur once governance details have been finalised and precinct partners provide their proprietary data. Once a PIM has been established, it becomes the framework for the next stages of retrofit design and construction and then operations. The integrated data model provides a 'life-time model' repository to enable the transition to a more sustainable operation. PIM becomes the foundation for more sophisticated eco-efficiency-oriented asset management system.

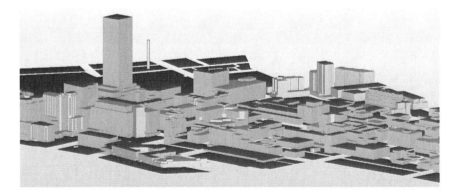

Figure 7.10 BIM model derived from City of Sydney 2012 FES data (view looking south-east from Darling Harbour across to Central Railway).

Case Study 2: brownfields precinct modelling

The precinct modelling illustrated in the following case study employs the latest generation of object-oriented, web-enabled, open-source software designed for application to a sustainability assessment of alternative precinct redevelopment scenarios within a rapidly evolving PIM environment. The software application (Figure 7.6) involved is MUtopia (Ngo et al., 2014). Here, integrated domain models (for energy, water, waste, transport, green space and demographics, among other data) inform a 3D spatial platform supporting urban infrastructure modelling for sustainable design. Models are simulated under alternative scenarios to generate customisable 2D and 3D reports. Outputs are provided for a wide spectrum of themes and indicators such as liveability, carbon emissions, water consumption, travel time, waste generation and life-cycle cost. The MUtopia tool is informed by the Australian Green Building Council's Green Star tools, One Planet Living and LEED frameworks, and sustainability principles more generally. These three rating systems help specify the data required in the fields of energy, water, waste, transport, food, liveability and governance. This forms the basis for the MUtopia tool. Its key features are: open, scalable and adaptable, cloud-based architecture; integrated GIS + BIM using PIM; advanced visualisation capabilities for rendering and reporting; predictive modelling capabilities, what-if scenario simulation; multi-user architecture, collaborative design and simulation platform; and public engagement capabilities via a web portal for community consultation. The tool can be used in multiple phases of a development project such as preliminary planning, stakeholder communication, master planning, community consultation, design and monitoring.

The value of a PIM-enabled precinct assessment and visualisation tool at preliminary planning stage was demonstrated by the application of MUtopia to a rapid one-month envisioning exercise focusing on a low carbon future for Fishermans Bend. Fishermans Bend is a 250-hectare brownfield precinct adjacent to the CBD in Melbourne required to accommodate around 120,000 residents and 60,000 commercial jobs over the next 40 years. It sought an urban design response for its buildings that aspired to be:

- low carbon: ultra-energy efficient buildings with maximum use of renewable energy;
- biophillic: optimising the exposure of buildings and their occupants to natural elements through, for example, green walls;
- water sensitive: minimising the import of potable water into and export of wastewater from the precinct by maximising the use of rainwater harvesting by buildings and greywater recycling within buildings.

For MUtopia modelling of the future Fishermans Bend precinct, a spectrum of building star ratings ranging from current practice to international best practice were employed. This was done to provide an estimate of the total amount of

128   *Newton et al.*

energy, water and CO_2 emissions associated with the required stock of residential and commercial buildings. This enabled a comparative performance assessment to be made as to the scale of environmental benefits to be achieved from adopting current versus best-practice performance targets in design briefs for developers. The latter represents what should be prescribed as a target for those developers wanting to be involved in the creation of Melbourne's largest inner city precinct. A full report on 'Ideas for Fishermans Bend' is to be found in CRC for Water Sensitive Cities and CRC for Low Carbon Living (2015); these documents report only on the MUtopia modelling undertaken for the residential apartments.

Using world leadership performance, a 10-star NatHERS[3] residence is expected to reduce operational energy by 27 per cent when compared to the currently mandated six-star dwelling. This is principally the result of the reduction of space heating and cooling requirements. Further reductions in energy demand of approximately 21 per cent can be gained from the installation of more efficient appliances and lighting. Energy efficiency gains beyond the building envelope, lighting and appliances can be derived from a decrease in hot water usage. Hot water usage reductions are obtainable with the installation of water-efficient appliances such as taps, shower heads, dishwashers and washing machines; delivering a 33 per cent reduction in hot water usage when compared to business-as-usual (BAU) practice (Australian Government, 2016). Further decreases in hot water usage could be obtained through behavioural adjustments such as reduction of shower length (Athuraliya et al., 2012) and, when combined with water-efficient appliances, can further reduce residential energy demand by 20 per cent.

Finally, local distributed energy systems were explored in order to determine the possibility of delivering carbon neutral or carbon negative precincts. Solar photovoltaics and storage appear most prospective at present. Given the likely development of approximately 180 new apartment buildings at Fishermans Bend, it is anticipated that approximately 143,000m² of roof area will exist. Accounting for services, walkways and a packing factor of 70 per cent for minimal shading it is therefore anticipated that a maximum of 14.4megawatts (MW) of solar photovoltaics could be installed. This solar photovoltaic (PV) installation will produce in the order of 60,800GJ of energy per year. For the BAU residential 6-star apartments this accounts for approximately 3 per cent of annual residential energy requirements. However, with the 10-Star building, water and energy efficient fixtures and behaviour adaptation scenario, a more sizeable 10 per cent of the residential energy demand can be accommodated by renewables. Combining all energy reduction measures (10-star NatHERS, energy efficient built in and plug in appliances, conservation behaviours by households and rooftop PV) a potential reduction in residential energy demand of 70 per cent could be achieved (Figure 7.11).

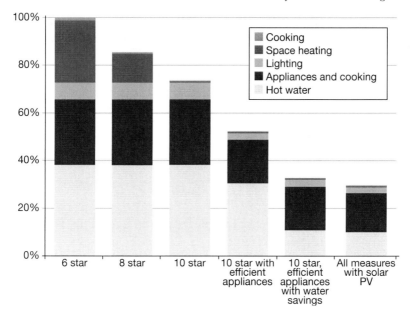

Figure 7.11 Alternatives to BAU energy scenario for new residential building at Fishermans Bend.

Conclusions

PIM is in its infancy, perhaps equivalent to where BIM was 20 years ago, but the frameworks and initiatives outlined in this chapter suggest significant promise for accelerated activity once its full potential has been recognised. Reflecting on the drivers that have led to the present international endorsement of BIM in the building and construction sector, we can identify several factors that will be needed to drive the widespread adoption of PIM. Academic research will provide the underpinning rationale for schemas and object libraries, among others. Prototype implementations and applications as discussed in this chapter will raise its profile by demonstrating the value that can be added to an urban design project with PIM-enabled databases and software.

The development of information exchange standards, especially those that cross the current divide between the urban planning and building/construction domains, will provide the platform for innovation using PIM. At the same time, PIM must be seen and applied as a collaborative tool that engages multiple disciplines, addressing different views of the information throughout the entire life-cycle development of streetscapes, city blocks, neighbourhoods and larger urban districts. This will include multiple stakeholders, such as property owners and managers, the design professions (architects, planners, landscape architects and engineers), local and state government agencies and end-users of the built environment. Positive support needs to emerge for the move to PIM, endorsed and even mandated by major public and private stakeholders.

130 *Newton et al.*

All this needs to be encouraged by software developers who see the opportunity to innovate and build new tools that employ precinct scale models that link into the emerging sources of open data to inform all aspects of our interaction with the built environment. Finally, there must be a willingness across all stakeholders to make a break with the past and adopt a new way of working. Indeed, the shift from current ways of representing, modelling and managing the built environment is as radical as the shift from 2D drawing to 3D modelling during the late twentieth century. Precinct Information Modelling offers a new and radical way of conceptualising how we manage our world. The opportunities are enormous if we can accelerate and scale up this transition.

Notes

1 GML stands for Geography Markup Language and is an XML grammar developed by OGC to describe geographical features.
2 A trigeneration (trigen) plant is a facility that uses the one energy source, typically natural gas or biomass, to simultaneously generate electrical energy, while using the waste heat for both direct heating and cooling.
3 NatHERS (Nationwide House Energy Rating Scheme) is Australia's energy rating system for residential buildings.

References

Athuraliya, A., Roberts, P. and Brown, A., 2012. *Yarra Valley Future Water: Residential Water Use Study Volume 2-Summer 2012*, Victoria: Yarra Valley Water.

Australian Building Codes Board (ABCB), 2016. *National Construction Code*, Australian Building Codes Board. Available at: https://services.abcb.gov.au/NCCOnline/ [Accessed 9 February 2016].

Australian Government, 2016. *Water Efficiency Labelling and Standards Scheme*, Canberra: Australian Government.

bsDD, 2015. *buildingSMART Data Dictionary*. Available at: http://bsdd.buildingsmart.org/ [Accessed 7 February 2016].

buildingSMART, 2016. *buildingSMART: International Home of OpenBIM*. Available at: www.buildingsmart.org/ [Accessed 15 February 2016].

Codoban, N. and Kennedy, C. A., 2008. The metabolism of neighbourhoods. *Journal of Urban Planning and Development*, 134(1), 21–31.

Crawford, R. and Fuller, R., 2011. *Energy and Greenhouse Gas Emissions Implications of Alternative Housing Types for Australia*, paper presented at State of Australian Cities National Conference, Melbourne, Australia, 29 November–2 December.

Cooperative Research Centre for Low Carbon Living (CRC LCL), 2015. *Retrofitting Urban Precincts to Create Low Carbon Communities*, project RP2018, Low Carbon Living CRC. Available at: http://lowcarbonlivingcrc.com.au/research/program-2-low-carbon-precincts/rp2018-retrofitting-urban-precincts-create-low-carbon [Accessed 15 February 2016].

Department of Infrastructure and Regional Development, 2011. *Creating Places for People: An Urban Design Protocol for Australian Cities*, INFRA 1219. Available at: www.gbca.org.au/uploads/244/36084/Creating%20Places%20for%20People%20-%20An%20

Urban%20Design%20Protocol%20for%20Australian%20Cities.pdf [Accessed 2 March 2016].

Galbraith, I., 2015. *Digital Infrastructure – The Future of Engineering*. Online. Available at: www.infrastructure-intelligence.com/article/may-2015/digital-infrastructure-%E2%80%93-future-engineering?utm_medium=email&utm_source=transactional&utm_campaign=weekly-email [Accessed 15 June 2015].

Gomez Zamora, P. and Swarts, M., 2014. Campus landscape information modeling: Intermediate scale model that embeds information and multidisciplinary knowledge for landscape planning, *Proceedings of the XVII Conference of the Iberoamerican Society of Digital Graphics: Knowledge-based Design*, Sao Paulo, Blucher Design Proceedings, 1(7), 61–65.

Hobson, A., 2013. *Technology, Innovation, and Collaboration in Project Delivery*. Network: Pathways to Innovation, A technical journal by Parsons Brinckerhoff employees and colleagues, 76, 21–23, Available at: www.pbworld.com/news/publications.aspx [Accessed 11 July 2015].

IFC4, 2013. *Industry Foundation Classes IFC4 Official Release*. Available at: www.buildingsmart-tech.org/ifc/IFC4/final/html/index.htm [Accessed 7 February 2016].

Ichikawa, Y., 2013. *ISO/TC268/SC1 – Smart Community Infrastructures*, PowerPoint Presentation, 05 February 2013, International Telecomunication Union. Available at: www.itu.int/en/ITU-T/jca/ictcc/Documents/docs-2013/YoshiakiIchikawa_JCA_Feb2013.pdf [Accessed 6 March 2017].

JBIM, 2010. *Journal of Building Information Modelling, Fall 2010*, Washington: National Institute of Building Sciences, buildingSMART Alliance. Available at: www.scribd.com/document/43227845/Journal-of-Building-Information-Modeling-Fall-2010 [Accessed 6 March 2017].

Mommers, B., 2014. The crossover revolution, *Geospatial World*, March,62–63. Available at: www.bimtaskgroup.org/wp-content/uploads/2014/04/crossover.pdf [Accessed 10 November 2016].

New South Wales (NSW) Government, 2015. *NSW Planning Legislation*. Available at: www.legislation.nsw.gov.au [Accessed 07 February 2016].

Newton, P., 1995. Virtual project teams. In: M. A. Fisher, K. H. Law and B. Luiten (eds) *Modeling of Buildings Through their Life Cycle*, CIB Publication 180, California: Stanford University.

Newton, P., 2008. *Transitions: Pathways Towards Sustainable Urban Development in Australia*, Dordrecht, The Netherlands: Springer.

Newton, P., 2012. Liveable and sustainable? Socio-technical challenges for 21st century cities, *Journal of Urban Technology*, 19(1), 81–102.

Newton, P., 2016. Framing new retrofit models for regenerating Australia's fast growing cities. In: T. Dixon and M. Eames (eds) *Retrofitting Cities for Tomorrow's World*, London: Wiley-Blackwell.

Newton, P., Hampson, K. and Drogemuller, R., 2009. *Technology, Design and Process Innovation in the Built Environment*, Spon Research Series, London: Taylor & Francis.

Newton, P., Marchant, D., Mitchell, J., Plume, J., Seo, S. and Roggema, R., 2013. *Performance Assessment of Urban Precinct Design: A Scoping Study*, Low Carbon Living CRC, August 2013. Available at: www.lowcarbonlivingcrc.com.au/resources/crc-publications/reports/performance-assessment-urban-precinct-design-scoping-study [Accessed 15 February 2016].

Newton, P., Pears, A., Whiteman, J. and Astle, R., 2012. The energy and carbon footprints of housing and transport in Australian urban development: Current trends and future

132 *Newton et al.*

prospects. In: R. Tomlinson (ed.) *Australia's Unintended Cities*, Melbourne: CSIRO Publishing.

Newton, P., Wilson, B. G., Crawford, J. R. and Tucker, S. N., 1993. Networking construction: Electronic integration of distributed information. In: K. Mathur, M. Betts and K. Tham (eds) *Management of Information Technology for Construction*, Singapore: World Scientific Publishing Co.

Ngo, T., Aye, L., Arora, M., Mendis, P. and Malano, H., 2014. M*Utopia: Platform for Envisioning Future and Transition Scenarios of Sustainable Urban Development*, paper presented at Practical Responses to Climate Change Conference, Melbourne, Australia, 23 May.

Open Geospatial Consortium (OGC), 2016. OGC: *Making Location Count*. Available at: www.opengeospatial.org/ [Accessed 15 February 2016].

Plume, J., 2015. Integrated digitally-enabled environment: The internet of places. In: A. Kemp (ed.) *AGI Foresight Report 2020*, London: The Association for Geographic Information, pp. 207–209.

Plume, J., Simpson, R., Owen, R. and Hobson, A., 2015. *Integration of Geospatial and Built Environment – National Data Policy*, Joint buildingSMART – SIBA Position Paper, Version 2, Canberra: SIBA.

Rawlinsons, 2011. *Australian Construction Handbook*, 29th edn, Perth: Rawlinsons Publishing.

Sharifi, A. and Murayama, A., 2013. A critical review of seven selected neighbourhood sustainability assessment tools. *Environmental Impact Assessment Review*, 38, 73–87

Trubka, R., Glackin, S., Lade, O. and Pettit, C., 2016. A web based 3D visualisation and assessment system for urban precinct scenario modelling. *ISPRS Journal of Photogrammetry and Remote Sensing*, 117, 175–186.

van Nederveen, S., Luiten, B. and Böhms, M., 2015. *Linked Data for Road Authorities*, paper presented at 32nd CIB W78 Conference 2015, Eindhoven, The Netherlands, 27–29 October.

8 Information integration for asset and maintenance management

Sonia Lupica Spagnolo

Introduction

Information integration along the entire life-cycle of built environment assets requires effective interoperability. This is described as a complete and efficient data exchange among information and communications technologies (ICT) and stakeholders. Interoperability has to be achieved from both a technology and process point of view and has to allow one application to read data from another application as well as share data effectively throughout the asset's life-cycle. The ultimate objective is to improve communication, making data sharing easy, reliable and fast. This is possible only if data duplication and conflicting datasets are completely avoided. The entire building process may be optimised if every stakeholder along the life-cycle of a built asset is able to find needed information and share existing or new datasets in a straightforward and conflict-free manner.

Since the early 1990s, improving information integration during the design phase has been one of the main areas of focus of academic research. According to the Royal Institute of British Architects Plan of Work (RIBA, 2013), achieving high levels of information integration during the design phase requires optimising information flows from concept design to technical design. This necessity is already evident when a single person is in charge of the design and has to develop a project that can be complex and vary significantly from concept to delivery. Such a requirement becomes more palpable when different designers work together. If a structural engineer could, for example, avoid re-drawing the architectural model to study load distribution for defining pillar size and position, it could prevent a significant amount of rework. If, then, an electronic application could automatically check possible clashes or conflicts between the two designs and immediately inform both structural and architectural designers, this could avoid variations after delivery, with a significant reduction of wasted time and resources.

This chapter will first explore the use of ICT tools for more effective life-cycle management and then present two examples from the Polytechnic University of Milan (Politecnico di Milano). The first case study aimed to develop an international platform for durability management, supported by reliable methods for service-life estimation. The second worked towards creating a national

134 *Lupica Spagnolo*

prototype database for an efficient data exchange among different stakeholders of the architecture, engineering and construction (AEC) industry. The adoption of the same approach to the entire construction field (not only buildings, but also infrastructure, services, processes and anthropomorphic environments), together with information sharing that is independent of the model, allowed the creation of a construction information management (CIM) system. These two tools may also be later merged to build a unique and interoperable platform for innovative life-cycle management.

ICT tools for more effective life-cycle management

As underlined by Steel et al. (2012), the information models now in use are large, complex and highly interdependent. They include architectural drawings; engineering schematics for structural, electrical and mechanical services; heating, ventilation and air-conditioning (HVAC) services, as well as cross-cutting concerns such as project management, scheduling and cost planning/estimation.

The introduction of Building Information Modelling (BIM) has been found to be a reliable way to face this problem (Volk et al., 2014). It allows working on the same 3D representation, adding more information about asset element properties with the use of attributes of BIM objects that together form the complete model of the built asset (Eastman et al., 2011). BIM has certainly brought benefits to the construction industry (see for example Hardi and Pittard, 2015; Love et al., 2014; Sanchez et al., 2016). Interoperability protocols for BIM software lay the foundation for a more efficient collaboration among designers, allowing them to exchange models across software platforms (Young et al., 2007; von Both et al., 2012). These protocols were launched by the International Alliance for Interoperability, now buildingSMART International (National Institute of Building Sciences, 2015). This organisation developed the open and freely available Industry Foundation Classes (IFC) model specifications that were also registered by the International Organization for Standardization through the official standard ISO 16739:2013 (ISO, 2013).

Despite this important contribution, achieving complete interoperability across different software packages is still challenging (Ciribini et al., 2015; Zanchetta et al., 2015). Several studies have shown that, for example, if IFC files are exported from IFC compliant software into another IFC compliant application, many attributes are lost (Jeongwon et al., 2015). Therefore, this is still an area of improvement that has the potential to reduce rework by promoting the widespread use of IFC or IFCxml[1] files among different stakeholders.

Another challenge faced by the industry is the use of BIM not only as a tool in the design process, but also as interface for the exchange of information between the different parties involved in the building process (Bianchi et al., 2014). The development of new and improved ICT tools coupled with cultural changes may enable a more universal interoperability. This would also allow full advantage of BIM to be taken as a new approach for built asset planning, design and delivery. This has been underlined by the US General Services Administration (GSA), that asserted that considerations about information are critical to successful integration of computer models for project coordination, simulation and optimisation, as well as asset and facility management (GSA, 2007). BIM has been used to improve a wide range of functions including design analysis, quantity surveying, cost estimation, environmental and performance assessment, scheduling and compliance with building codes and regulations.

The next big challenge is achieving more effective information integration through a single, open, web-enabled and free platform, which efficiently stores and integrates a wide range of data related to the built asset. This becomes particularly crucial when aiming to optimise asset management and maintenance. The possibility of collecting and storing data from the first life-cycle phases encourages a reliable maintenance plan to be prepared. Accurate tracking of maintenance interventions and asset variations during the service-life of constructed facilities additionally permits optimising facility management. This kind of information can also be added to existing BIM objects, allowing the use of IFC compliant software for maintenance planning or facility management.

Moreover, this platform could collect the reference service-life (RSL) of buildings, components, assembly and materials. RSL is defined in ISO 15686-1:2011 (ISO, 2011) as 'service life that a building or parts of a building would expect (or is predicted to have) in a certain set (reference set) of in-use conditions'. Having the RSL information, designers and maintenance planners can then calculate the estimated service-life (ESL), applying one of the methodologies proposed by ISO 15686-2 (ISO, 2001a). This allows forecasting the service-life and estimating the timing of necessary maintenance and replacement of components. Durability management increases constructed facilities' quality and sustainability because it encourages adopting the best maintenance strategy. It also provides a reliable performance control process over time (Daniotti and Lupica Spagnolo, 2007).

Real estate, maintenance managers, building contractors and manufacturers are usually very knowledgeable about asset durability. Their know-how is based on the estimated or measured service-life and reliability, and is centred on their experience relating to the number and type of maintenance interventions. However, such information and knowledge is only occasionally shared. This is mainly due to the lack of appropriate tools for data storage and dissemination (Re Cecconi and Lupica Spagnolo, 2009).

Service-life planning (SLP) and data collection from asset management are only the first steps towards an efficient asset management and maintenance process. BIM can be exploited to optimise SLP (ISO, 2014) and can also be used to better organise facility management operations (see for example Lo Turco, 2015). Nevertheless, it is still necessary to develop specific ICT tools for life-cycle data use and sharing (Daniotti and Lupica Spagnolo, 2015). It is difficult to gain the required input data without an integrated information managing system. This shortfall may also negatively affect the accuracy of maintenance planning and management. The gap also results in wasted time and resources as well as in less effective control of the sustainability of building processes. For example, without

136 *Lupica Spagnolo*

knowledge about site work procedures, in-use conditions or end-of-life interventions for specific construction sites, the choice of sustainable products could be sub-optimal and life-cycle assessment (LCA) could be incomplete.

Therefore, the creation of a unique web-based platform for data storage and sharing, where information about service-life and performance decay over time is also available, is the first step towards more effective asset planning, construction and management. It is also important to develop specific ICT tools for service-life appraisal, maintenance planning and durability management. Moreover, talking the 'same language' with a common classification and coding system avoids communication errors, especially during design and construction.

The international platform for service-life appraisal, data storage and sharing

As mentioned earlier, SLP is an important step during the design phase because choices made will influence the entire LCA of a building process. Moreover, the choice of the most suitable building components and materials has to consider not only their duration but also their failure-rate curve over time. An effective and sustainable process depends on ensuring optimum maintenance planning. This is only possible if the knowledge of building components' durability is available and applied. The need for more accurate service-life data is therefore essential and compounded by the necessity for a wider availability and an easier accessibility of such information. In response to this need, the French Scientific and Technical Centre for the Building Industry (Centre Scientifique et Technique du Bâtiment, CSTB) and Politecnico di Milano structured the French RSL database (Hans et al., 2008). It was accomplished through the analysis of the data needed by designers to evaluate the planning and duration of asset maintenance processes.

This database contained some of the input data required by ICT tools that include service-life management applications. It became the starting point for integrating results from past research activities and developing new research programmes that aim to enhance the database and make it useful for the international community (Daniotti et al., 2010).

The development of an international database requires using a standardised international component and materials classification to ensure efficient communication across tools and actors. In the construction field, this level of standardisation and interoperability can facilitate a faster individualisation of characteristics, a proper organisation of each set of documents and, eventually, reliable costs evaluation, time scheduling and sustainability assessment.

This classification system must have two essential characteristics in order to meet this purpose:

- to individualise unequivocally each family of objects and to let additional information be added to allow more precise descriptions of individual objects;
- to be applicable to every class of objects, independently of their complexity.

Asset and maintenance management 137

Classification systems are based on separating elements into classes. These can be grouped according to two main methods:

- direct grouping: elements are identified as belonging to a class. Classes are organised according to a hierarchical order. Some primary classes include sets of sub-classes and so on. For instance, parts of a building such as walls, floors, foundations and roofs are macro-components formed by other elements, which are contained within the main class. MasterFormat (CSI and CSC, 2016), developed in the United States (US) and Canada, is an example of such a classification based on direct grouping; and
- combined grouping: elements are identified by the free aggregation of different informative attributes. Examples of this classification include the Swedish standard SfB (Ray-Jones and Clegg, 1976) and the Italian standard UNI 8290 (UNI, 1981).

As described in Lupica Spagnolo (2016), the first activity undertaken to develop this classification was to identify a suitable system for adoption. This was done based on an extensive review of the most common internationally available classification systems (Figure 8.1).

The SfB classification initiative (Swedish acronym for Samarbetskommittén för Byggnadsfrågor, Cooperative Committee for Construction Issues) was born in 1956 with the objective of becoming an international classification system. Developed in Sweden, it was later presented in the Netherlands to a community of European experts. The first English version was later translated into German and Italian. In spite of its transnational origin, the SfB classification has not been widely used (Lupica Spagnolo, 2016).

Another international classification system is the Universal Decimal Classification (UDC). This is not focused on the construction field, but is aimed at organising a wide range of information types. The Italian standard UNI 8290, published in September 1981, standardised the terminology employed,

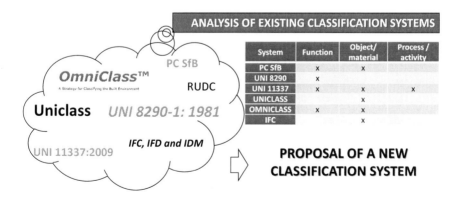

Figure 8.1 Comparison between existing classification systems.

138 *Lupica Spagnolo*

representing an important step towards supplying a national classification model. This standard provides residential buildings with a classification and language to articulate technological units and technical elements into which the system is divided (UDC Consortium, 2005).

The Construction Project Information Committee (CPIC) introduced the Uniclass (Unified Classification for the Construction Industry) system in the United Kingdom in 1997 and suggested a method for classifying the building sector based on 15 tables (Crawford et al., 1997). Each chart looks at a specific information aspect and it can be used separately or linked together with other tables to express complex concepts.

MasterFormat represents the most frequently used standard of communication in the US and Canada to organise design content and documents. The project, through the Construction Specification Institute (CSI) proposal, can be split into divisions and sections, with a standardised procedure to manage every type of project information (CSI and CSC, 2016). OmniClass on the other hand, is a classification system that was accepted by the US building industry in 2006. It is a freely available standard based on a common language for the whole building sector and a standardised system to classify information regarding the entire building life-cycle, from design and planning to demolition (OmniClass, 2016).

According to ISO 12006-2, classifications of designed elements, work results and products are useful (ISO, 2001b). The Italian standard UNI 11337:2009 (UNI, 2009) introduced a new framework to classify contemporary functions, objects and processes/activities. It organises a list of items according to the sequence of the work to be carried out and permits analytical evaluations such as the cost per measurement unit for each intervention required by each task. It also allows elementary evaluations to be conducted such as the cost per unit of technological objects.

It was decided to adopt a new classification system based mainly on the SfB, UNI 8290 and UNI 11337 (UNI, 1981; 2009) provisions for the international RSL database. As the aim was to encourage the assembly of RSL data covering building objects, this proposal does not contain a classification of activities, vehicles, tools, human resources or environments. For example, it can be used to individualise unequivocally a pillar, but not an activity like daubing, a vehicle like a crane, a tool like a trowel, a person like a bricklayer or a place like a kitchen.

The Polytechnic University of Milan contributed to the:

- RSL definition, through research activities on different building materials and components and by proposing a new approach to define accelerated ageing cycles in laboratory (Daniotti et al., 2008);
- ESL appraisal, creating 'driving grids' to enhance the application of the Factor method (Daniotti et al., 2010).

The international standard ISO 15686-2 (ISO, 2001a) describes some methodologies to evaluate service-life. Among these, the factor method emerges as the most exploitable due to its simplicity of use. Service-life in design conditions

Table 8.1 List of noteworthy factors used in the factor method

Agents		Noteworthy factors	
Agent related to the inherent quality characteristics	A	Quality of components	Manufacture, storage, transport, materials, protective coatings (factory-applied)
	B	Design level	Incorporation, sheltering by rest of structure
	C	Work execution level	Site management, level of workmanship, climatic conditions during execution of the work
Environment	D	Indoor environment	Aggressiveness of environment, ventilation, condensation
	E	Outdoor environment	Elevation of the building, microenvironment conditions, traffic emissions, weathering factors
Operation conditions	F	In-use conditions	Mechanical impact, category of users, wear and tear
	G	Maintenance level	Quality and frequency of maintenance, accessibility for maintenance

is calculated by correcting the RSL with seven multiplicative factors, as listed in Table 8.1.

The factor method generates the following formula for the calculation of the ESL:

$$ESL = RSL * A * B * C * D * E * F * G$$

These correction factors are generally weighted between 0.8 and 1.2 and are used to take into account the particular conditions in which the component is utilised. This method considers characteristics like the quality of production, design and execution, the presence of stressing agents, the frequency of maintenance interventions and attention to executive details, among other factors. The biggest advantage of such a method, which at the same time is its greatest limitation, is its simplicity. This is why it can be applied to both small and large projects, even if it cannot model complex phenomena.

Each single building component is associated with a specific factor grid, which translates all the variables affecting service-life into different sub-factors belonging to the seven main factors proposed by ISO 15686 factor method (ISO, 2001a; 2011). Thus, the construction of the database consists of four main steps, namely:

1   A grid is built for a given building component by a panel of experts.
2   The grid is shared among the stakeholders of the construction sector.
3   Information regarding properties and service-lives is collected by the stakeholders for the comprised sub-factors.
4   The platform administrator validates the data.

140 *Lupica Spagnolo*

As a consequence, the RSL database not only permits the convergence of all the information coming from experimental research, but is also an indispensable tool for the application of existing methods for SLP, namely ISO 15686-2 (ISO, 2001a), UNI 11156-3 (UNI, 2006) and, in particular, the factor method.

INNOVance, the Italian database for construction information management

Another step forward in advancing the BIM approach, through Building Information Modelling and Management (BIM&M), towards holistic construction information management (CIM) was made during a three-year research project called INNOVance. This project has seen the participation of a wide range of stakeholders in the building industry (Figure 8.2) across public procurement, contractors, manufacturers, designers, research centres, ICT developers and end-users.

This research project, funded by the Italian Ministry of Economic Development, started in July 2011 with the aim of introducing radical innovation into the Italian construction sector by creating the first national construction database and a prototype of an interoperable BIM&M platform (Pavan et al., 2014).

To overcome possible inefficiencies due to incorrect or redundant exchange of information among actors of the construction chain throughout the entire life-cycle of a project, the Polytechnic University of Milan developed a new classification, naming and coding system. It allows an unambiguous language to be spoken where words and their meanings are clearly defined (Daniotti et al., 2013). The naming and coding system developed with INNOVance considers the entire built asset as the last output of the construction industry, analysing it from five points of view (Figure 8.3):

- a functional-spatial logic which highlights environment and functional areas into which the built asset is divided, for example, commercial areas inside a building;
- a technological logic for construction which encourages disassembly into its elements, from entire components to single products;
- a technological logic for equipment which encourages disassembly into its parts, from entire systems to single products;
- an anthropomorphic logic to represent all the changes in the environment including natural aspects, for example, excavation and embankments;
- the processual logic to individualise and codify activities, vehicles, tools and human resources.

Each of these logics contains three levels of objects, according to their complexity.

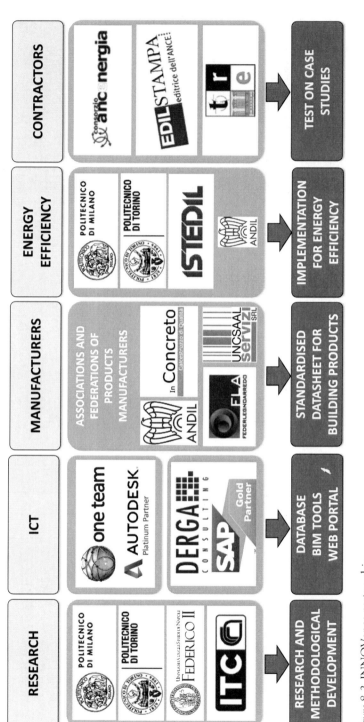

Figure 8.2 INNOVance partnership.

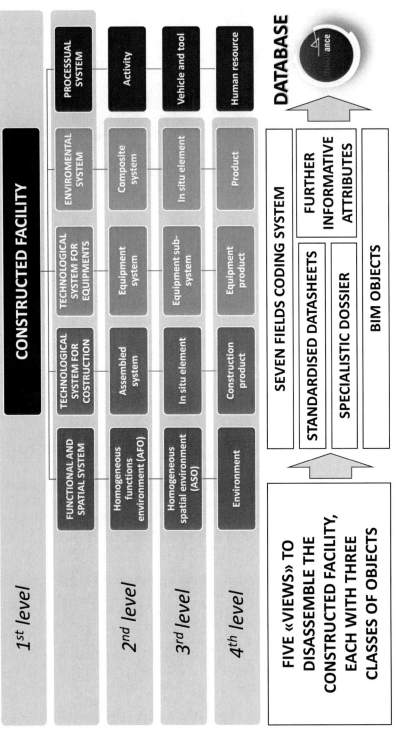

Figure 8.3 INNOVance hierarchical structure.

Asset and maintenance management   143

The functional-spatial system

This logic splits the constructed facility into a set of spatial elements defined by their functions, dimensions, shape and position with respect to themselves and the external environment. It is divided into the following three levels:

- homogeneous functional environments – AFO, using the Italian acronym;
- homogeneous spatial environments – ASO, using the Italian acronym;
- single environments.

Homogeneous functional areas make it possible to highlight each function, breaking down the built asset into commercial AFO, residential AFO and administrative AFO, among other categories. Inside a commercial AFO it is then possible to individualise, for example, the sales area or the incoming goods area, which are considered homogeneous spatial environments (ASO). In each ASO, it is possible to find single rooms, such as single shops, bathrooms, kitchens, bedrooms and living rooms.

The technological system for construction

This system is a structured set of technological units and/or technical elements, defined with technological requirements and performance specifications. It is divided into three levels:

- assembled systems;
- on-site elements;
- construction products.

For instance, each constructed facility is formed by assembled systems such as walls, floors and coverings. These assembled systems are composed of different on-site elements, such as the layers of finishing, tiles, bricks, thermal and acoustic insulation. Moreover, each single layer (the on-site element) consists of several construction materials, such as cement, mortar, insulation panels and anchors.

The technological system for equipment

The above-described approach for construction can be applied to services so that HVAC, for example, is considered the equipment system. The distribution system becomes the equipment sub-system, whereas each single pipe is the equipment product.

The technological system for equipment is divided into the following three levels:

- equipment systems;
- equipment sub-systems;
- equipment.

144   *Lupica Spagnolo*

The environmental system

This system is the structured set of elements which belong to the natural environment with or without anthropomorphic intervention. It is divided into three levels:

- a composite system;
- on-site elements;
- products.

For example, a road can be considered as a composite system which consists of different on-site elements (such as a surface course, shoulders, base and bed) and natural products such as the earth foundation and trees along the sides.

The processual system

This system comprises every activity, vehicle, tool and human resource. Considering the above-described hierarchy, the adopted coding system then defines a structure with an open compilation. There are seven fields, which can describe every element to be encoded, independently from object complexity and the set system to which it belongs (Figure 8.4):

1   Category.
2   Type.
3   Function and in-use characteristics or, for construction products, harmonised technical standards.
4   Performance characteristics.
5   Geometrical features such as shape, aesthetics and construction traits.
6   Dimensional features.
7   Physical/chemical properties.

A fixed seven-field coding system was chosen to standardise the output code and provide a coherent individualisation of each item of information inside the code itself. The second step was to create a tool with a standardised procedure for data sharing to avoid misunderstanding, errors, over-costs and delays, mainly due to documents and drawings that are not correctly interpreted and updated.

Exploiting the unambiguous classification system for every object and the defined hierarchical structure, the Polytechnic University of Milan, in

SEVEN FIELDS CODE, INDEPENDENTLY FROM THE OBJECT COMPLEXITY						
				Technical characteristics		
Category	Type	Functional characteristics/ Harmonised technical standards	Performance characteristics	Geometrical features	Dimensional features	Physical/ chemical properties

Figure 8.4 INNOVance coding system.

collaboration with Derga Consulting S.r.l.[2] created a unique database to store and share information. This ICT tool allows each object individualised in the above-mentioned hierarchical structure to be collected:

- standardised datasheets and specialised dossiers, covering, for example, images, videos and drawings;
- further informative attributes which cannot take place in the defined datasheets and dossiers;
- BIM objects in different file formats; including the interoperable IFC format.

All the structured data is contained within an enterprise resource planning (ERP) software package by Systems, Applications and Products in Data Processing (SAP). This is organised in the INNOVance database through the use of unique names, technical datasheets, specialised dossiers, complementary data and BIM objects. All the data is accessible through a user-friendly web portal and each user can manage their projects from design brief through all further phases.

Many operations are possible through this website. For example, manufacturers can create, upload and modify the standardised datasheets of their construction products as well as add BIM objects. Designers can describe the technical solutions designed and share their projects. Companies can consult data sheets established by designers and manufacturers and check the correspondence between ordering and arrival of the goods.

The web portal constitutes an exchange platform with

- a public section open to any user where it is possible, for example, to find a BIM library where information (graphical, alphanumerical or multimedia) is embedded into generic BIM objects;
- a private section open only to logged-in users where it is possible to store a personal library of datasheets and dossiers and where, thanks to the BIM&M server, they can convert generic BIM objects from the BIM&M library into specific BIM objects.

This web platform aims to become an interactive virtual bridge between different stakeholders. To do that, while dealing with the large amount of information to be collected, there are different views of the database according to the user's profile. This permits only useful data to be filtered for the relevant stakeholder, but leaves open the possibility of searching, visualising and possibly modifying other information in the database.

Conclusions

Effective asset and facility management requires gathering, storing and analysing large volumes of information about the behaviour of different building materials, products and components. Service-life planning and data collection from facility

146 *Lupica Spagnolo*

management are only the first steps towards more efficient asset management driven by the need to develop ICT tools that are specific to life-cycle data use and sharing.

Rationalising the information flows connecting construction process stages (planning, design, construction, use, management, maintenance and disposal or reuse) and the various actors involved (clients, users, designers, contractors and components manufacturers, among others) is a strategic step towards the optimisation of the building process and the whole construction sector.

This chapter explored the experience of an international reference service-life database and the INNOVance platform, a complete construction information management system. The latter constitutes a unique collaborative tool whose purpose is to store, elaborate and share useful data across building practices and to support all actors involved in the process. Potential impacts from using such information integration tools are expected across the entire building industry. These tools help avoid communications misunderstandings and data redundancies, thereby increasing the efficiency of the building process across the entire life-cycle of each built asset. The exchange of information is associated with unique codes and can be done in a fast, interoperable and reliable way. Information integration and knowledge exchange for asset and maintenance management enables attaining higher quality, which can mean high-performance in safety, security, sustainability or aesthetic value; ideally a combination of these attributes. Moreover, optimised access to information allows the user to choose the safest technologies and materials and to properly check the achievement of the seven basic requirements. Those fundamental requirements cover mechanical stability, fire safety, hygiene accessibility in use, protection against noise, energy efficiency and sustainable use of resources.

The use of ICT tools, such as the above-described examples, can profoundly reshape the way industrial companies interact with customers, creating additional opportunities for substantial gains along the entire industry value chain. Easier access to information can lead to more aware clients who participate more actively during the design stage and make better-informed decisions. Use of the tools can also enable a more comprehensive analysis of the impact of different materials and products across the entire life-cycle of the asset as well as influencing operational and maintenance costs. This will, however, require further research before the untapped potential of ICT for more effective life-cycle management can be realised.

Notes

1 IFCxml is the Industry Foundation Classes (IFC) format using the eXtensible Mark-up Language (XML).
2 Derga Consulting S.r.l. is a partner of SAP, acronym for Systemanalyse und Programmentwicklung or Systems, Applications and Products in Data Processing, a software company which has developed some well-known enterprise resource planning and data management programs. More information is available at www.derga.it/home.

References

Bianchi, L., Chiozzi, M., D'Alessandro, R., Daniotti, B., Di Fusco, A., Galli, M., Giorno, C., Gulino R., Lupica Spagnolo, S., Pasini, D., Pavan, A., Pola, M. and Rigone, P., 2014. *Building Process Optimization Through an Efficient Data Management,* paper presented at the 13th ISTeA Conference on Energy, sustainability and Building Information Modelling and Management, Bari, Italy, 10–11 July.

Ciribini, A. L. C., De Angelis, E., Tagliabue, L. C., Paneroni, M., Mastrolembo Ventura, S. and Caratozzolo, G., 2015. *Workflow of Interoperability Toward Energy Management of the Building,* paper presented at the 14th ISTeA Conference on Environmental Sustainability, Circular Economy and Building Production, Milan, Italy, 24–25 September.

Crawford, M., Cann, J. and O'Leary, R., 1997. *Uniclass – Unified Classification for the Construction Industry.* London: RIBA Publishing.

CSI and CSC, 2016. *MasterFormat, Master List of Numbers and Titles for the Construction Industry.* Alexandria: The Construction Specifications Institute and Construction Specifications Canada.

Daniotti, B. and Lupica Spagnolo, S., 2007. *Service Life Prediction for Buildings' Design to Plan a Sustainable Building Maintenance,* paper presented at the conference on Sustainable construction, materials and practices, Lisbon, Portugal, 12–14 September.

Daniotti, B. and Lupica Spagnolo, S., 2015. *An Interoperable ICT Tool for Asset and Maintenance Management – Advances in Research,* paper presented at the 1st International Symposium on Building Pathology (ISBP2015), Porto, Portugal, 24–27 March.

Daniotti, B., Lupica Spagnolo, S. and Paolini, R., 2008. *Climatic Data Analysis to Define Accelerated Ageing for Reference Service Life Evaluation,* paper presented at the 11th International conference on Durability of Building Materials and Components, Istanbul, Turkey, 11–14 May.

Daniotti, B., Lupica Spagnolo, S., Chevalier, J. L., Hans, J. and Chorier, J., 2010. *An International Service Life Database: The Grid Definition for an Actual Implementation of Factor Methods and Service Life Prediction,* paper presented at the CIB World Congress 2010, Salford Quays, UK, 10–13 May.

Daniotti, B., Lupica Spagnolo, S. and Pavan, A., 2013. *The Unambiguous Language for Construction,* paper presented at the 12th ISTeA Conference on ICT, Automation and the Industry of the Built Environment: From the Automation Exchange to the Field Management, Milan, Italy, 3–4 October.

Eastman, C., Teicholz, P., Sacks, R. and Liston, K., 2011. *BIM Handbook: A Guide to Building Information Modeling for Owners, Managers, Designers, Engineers and Contractors,* 2nd edn, Hoboken: Wiley.

General Services Administration (GSA), 2007. *GSA Building Information Modeling Guide Series 01 – Overview,* Washington, DC: US General Services Administration.

Hans, J., Chorier, J., Chevalier, J. L. and Lupica, S., 2008. *French National Service Life Information Platform,* paper presented at the 11th International Conference on Durability of Building Materials and Components (11DBMC), Istanbul, Turkey, 11–14 May.

Hardi, J. and Pittard, S., 2015. If BIM is the solution, what is the problem? A review of the benefits, challenges and key drivers in BIM implementation within the UK construction industry. *Journal of Building Survey, Appraisal & Valuation,* 3(4), 366–373.

148 Lupica Spagnolo

ISO, 2001a. 15686-2:2001, *Building and Constructed Assets – Service Life Planning – Part 2: Service Life Prediction Procedures*. Geneva: International Organization for Standardization.

ISO, 2001b. *12006-2:2001, Building Construction – Organization of Information about Construction Works – Part 2: Framework for Classification of Information*. Geneva: International Organization for Standardization.

ISO, 2011. *15686-1:2011, Buildings and Constructed Assets – Service Life Planning – Part 1: General Principles and Framework*. Geneva: International Organization for Standardization.

ISO, 2013. *16739:2013, Industry Foundation Classes (IFC) for Data Sharing in the Construction and Facility Management Industries*. Geneva: International Organization for Standardization.

ISO, 2014. *15686-4:2014 Buildings and Constructed Assets – Service Life Planning – Part 4: Service Life Planning using Building Information Modelling*. Geneva: International Organization for Standardization.

Jeongwon, R., Jiyong, L. and Jungsik, C., 2015. IFC-based information interoperability process for freeform building, *Advanced Science and Technology Letters*, 120, 449–452.

Lo Turco, M., 2015. Representing and managing real estate: BIM for facility management. *Territorio Italia*, 2, 31–45.

Love, P. E., Matthews, J., Simpson, I., Hill, A. and Olatunji, O. A., 2014. A benefits realization management Building Information Modeling framework for asset owners. *Automation in Construction*, 37, 1–10.

Lupica Spagnolo, S., 2016. *From BIM to CIM to Innovate Life Cycle Management of Entire Works and their Parts*, paper presented at the CIB World Congress 2016 (WBC16), Tampere, Finland, 30 May–3 June, 5, 880–891.

National Institute of Building Sciences, 2015. *buildingSMART Alliance 2015 Strategic Plan*. Available at: www.nibs.org/resource/resmgr/BSA/bSa2015_StrategicPlan.pdf [Accessed 4 July 2016].

OmniClass, 2016. *Omniclass Home*. Available at: www.omniclass.org/ [Accessed 4 July 2016].

Pavan, A., Daniotti, B., Re Cecconi, F., Maltese, S., Lupica Spagnolo, S., Caffi, V., Chiozzi, M. and Pasini, D., 2014. *INNOVance: Italian BIM Database for Construction Process Management*, paper presented at the International Society for Computing in Civil and Building Engineering (ICCCBE) and CIB W078 Conference, Orlando, Florida, USA, 23–25 June.

Ray-Jones, A. and Clegg D., 1976. *CI/SfB Construction Indexing Manual*. London: RIBA Publishing.

Re Cecconi, F. and Lupica Spagnolo, S., 2009. *Service Life Data in Life Cycle Management of Building*, paper presented at the International Conference on Construction and Building Research, Madrid, Spain, 14–26 June.

RIBA, 2013. *Plan of Work*. Available at: www.ribaplanofwork.com/Download.aspx [Accessed 6 December 2015].

Sanchez, A. X., Hampson, K. D. and Vaux, S., 2016. *Delivering Value with BIM: A Whole-of-life Approach*. Oxford: Routledge.

Steel, J., Drogemuller, R. and Toth, B., 2012. Model interoperability in Building Information Modelling. *Software & Systems Modeling*, 11(1), 99–109.

UDC Consortium, (2005). *UDC: Universal Decimal Classification. Standard Edition*. London: British Standards Institution.

UNI, 1981. *8290-1:1981, Residential Building – Building Elements – Part 1: Classification and Terminology*. Milan: UNI (Italian Organization for Standardization).

UNI, 2006. *11156-3:2006. Evaluation of Durability of Building Component. Part 3: Evaluation Method for Service Life*. Milan: UNI (Italian Organization for Standardization).

UNI, 2009. *11337:2009, Building and Engineering Works – Coding Criteria of Works, Activities and Resources – Identification, Description and Interoperability*. Milan: UNI (Italian Organization for Standardization).

Volk, R., Stengel, J. and Schultmann, F., 2014. Building Information Modeling (BIM) for existing buildings – Literature review and future needs. *Automation in Construction*, 38, 109–127.

Von Both, P., Koch, V. and Kindsvater, A., 2012. *Potential and Barriers for Implementing BIM in the German AEC Market*, paper presented at the 14th International Conference on Computing in Civil and Building Engineering, Moscow, Russia, 27–29 June.

Young, N. W., Jones, S. A. and Bernstein, H. M., 2007. *Interoperability in the Construction Industry. SmartMarket Report*. New York: McGraw Hill Construction.

Zanchetta, C., Paparella, R., Cecchini, C. and Alessio, G., 2015. *Performance Based Building Design through Building Energy Modeling (BEM)*, paper presented at the 14th ISTeA Conference on Environmental Sustainability, Circular Economy and Building Production, Milan, Italy, 24–25 September.

9 IT support for contractor monitoring of refurbishment projects

Jan Bröchner and Ahmet A. Sezer

Introduction

Refurbishment activities comprise a high proportion of construction industry output in most developed countries. There is no international consensus among statisticians as how to define refurbishment or renovation of buildings (Mansfield, 2002). The UK Office for National Statistics publishes data indicating that the volume of 'repair and maintenance' corresponded to about 60 per cent of new work during 2014 and 2015. 'Repair and maintenance' was roughly equally divided between housing and non-housing, and it is probable that much of this was refurbishment. There are no obvious reasons why ongoing investment in existing buildings should decline and the potential for increasing environmental sustainability by improving energy performance in the building stock remains considerable. The EU Energy Efficiency Directive (2012/27/EU) adopted in 2012 includes a requirement for member states to develop long-term renovation strategies for their national building stocks.

Efficient processes for the physical transformation of existing buildings are therefore needed. With few exceptions, refurbishment projects are characterised by information gaps and poor integration. Instead, there are numerous and often freestanding activities: design, production planning, waste management and site monitoring. Also, there is only a weak link, if any, to information systems supporting the future management of the refurbished building. This chapter focuses on how information technologies (IT) tools can support contractors when monitoring their refurbishment projects.

Refurbishment contractors face three challenges in information integration. First, there is the potential for integrated information flows across stages of the entire design, construction and facilities management process. The second challenge lies in integrating with existing enterprise systems, and the third is that of integration of information in the supply chain. Before discussing the three integration challenges, this chapter will explore current practices for monitoring on refurbishment sites.

Site monitoring

Typically, refurbishment contractors monitor adherence to budget, specifications and time schedules, just as for new construction. Managing refurbishment projects differs however from new construction in several respects (Egbu, 1997). It is possible to see refurbishment projects as subject to a long range of specific constraints with complex relations to broad performance indicators for cost, time, quality and safety; this may explain the degree of informality in monitoring such projects (Singh et al., 2014). Although new construction projects occasionally have to handle challenges such as geotechnical complications that were not foreseen at the design stage, unexpected discoveries are frequent in the process of refurbishment. A common problem found is the low quality of archival drawings; buildings were not built as in the original building permit documents, and later alterations were poorly recorded or not at all. Unexpected discoveries of asbestos and other hazardous materials, structural weaknesses and hidden damages from moisture are examples of what tends to complicate refurbishment processes. The difficulty of knowing what is behind wall surfaces for example, requires experienced site managers and workers. A need for coordinating refurbishment activities with client activities, perhaps in the same building, may introduce further uncertainty and complications. In comparison to new construction, refurbishment projects can be expected to need more flexible monitoring systems because of frequent deviations from original assumptions and schedules.

Worldwide, accounting regulations and the need for data to be used for tender estimates appear to be strong determinants of how projects are currently monitored. Data collected for traditional accounting purposes and as a basis for tendering cannot be used directly for comparing efficiency in resource use or productivity across projects. When a broader set of project data relating to resource use is obtained, benchmarking is more unreliable than for new construction because the context of refurbishment projects as well as prior conditions of buildings vary greatly. A more ambitious attempt to measure site productivity implies including not only input but also output monitoring. While it can be discussed whether new construction is to be viewed as a specific type of service, the importance of service process qualities in the view of customers is obvious in the case of refurbishment. Examples of process qualities are absence of dust and noise, time precision and two-way communication in projects. The fundamental issues that complicate the measurement of service productivity in comparison with the manufacturing industry are relevant (Sezer and Bröchner, 2014).

In highly decentralised, project-based organisations such as large refurbishment contractors, an absence of benchmarking initiatives can be explained through lack of proper definitions, methods and IT tools. There is also the possibility that site managers feel that it is more efficient to spend time on actually improving the outcome of projects than engaging in data collection. There are many reasons why managers might be unenthusiastic towards benchmarking (Williams et al.,

152 Bröchner & Sezer

2012). They might hesitate to provide details of how they solve complicated problems on site if they perceive personal risks. Current practices for monitoring projects are also dominated by a search for deviations from budgets and schedules rather than measuring the efficiency of resource use. Focus on deviations can be more related to a blame culture than providing a strong basis for improving efficiency. Also, since it is difficult to assess differences in prior conditions and project contexts for refurbishment, comparisons between projects rather than between project forecasts and outcomes appear to be unusual.

BIM in refurbishment

Compared to the number of published studies about Building Information Modelling (BIM) in new construction, there have been few contributions dealing specifically with refurbishment projects. In fact, research appears to be more developed, even if just slightly, in the field of BIM for facilities management than BIM for refurbishment (İlter and Ergen, 2015). There is however a rapidly growing body of research into methods for generating BIM models for existing buildings (Volk et al., 2014). In parallel with the successive introduction of BIM tools for new construction, recent advances in automatic recording of the geometry of existing structures promise greater use of object-oriented presentations that allow more precise monitoring of refurbishment projects (Sanchez and Joske, 2016).

The main tasks of 3D as-built modelling are geometric modelling, object recognition and relationship modelling (Tang et al., 2010). One option is to start from existing drawings, another and more widely researched, is based on using cameras or laser scanners in existing buildings to develop the 3D model. As to methods and tools for generating 3D building models from 2D drawings, it can be said that most of the earlier research has focused on particular steps in the process of generating 3D models. This is according to a review by Gimenez et al. (2015) who have developed a method which uses an optical imaging scanner on 2D paper floor plans (Gimenez et al., 2016). This method requires removing irrelevant information (noise) from the images and separating the text on the drawings from their geometrical information. Typical difficulties are the identification of openings in walls and dealing with the diversity of drawing conventions. Methods for analysing existing paper documents are important, as the primary function of BIM in actual projects concerns the geometry of the existing building. Unfortunately, old drawings are often unreliable; one cannot be sure that the building was erected originally in accordance with the available documentation or that later alterations have been recorded.

Working with existing buildings, laser scanning[1] and camera images allow the creation of point clouds, which then have to be interpreted so as to be included in a workable digital model (Pătrăucean et al., 2015). The clutter typical of building interiors, especially if they are in current use, is one of the problems, and, in general, the size and complexity of the resulting point clouds is difficult to handle (Hong et al., 2015). Current methods are at best semi-automatic; the

Contractor monitoring of refurbishment   153

need for manual modelling or at least correction is often great and has obvious cost consequences.

In contrast, the object-oriented data structure is already in place for new buildings that have used BIM for the design phase. Thus, the effort required to create as-built BIM models is significantly reduced. Regardless of the origin of the as-built digital model, eliminating unnecessary details that complicate contractor use of the model on site remains a vital task in the context of refurbishment. Determining the appropriate level of detail for purposes of monitoring deconstruction and waste management is one of the crucial issues to be addressed. In general, it must be said that although more efficient tools and methods are emerging, preparing accurate and useful 3D models of existing buildings is still costly.

A UK study of housing refurbishment has found that the two major obstacles to BIM use are the investment needed and the lack of demand from clients (Kim and Park, 2013). Insufficient client demand is undoubtedly a sign of lack of integration with data flows in the maintenance and operation of the refurbished facilities. This lack of integration may arise from ignorance of potential benefits, but it might also reflect a reluctance to invest heavily in new and stronger information and communication technology (ICT) support for running the client's stock of buildings. One strategy to minimise the cost of producing the 3D model for multi-family housing estates is to start from a single dwelling unit and then expand by adapting the model to varieties of dwelling units, leading to full models of the building complexes. There is always a risk of geometric anomalies. It is therefore important to have good procedures for checking all spaces that are affected by a refurbishment project.

BIM use may vary from country to country. Current use of BIM for refurbishment in Sweden among leading contractors is mostly found in large projects where there are advantages of scale. BIM use also depends on how projects are procured. The design-build route, as opposed to traditional contractual arrangements with design documents provided by the client, encourages the contractor to invest in integrated flows of data from design to the refurbishment site activities. The value of BIM in the design phase of refurbishment projects appears to be lower than for new construction because the range of design choices is typically narrower when faced with an existing structure. The potential for gaining efficiency through considering a number of design alternatives is thus lower.

What has been described now goes a long way towards explaining why BIM integration based on laser scanning has made little headway. Both contractors and clients need to see benefits that exceed the cost of preparing digital models of existing structures. Such benefits can arise during the refurbishment process and subsequently for operations and maintenance of the facility. There will be cultural heritage projects, where a government client may require great precision in documenting status of a building before and after refurbishment. The usefulness of BIM has been illustrated by the renovation of the Durham Cathedral Chapter House (Tapponi et al., 2015). In addition, for structures such as hospitals with a heavy concentration of building service systems and subject to frequent

154 Bröchner & Sezer

alterations, it will be easier to demonstrate that benefits of detailed as-built BIM outweigh the associated costs. Large projects where many similar structures are to be refurbished according to a uniform pattern are otherwise the obvious candidates for a BIM approach. Over the years, there will also be more cases of alterations to buildings that were originally designed in a digital 3D model environment, thus facilitating further use of the design model.

Contractors, clients, suppliers and their systems

Firms undertaking refurbishment projects are often small, as are their projects. Large-scale renovation projects can be thought to afford platform development including significant IT investments. Small firms can be expected to be unwilling to invest heavily in comprehensive digital models (Sun et al., 2015), which also might not be sufficiently appreciated by their clients, often small property owners themselves. Depending on national and local practices, most projects are likely to be traditional design-bid-build, whereas other projects are procured as design-build, with potential consequences for IT use linking design more strongly to process monitoring.

Many smaller contractors depend today on commercially available enterprise systems that fulfil a number of functions such as linking project planning to accounting, although primarily intended for new construction (Tatari and Skibniewski, 2010). Although these software packages fall short of being manufacturing-type enterprise resource planning (ERP) systems, performance indicators can be derived from them (Skibniewski and Ghosh, 2009). Integration of data flows within the organisation may then be of more obvious use than attempting to exchange digitally structured information with the client, either received from the design stage or provided by the contractor as input to facilities management. The available software packages have to be adapted to the specific needs and conditions of refurbishment projects. For larger contractors with their own internally integrated systems, there are similar needs for adaptation to refurbishment.

Linking to the refurbishment contractor's suppliers of goods and services may also be assigned a higher priority than directly tapping into client design information. Much refurbishment work is subcontracted. Sub-contractor involvement in the planning of refurbishment activities tends to be limited, mostly ad hoc in nature and according to their specialist contributions (Rahmat and Ali, 2010). Depending on the technology involved, sub-contractors such as those delivering piping and electrical services may have their own, subsector specific information structures and systems, which are able to handle the complexity of the components to be diagnosed, replaced or newly installed. For these sub-contractors, the main barrier to integration is not so much that their ICT skills and capabilities are lower than that of main contractors. The problem often lies in the lack of interoperability. There are many potential benefits of closer integration and information exchange across sub-contractors (Loosemore, 2014). However, although a main contractor might wish for a broader view of resource use and performance on the refurbishment site, sub-contractors tend to remain black boxes.

Case study – Swedish site managers

This section will discuss the results from a case study carried out in Sweden in 2015. Interviews with 19 site managers of Swedish refurbishment projects showed the importance of scale and repetition for the development of monitoring routines. Those interviewed belonged to two of the leading contractors and three smaller, specialised contractors. Many refurbishment projects are small and unique. Software use on site is dominated by simple spreadsheets. The large housing estates with blocks of dwellings built during the Swedish 1965–1974 *Million Homes* programme (Hall and Vidén, 2005) and now in need of modernisation allow repetition of standard solutions during the refurbishment process. In one of these large projects with renovation of several buildings which had similar structural and architectural features, the site manager explained that they had chosen one of the buildings as a pilot and then had created plans for waste management and other activities based on that particular building. The developed routines were then re-applied with small modifications throughout the estate. Despite the scale of the project, there was no on-site use of BIM.

Why did site managers choose particular ICT tools? Was it because of ease of use? Did these tools prove to be a more efficient substitute for more primitive technologies? Was there any added value from new functions? The interviews and a questionnaire survey of site managers shed some light on these questions. It was initially thought that site managers may not accept spending more time on using a given ICT tool to complete a task than they would when using traditional methods. On the contrary, one interviewee brought up an example of support for visualising project schedules. This relied on a software platform that required spending much time on producing and updating schedules but was still seen as being very useful for on-site management. The ability to enter project data directly on a portal, using mobile phones and (more seldom) tablets, was one of the features mentioned as beneficial. Nevertheless, not all innovative features were welcomed by site managers; one interviewee for example questioned a new application allowing them to print directly from mobile phones.

The questionnaire responses indicated that, out of eight functional features, there were three that corresponded to requirements shared by a majority of site managers:

- my ICT support must allow easy updating of project information;
- it must be easy to interpret what I see on my screen;
- my ICT support must allow easy entry of data.

At the other end of the scale, there were three clearly less important requirements, according to the site managers:

156 *Bröchner & Sezer*

- my ICT support must be possible to link to the client's IT system;
- my ICT support must allow simple linking to a BIM model;
- my ICT support questions must be answered within two minutes, also outside ordinary work hours.

Two requirements received intermediate responses:

- the IT system must be accessible without failures lasting longer than 30 seconds;
- information in the IT system must be accessible on my smartphone.

Site managers thus seemed primarily concerned with ease of use of ICT tools for performing simple everyday tasks. The reason why they did not prioritise quick access to a helpdesk might be that they were unlikely to use complicated ICT tools on site. While linking to a client system and BIM implies a potential for a higher level of usefulness, it was not part of the site managers' world. For purposes of monitoring, the perceived needs of integration with external (client) information and BIM are thus less prominent than the immediate tool usability features.

Potential drivers for higher uptake rates – sustainability and integration

Growing concerns with environmental sustainability may turn out to be a strong driver that increases the uptake of new technologies. Leadership in Energy and Environmental Design (LEED), the Building Research Establishment Environment Assessment Method (BREEAM) and other building certification systems (Cole and Valdebenito, 2013) encourage clients and contractors to systematise the monitoring of materials used, not least because certification procedures are facilitated by BIM. Hammond et al. (2014) have presented a case of a university office building that could be renovated in order to achieve a major reduction in energy use and analysed how the project would benefit from BIM linked to LEED certification. A conceptual model for bridging BIM and an asset information model based on certification systems has been proposed by Alwan and Gledson (2015).

BREEAM Refurbishment Domestic Buildings and the LEED for New Construction and Major Renovations are two of the commonly used client-oriented building certification schemes. These are special versions of more general sustainability rating schemes that recognise refurbishment of existing buildings and have a simplified approach to the ecological sustainability of refurbishment as a physical process (Sezer, 2015). In the BREEAM Refurbishment Domestic Buildings, the energy section is given more weight than for new construction, while the sections for waste, materials and pollution are given less weight. LEED for New Construction and Major Renovations has a long history, but there is little of it that is specific to refurbishment processes. The Comprehensive

Assessment System for Built Environment Efficiency (CASBEE), the Japanese system, does include CASBEE for Renovation in addition to an economic efficiency evaluation tool. The German Association for Sustainable Buildings now provides a tool for evaluating existing buildings and renovations which does cover all three aspects of sustainability and includes the categories of sociocultural quality, functional quality and process quality. Also, the Australian Green Star Certification includes a version for major refurbishments.

With an increased emphasis on sustainability, more detailed waste management on refurbishment sites needs better support from integrated digital models. These can include semantic information such as attributes of building materials that are to be removed or demolished, as well as replacement materials and components which are environmentally friendly, recyclable or otherwise. In their analysis of office retrofit projects, Li and Yang (2014) mention the potential role of BIM in waste analysis. In the case of Sweden, linking BIM-based refurbishment planning to a sustainable materials database such as the SundaHus, which provides a system of standards for assessing construction materials and products from environmental and health perspectives, appears to be desirable for both contractors and clients. Green public procurement initiatives can raise the level of awareness here.

Conclusions – slow progress, but sustainability is a driver

Modern ICT tools often replace the corresponding pen-and-paper operations. For contractors that pursue a wide range of project types and sizes, there is a challenge in downscaling IT systems from large new construction settings to refurbishment projects. In principle, the BIM database will support detailed budgets and production schedules as well as on-site logistics. For most refurbishment projects, waste generation is a serious challenge and can be facilitated by access to BIM data. However, the production of an accurate and useful geometrical model remains costly and applications remain limited to large, repetitive projects or those where the client can see immediate benefits from investing in a 3D digital representation.

Information technology may not have developed as rapidly in this field as expected by specialist proponents in the 1990s, although there is steady progress through software that is becoming more efficient through the accumulation of international experiences. Client demand for integration throughout the process starting with design and ending with further management of the refurbished building is crucial for progress. Given the typically high dependence on specialist sub-contractors in refurbishment projects, the main contractor faces another challenge of information integration with the supply chain. More widespread use of BIM in refurbishment requires distributing benefits and incentives throughout networks of firms.

Although main contractors today might identify advantages more readily, they must interact with both clients and sub-contractors whose IT capabilities vary greatly. Downscaling methods and experiences from large refurbishment projects

158   *Bröchner & Sezer*

with skilled participants is an urgent task. More emphasis on documented environmental sustainability of refurbishment processes is to be expected in the future, and this will be increasingly important for BIM-based integration of information flows in refurbishment projects.

Acknowledgements

Support from the Swedish Construction Federation (BI) and the Centre for Management of the Built Environment (CMB) at Chalmers University of Technology is gratefully acknowledged.

Note

1   Laser scanning refers to laser beams being steered, usually through moveable mirrors, also measuring the distance at every pointing direction. This captures 3D shapes of objects and facilities. A point cloud of geometric samples of surfaces is created. From these points, the shape of the object is reconstructed. Ideally, this process should generate a 'semantically rich 3D model of the facility, composed of objects characterised by geometry, relations, and attributes' (Pătrăucean et al., 2015), and the result would be a BIM file.

References

Alwan, Z. and Gledson, B. J., 2015. Towards green building performance evaluation using asset information modelling. *Built Environment Project and Asset Management*, 5(3), 290–303.

Cole, R. J. and Valdebenito, M. J., 2013. The importation of building environmental certification systems: International usages of BREEAM and LEED. *Building Research and Information*, 41(6), 662–676.

Egbu, C. O., 1997. Refurbishment management: Challenges and opportunities. *Building Research and Information*, 25(6), 338–347.

Gimenez, L., Hippolyte, J. L., Robert, S., Suard, F. and Zreik, K., 2015. Review: Reconstruction of 3D building information models from 2D scanned plans. *Journal of Building Engineering*, 2, 24–35.

Gimenez, L., Robert, S., Suard, F. and Zreik, K., 2016. Automatic reconstruction of 3D building models from scanned 2D floor plans. *Automation in Construction*, 63, 48–56.

Hall, T. and Vidén, S., 2005. The million homes programme: A review of the great Swedish planning project. *Planning Perspectives*, 20(3), 301–328.

Hammond, R., Nawari, N. O. and Walters, B., 2014. BIM in sustainable design: Strategies for retrofitting/renovation, in *Proceedinds of the 2014 Computing in Civil and Building Engineering (ASCE)*, Orlando, 23–25 June, pp. 1969–1977.

Hong, S., Jung, J., Kim, S., Cho, H., Lee, J. and Heo, J., 2015. Semi-automated approach to indoor mapping for 3D as-built Building Information Modeling. *Computers, Environment and Urban Systems*, 51, 34–46.

İlter, D. and Ergen, E., 2015. BIM for building refurbishment and maintenance: Current status and research directions. *Structural Survey*, 33(3), 228–256.

Kim, K. P. and Park, K. S., 2013. BIM feasibility study for housing refurbishment projects in the UK. *Organization, Technology and Management in Construction*, 6(2), 765–774.

Li, M. and Yang, J., 2014. Critical factors for waste management in office building retrofit projects in Australia. *Resources, Conservation and Recycling*, 93, 85–98.

Loosemore, M., 2014. Improving construction productivity: A subcontractor's perspective. *Engineering, Construction and Architectural Management*, 21(3), 245–260.

Mansfield, J. R., 2002. What's in a name? Complexities in the definition of 'refurbishment'. *Property Management*, 20(1), 23–30.

Pătrăucean, V., Armeni, I., Nahangi, M., Yeung, J., Brilakis, I. and Haas, C., 2015. State of research in automatic as-built modelling. *Advanced Engineering Informatics*, 29(2), 162–171.

Rahmat, I. and Ali, A. S., 2010. The involvement of the key participants in the production of project plans and the planning performance of refurbishment projects. *Journal of Building Appraisal*, 5(3), 273–288.

Sanchez, A. X. and Joske, W., 2016. Benefits dictionary, in A. X. Sanchez, K. D. Hampson and S. Vaux (eds) *Delivering Value with BIM: A Whole-of-life Approach*, London: Routledge, 103–204.

Sezer, A. A., 2015. Contractor use of productivity and sustainability indicators for building refurbishment. *Built Environment Project and Asset Management*, 5(2), 141–153.

Sezer, A. A. and Bröchner, J., 2014. The construction productivity debate and the measurement of service qualities. *Construction Management and Economics*, 32(6), 565–574.

Singh, Y., Abdelhamid, T., Mrozowski, T. and El-Gafy, M., 2014. Investigation of contemporary performance measurement systems for production management of renovation projects. *Journal of Construction Engineering*, 417853, 9 pages.

Skibniewski, M. J. and Ghosh, S., 2009. Determination of key performance indicators with enterprise resource planning systems in engineering construction firms. *Journal of Construction Engineering and Management*, 135(10), 965–978.

Sun, C., Jiang, S., Skibniewski, M. J., Man, Q. and Shen, L., 2015. A literature review of the factors limiting the application of BIM in the construction industry. *Technological and Economic Development of Economy*, doi:10.3846/20294913.2015.1087071.

Tang, P., Huber, D., Akinci, B., Lipman, R. and Lytle, A., 2010. Automatic reconstruction of as-built building information models from laser-scanned point clouds: A review of related techniques. *Automation in Construction*, 19(7), 829–843.

Tapponi, O., Kassem, M., Kelly, G., Dawood, N. and White, B., 2015. Renovation of Heritage Assets using BIM: A case study of the Durham Cathedral, in *Proceedings of 32nd CIB W78 Conference, Eindhoven, October 27–29*, pp. 706–715.

Tatari, O. and Skibniewski, M. J., 2010. Empirical analysis of construction enterprise information systems: Assessing system integration, critical factors, and benefits. *Journal of Computing in Civil Engineering*, 25(5), 347–356.

Volk, R., Stengel, J. and Schultmann, F., 2014. Building Information Modeling (BIM) for existing buildings – Literature review and future needs. *Automation in Construction*, 38, 109–127.

Williams, J., Brown, C. and Springer, A., 2012. Overcoming benchmarking reluctance: A literature review. *Benchmarking: An International Journal*, 19(2), 255–276.

10 Experience with the use of commissioning advisers in design

A Danish context

Marianne Forman

Introduction

After several years of increasing demands for higher energy performance in Danish buildings, the experience is that there appears to be a large gap between the estimated energy consumption from the design phase and the actual energy consumption experienced during operations. The root of this situation can be traced back to the need for information integration across construction projects and large operating organisations (see also Chapter 11). Finding new approaches to integrating information between construction projects and large operating organisations should be a major priority in Denmark and many other countries, due to increased pressures for energy reduction. In Denmark, what is referred to as commissioning is one of the up-coming methods for handling this problem. Commissioning is a quality assurance (QA) system that is designed as a component of project delivery. The development of the concept of 'commissioning', which has taken place in the Danish building industry, is strongly inspired by the well-established ASHRAE 0-2005 guideline from the United States (US) (Ágústsson and Jensen, 2012). This chapter will draw on the Danish experience of adapting and using this QA model.

In the related Danish standard, a commissioning process is described as follows:

> The commissioning process for a building is a quality-assurance process that is performed to obtain, verify and document that a building's technical facilities, installations and systems are planned, designed, installed and tested, and that they are operated and maintained so that they meet the requirements for the total economy of the building, requirements in building regulations and other legislation as well as the client's clearly defined requirements.
>
> (DS 3090, 2014)

In 2006, the Danish Building Regulations (BR06) introduced a new requirement for the energy performance of buildings based on energy frames, representing a shift from previous descriptive regulations to a performance-based regulation. The use of energy frames as the main requirement of new construction was a result of the European Union (EU) Directive on Energy Performance of Buildings. The shift from descriptive to performance-based regulation has led to the need for new energy solutions in the construction industry.

In Denmark, new priorities in construction are often addressed through new standards, including new types of consulting services and coordination tools. The increased complexity of buildings' technical systems has led to a greater focus on commissioning and more advisers have started offering commissioning as a consulting service to their clients. In 2013, the Value Creating Construction Process published instructions on the commissioning process. The new Danish standard DS 3090 (Danish Standards, 2014) was released the following year and introduced as a way of handling information integration across the construction project and with the client. This has been followed by an ongoing debate about whether commissioning should be introduced as a legal requirement for large buildings in Denmark. However, there is insufficient practical experience and knowledge about the commissioning process and how it may impact construction processes and building performance.

This chapter will illustrate the role of commissioning in construction and its effect on coordination, knowledge flows and practice through a case study about the design phase of a large municipal school in Copenhagen. Previous municipal projects have identified a large gap between the energy consumption calculated in the design phase and the actual energy consumption in operation. This realisation has led to the case study project having strict requirements for the building's energy performance. Copenhagen Property, which manages and services the properties owned by the City of Copenhagen, has chosen to test whether the use of a commissioning adviser can contribute to a process which ensures that the operation and use of the building is mainstreamed from the design stage of the building's technical systems. This project used the new Danish standard for commissioning.

This case study will show that the design and operation of high-performance buildings is more than information integration across life-cycle phases, existing knowledge and a project-based quality-assurance system. Instead, it is also about developing and adapting practices of the involved communities based on a mutual learning process. Integration of information across different professions and users in relation to new priorities is thus not only about how data and information flow between the various actors. It is also about the opportunities those actors have to use the information locally in a meaningful way, for example, as a basis for architects and engineers through the design process and for operations departments to formulate their requirements for the operational phase before construction starts.

Role of information integration in the context of design and operation of high-performance buildings

Commissioning is introduced by consulting firms and Danish Standards as a way to handle information integration across the construction project and the client

162 *Forman*

organisation. According to Ágústsson and Jensen (2012), commissioning is 'not a replacement for existing quality inspection process, but an addition to that process'. The methodology, which is developed to achieve quality assurance of the process, consists of a series of procedures and templates that are used throughout the commissioning process.

There is, however, a lack of clarity about the reasons behind the difficulties of achieving high energy performance buildings and what commissioning is. A French–Danish research project on how the construction industry adapts to new societal demands found that

> The construction sector's innovation is characterized by a continuous translation of societal demands, project-related requirements, requirements from the operational system and user requirements in the individual construction project, and that this translation is done in many places by many different actors in the construction sector.
>
> (Bougrain et al., 2014)

In order to transition towards the new energy requirements, it is here argued that the issue must be clearly understood within the context of new challenges such as:

- technical complexity – composition of technical systems, run-in technical solutions and troubleshooting, and change of operational systems;
- organisational complexity – new partnerships between private and public actors;
- process complexity – new types of contracts and performance goals;
- user complexity – new types of interactions within the use phase;
- service and operational complexity – long lifespan and involvement of new stakeholders.

The integration of information across construction projects and large operating organisations required to handle these new challenges is not limited to sharing new types of information across actors. It also includes developing new relationships between practices of information integration within a project and an operating organisational setting. Vertically within the project, there is a need to develop and articulate the client's technical requirements so that these can serve as input for the design team throughout the design process. This is a knowledge-sharing process across architects, engineers and operation teams in construction projects. Longitudinally, across portfolios of buildings in major operating organisations, there is also a need to integrate operational experience with technical systems as a platform for articulating technical requirements for new buildings and to qualify operating organisations for performance-based operation.

The case study presented in later sections of this chapter draws on ethnographic methods and theories and will use the theory of boundary objects to analyse the ways in which commissioning supports the need for knowledge sharing across the construction and operation phases. The key concepts from the ethnographic approach which are relevant for this case study and the way in which they are used will be described in the following section. The ethnographic approach was chosen to focus on the analysis of information integration as a knowledge-sharing process across knowledge boundaries.

Successful collaboration across multiple groups in different communities is often perceived as being based on consensus, but Star and Griesemer (1989) found in their studies that groups collaborate without necessarily reaching consensus. They have developed the concept of boundary objects as a way of explaining how cooperation can occur without consensus across different groups. Boundary objects are present on organisational boundaries, where they are used as tools to communicate and coordinate knowledge across knowledge boundaries. Boundary objects are developed by groups who want to collaborate towards a particular goal. These structures have different physical and organisational characteristics, and meet the specific information and coordination needs of each group. Over time, the objects and associated organisational practices will be taken for granted by the groups and be stabilised as part of their infrastructure. Interpretative flexibility describes the fact that different groups interpret the same object in different ways making cross-group collaboration essential so that it makes sense for the different groups in their own practice (Star, 1999; 2010). 'Boundary objects are at once temporal, based in action, subject to reflection and local tailoring, and distributed throughout all of these dimensions. In this sense, they are n-dimensional' (Star, 2010). Star and Griesemer (1989) developed four different types of boundary objects from their empirical analysis. They are:

- repositories: 'these are ordered piles of objects which are indexed in a standardised fashion' (e.g. a library or museum). Individuals can use or borrow from the 'pile' for their own purpose;
- ideal types: 'this is an object such as a diagram, atlas or other description which in fact does not accurately describe the details of any one locality or thing. It is abstracted from all domains, and may be fairly vague. However, it is adaptable to a local site precisely because it is fairly vague';
- coincident boundaries: 'these are common objects which have the same boundaries but different internal contents' (e.g. a map). 'Work in different sites and with different perspectives can be conducted autonomously, while cooperating parties share a common referent';
- standardised forms: 'these are boundary objects devised as methods of common communication across dispersed work groups'. 'The advantages of such objects are that local uncertainties are deleted'.

Gal, Lyytinen and Yoo (2008) have found that boundary objects, information infrastructures and organisation identities are intricately interwoven. If boundary objects that are shared by multiple organisations are changed, the boundary practices that interconnect the organisations and the identity of the organisations will also change (Gal et al., 2008).

164 *Forman*

The essential knowledge boundaries that the new commissioning objects operate in are defined as being between clients, energy consultants, the operation team and the commissioning adviser. Since the focus is on the design phase, suppliers and contractors are not currently involved. Each organisation is described in terms of how knowledge is localised, embedded and invested (Carlile, 2002). Knowledge is invested in practices such as methods and ways of doing things. Success demonstrates the value of the knowledge developed, which means that success will motivate maintaining the same practice. The way the different communities use the different objects at micro level depends on how the various objects are able to transfer, translate or transform information across the knowledge boundaries, as well as the various communities' ability to use the information locally in a meaningful way.

Carlile (2002) found that existing knowledge needs to be transformed in connection with product development. He states that only objects, models and maps with coincident boundaries are capable of transforming knowledge across knowledge boundaries, which in turn will change understanding within each knowledge domain. However, it is not enough simply to transfer or translate information across knowledge boundaries; knowledge needs to be transformed within each knowledge domain. Although it might seem to be parallel to product-development processes, new information infrastructure has to be established across the different boundaries to handle the new information requirements.

The different objects described in the commissioning standard are characterised according to Star and Griesemer's (1989) and Star's typologies (1999; 2010). The key objects are: commissioning plan (Cx plan), commissioning meetings (Cx meetings), commissioning log (Cx log) and commissioning scrutiny (Cx scrutiny). Each of these objects has different structures that allow them to coordinate knowledge sharing in different ways across communities. In the case study, the different objects are described with regard to whether they act as boundary objects across the communities or have some other function. In the context of construction projects, different types of boundary objects are historically created by the various parties to ensure communication and coordination in the projects, such as contracts and agreements. They are all part of the project's infrastructure that guides the project's organisation. In the case study, the commissioning objects are described based on how they affect existing boundary objects and boundary practice in the design phase and their impact on long-term information integration structures between the design team and the operating organisation.

Case study

The municipal council of Copenhagen aims to make the city carbon neutral by 2025. This places special demands on the energy performance of municipal buildings. Prior experience indicates that there is often a gap between the designed (calculated energy demand) and the actual energy consumption of buildings. The lack of accurate energy forecasts means that it can be difficult to

Commissioning advisers in design 165

plan effectively and achieve the desired CO_2 reductions. This section explores a case study about the design phase of a large municipal school in Copenhagen. In this case, the design phase was divided into four components: the overall programme, outline, project and main project.

The aim of the school project was twofold: to build a new school that met the energy performance requirements for the buildings, both in the project design phase as well as the operational phase, and to test commissioning as a method for achieving the energy performance objective. The central actors were the City of Copenhagen, a Danish energy engineering company that works closely with a Danish architectural firm in the design team, and an engineering company functioning as client adviser on the project, including the commissioning adviser function.

Data was collected through observations, interviews and building documentation. The construction project was tracked during 2014–2015, when the school was in the design phase. Due to the focus on commissioning, Cx meetings and special working meetings organised as part of the commissioning process were continuously observed throughout the design period. The research team focused on:

1　Which actors used the various boundary objects and what they used the objects for.
2　What topics were brought up at the meetings and what decisions were taken.

In addition, all documents and minutes of meetings were collected and the research team had access to the project online database including all project documents, project proposals and options, and all examination reports. Qualitative interviews were conducted twice with the client, the commissioning adviser, the architect, the energy consultant responsible for the design of all the technical facilities, the operation coordinator, the operation officer working with energy systems, and an employee from the department which ordered the school. The interviews were conducted as semi-structured interviews, in which the central themes were local work tasks, local work processes, the use of methods, theories and tools in the work processes and parameters for a successfully completed project. Interviews also sought stakeholders' perceptions of the contribution of the commissioning process with respect to the improvement of information integration between the various actors.

City of Copenhagen

Four government units are relevant in connection with the construction of schools in the City of Copenhagen:

* Construction Copenhagen is the client organisation of Copenhagen Municipality and responsible for carrying out the design and construction of the school;

166 *Forman*

- Copenhagen Property manages and services the properties owned by the City of Copenhagen and is subsequently responsible for the operation of the school;
- Children and Youth Administration manages the schools of the City of Copenhagen and is responsible for ordering the new school and user involvement during the project;
- Culture and Sport Administration manages the sports halls of the City of Copenhagen and is responsible for ordering the new sports hall for the school.

At the beginning of the school project, Copenhagen Construction was part of Copenhagen Property, but later became a separate unit. In this case, the Children and Youth Administration ordered the school and the Culture and Sport Administration ordered the sports hall. They were responsible for formulating a vision as well as specifying functional requirements for the project. Historically, both administrations have developed general functional requirements for buildings which, also in this project, were input to the initial programme phase in the design phase.

Operational requirements that have emerged as a result of new societal demands for enhanced energy performance in buildings have represented a challenge for the City of Copenhagen. There have been major problems with the delivery of new buildings where the users have not been able to establish how to properly operate the building, which in turn has cost Copenhagen Property time and money. To meet these challenges, Copenhagen Property has initiated an internal project 'From Build to Operation' (BtO) to improve communication of expectations between Copenhagen Construction and Copenhagen Property. In 2013, a task force was set up consisting of four members from Copenhagen Property and Copenhagen Construction to develop a concept for BtO with a special focus on early inclusion of operational requirements in the design phase of construction projects. The task force developed an operational log in which they gathered experience from operations as well as a process for participation in the construction project. The process is structured so that operating staff are involved in the development of the conceptual design, project proposal and commissioning training just before final delivery. In theory, operating personnel should be fully prepared for the take-over of the building's operational phase. The aim of this process is to ensure alignment around the type of building desired, level of technical complexity and type of skills required by the local operating staff. The task force followed a number of projects, arranged workshops for colleagues and are continuing their work on the operational log and the process participation concept.

Project participants

The following sections will describe the four central organisations in terms of tasks, responsibility, tools and key success factors as well as those key objects that influenced the design phase across the actors.

Client

Construction Copenhagen is responsible for carrying out the construction of the school. The client is an architect and has many years of experience. The client's task is to organise the project in order to achieve the project objective and to comply with the City of Copenhagen's policies. This includes preparation and conclusion of contracts with architects, engineers, specialist consultants, contractors and client advisers, all of whom are needed to help carry out a construction project. Furthermore, it is the client's responsibility to follow up on all agreements that are made. These relate to managing the process so that the ordering body (Children and Youth Administration and Culture and Sport Administration) and all the various construction companies' commitments are on schedule. Additionally, the client is required to document all decisions made during the project to address any potential conflict. The key personnel within Copenhagen Construction that the client collaborates with in relation to the tender documents and contracts are a lawyer and the manager of the department. In addition, the client adviser is the client's main 'sparring partner'. The client expressed that the key success factors of their work are to deliver a construction project on time, within budget and to the expected quality, including requirements for energy performance.

The operating organisation

Copenhagen Property is responsible for the operation of the school. The BtO coordinator is an experienced building constructor whose task is to involve the 'right people' from Copenhagen Property at the 'right time' in the construction process in relation to reviewing project documents. This includes ensuring that they receive information and submit comments for scrutiny on time. The Cx process in this project is an extended procedure in relation to BtO. Here, the coordinator engaged technical staff from four different areas: emergency operations, ventilation, electrical installations and energy management. The coordinator perceived that the key success criteria for their work were to ensure the school functioned for a variety of users and to strengthen the BtO concept.

The energy consultant

The energy consultant, responsible for the design of the technical system, is a medium-sized company and forms part of the team set up by the main consultant for the design process. The energy consultant is an energy engineer with many years of experience. In design work, industry performance specifications are important guidelines for the successful delivery of project material in the various phases of the design process. The energy consultant distinguished between integrated energy design, traditional design and follow-up, including the commissioning of the various phases of the construction project. Integrated energy design was applied from the beginning of the project and refers to the

168　*Forman*

process of establishing a dialogue with the architect from the start of the design process as a way of integrating energy considerations into the geometric design, choice of materials and orientation of the building. When technicians design technical systems they use different calculation methods, simulation tools, standards and regulations, which represent the traditional design phase. The energy consultant perceived that their key success criterion was to deliver projects that clients were subsequently satisfied with.

Commissioning adviser

The company, functioning as client adviser on the project, incorporated the commissioning adviser function and is a large engineering company. This organisation is experienced in the engineering field and works with both technical design and commissioning. The commissioning adviser coordinates incoming client data regarding requirements for technical facilities and monitors compliance. The contract refers to the Danish Standard about commissioning, but the consultant's services have been adapted to the needs of Copenhagen Property. This means that the role of the commissioning adviser is to support increased contact and dialogue between Copenhagen Property and Copenhagen Construction. The commissioning adviser sees the commissioning process as a way to use the various standard templates to achieve a good result from a process that is highly dependent on input from operations. The commissioning adviser perceived that the key success criteria for their work were that a building is fully operational from the moment of hand-over and that the building's servicing needs have been thoroughly established.

Objects that influence the design phase across actors

Table 10.1 summarises existing boundary objects, new objects from the operating organisation and new Cx objects which all affect the coordination of information and cooperation between the actors in the design phase.

Significance of using commissioning in the design phase for boundary practice

The following sections will describe the commissioning processes in the design phase. It will then explore the way the new objects influenced boundary practice, the consequences for the existing boundary objects and finally the dilemmas that the use of commissioning causes for the various parties.

Cx plans

The Cx plan, among other things, specifies the commissioning organisation. The members of the organisation were the commissioning adviser, client, architect, energy consultant, and representatives from the operating organisation, the Children and Youth and the Culture and Sport administrations.

Commissioning advisers in design 169

Table 10.1 List of existing objects from the design phase, new objects from the operating organisation and new Cx objects

---

Existing project boundary object in the design phase:

- contracts (roles and responsibilities);
- industry performance specification of professional services (expectation to deliveries in defined phases);
- process documentation in relation to ongoing decisions (legal position of decisions and placement of responsibility);
- project material in relation to the phases: program, outline, project and main project (the ongoing representation of the building);
- simulation tools;
- general functional requirement for buildings (requirements on which building programmes are established).

New objects from the operating organisation:

- operational log (first version of the operating organisation's requirements to buildings);
- BtO concept.

New Cx objects:

- Cx plan;
- Cx meetings;
- Cx log;
- Cx scrutiny.

A common understanding of the commissioning process was that the process should ensure knowledge sharing between the design team and the operating organisation. The commissioning organisation was perceived as appropriate because both sides were represented and made aware of problems and challenges related to the design and use of technical systems. In addition, it was seen as an advantage that there were many users of operating activities, as it gave them the opportunity to understand all the choices that needed to be made in connection with the construction project.

Cx meetings

Cx meetings were held regularly. Additional participants are included based on the agenda items. The meetings have pushed the focus on user needs and experience in operating the building to an earlier stage in the design process. Solutions that users could relate to have been presented early in the different phases and this gave rise to a number of useful discussions early in the design phase. A joint assessment was that the Cx meetings strengthened this process as they provided the opportunity to discuss operation, maintenance and user-oriented features of the building. Furthermore, they all found that it was important to have someone who was responsible for arranging meetings.

170 *Forman*

Cx log

According to the standard, the Cx log represents the client's defined requirements for technical facilities, installations and systems that the finished building must provide. In this project, the task of defining requirements was placed with the operating organisation and the other users, and the operational log was split into a Cx log and an operational log. The commissioning adviser believes it is necessary to separate the two logs as many items in the operational log are outside the scope of the technical systems that belong to the commissioning task. Experience shows that the Cx log is not established from the beginning with articulated requirement to the technical system, but is built through the design process. However, the Cx log also had the function to uphold the requirements of the design process as soon as they were articulated.

Scrutiny processes

The design process was divided into four phases: programme, outline, project and main project. The Cx scrutiny is a standardised schema where all the different communities have the opportunity to comment on project documentation at the different phases. The various communities raise questions that need to be answered by the design team and in turn must be accepted by the questioner. The commissioning function and therefore the scrutiny process was handled by the client adviser.

The results show that, where the users and the operating organisation are able to comment on different solutions and can ask clarifying questions, the client adviser's technical staff raise new technical solutions. In this way, the comments from the client adviser's staff act as a competing engineering paradigm on project solutions rather than an appraisal of the design team's proposals. The dialogue in the scrutiny schema could for example address the basis for calculations of expected energy consumption which is an input into the simulation tool. However, the Cx scrutiny process also had the function of identifying specific problems where further investigation was appropriate.

Commissioning influence boundary practices

This section will describe the changes in boundary practices initiated by the reinterpretation of the existing boundary objects and the significance of the design work. In construction projects, 'contracts', 'industry performance specification of professional services' and 'process documentation in relation to ongoing decisions' are all part of the information infrastructure that serves as a common reference. They guide the different roles, responsibilities and obligations of organisations in the design process. At the same time, they provide the framework for collaboration among organisations and form the basis for assessment of the various organisations if the collaboration succeeds or fails.

The regular Cx meetings and the Cx log make it possible for the client to make decisions about technical matters. By supporting participation of the technical staff in the construction project with a commissioning adviser, the client sees that the operating organisation has become a better project participant in the sense that the commissioning adviser has helped the department to qualify their input for the design process. In previous projects, the client had to spend time searching for the relevant staff in the operating organisation. It seems that the client used the process to optimise coordination tasks related to the operating organisation as well as maintaining documentation generated through processes that support the client's record-keeping needs. In this way, the client used the process as a coordinating tool to ensure progress.

The scrutiny process was used by the client as a new project tool for checking deliveries for the various design phases, which in turn changed boundary practices between the energy consultant and the client. The scrutiny processes mean that the client received detailed feedback from many different knowledge domains to assess project documentation at each phase. This in turn meant that the client could comment on project documents, as opposed to past experiences where the client often simply accepted them. From the design team's perspective, use of the scrutiny processes represented a shift in deliveries expectation. The energy consultant considered that the level of detail for the various design phases, as described in industry performance specifications of professional services, could be interpreted in several ways. They also thought that the scrutiny process raised expectations regarding the level of detail required for each phase of the design process. The consequence for the design team is that the representation of the building in the project material was perceived as more complete in each phase than it really was.

The contract system in the sector assigns roles and responsibilities across organisational boundaries and is a key coordination tool in construction projects. The commissioning process put pressure on the coordination of different roles and responsibilities in the design phase. The client was very careful to keep the various actors in their roles and found that it can be difficult at times to keep the support of the commissioning adviser and the team. There can be a tendency for technical staff to develop technical proposals, but if the commissioning adviser and the team behind them do so, they change roles and responsibilities in the project organisation. In this case, the client risks the design team disclaiming responsibility for a technical solution by referring to the commissioning adviser, who represents the client. In contrast to the client, the BtO coordinator will retain the commissioning adviser in a supporting expert role. However, this will push the commissioning adviser and the team behind him to assess solutions, thus creating ambiguous roles that the client wants to avoid.

From the consultant's perspective, the problem regarding roles and responsibility is that the commissioning adviser gathers and coordinates information for the various users, but they do so without responsibility for the design. Since they are the responsible parties, the energy consultant and his team therefore need to check all information. In continuation of this argument, the consultant believes

172 *Forman*

that the client will have to pay too much for commissioning as it is organised in this project. They were baffled by the fact that the consultants who designed the systems were not brought in to hand over the system to the operating staff, as it is they who know the system best. A commissioning adviser is now paid to follow the entire project.

The design work was also influenced by the commissioning process. The consultant's experience with scrutiny is that it has worked well with the various users, but it has given rise to problems with the client-adviser team; i.e. third parties. The user work activities take place in or are related to the finished building, and their issues are therefore related to its operation. The consultant considers the users' requirements in the review as a matter for dialogue and clarification by both sides. In contrast, the scrutiny comments made by other engineers (third parties) imply that things should be done differently. The two competing engineering paradigms represent a dialogue within the traditional design practice and can be explained by the fact that the two companies come from communities with quite similar knowledge. The consequence of using scrutiny in this way may be that the traditional design practice is not being challenged. Instead, it may challenge contractual liability, since placement of responsibility can be blurred and thus change boundary practice.

In the project, the operational log was split into a Cx log and an operational log. For the BtO coordinator and the technical staff, this meant that they left the responsibility for issues in the Cx log to the commissioning adviser. There is only one engineer with experience in control systems in Copenhagen Property, so he is often busy. He has been involved in the Cx process but has largely left his work to the commissioning adviser. In this way, the operations department uses the commissioning adviser as an additional resource.

Project material and operational log as transformative boundary objects

The energy consultant perceives the operational log as a tool to support dialogue and clarification from both sides. In this way, the operational log can be viewed as a coincident boundary object, which is characterised by having the same boundaries but different content. As a boundary object between the operating organisation and the energy consultant/design team, the list of requirements can be interpreted from different communities and represent local operating experience and input to a design process. The combination of Cx meetings and the use of the operational log has contributed to new understandings of design practice as well as operation practice and helped the operating organisation and design team to alter their own practices. It seems that the operational log as a boundary object between the operating organisation and the design team has the potential to contribute towards new information infrastructure based on transformed knowledge between the two communities.

In the energy consultant's opinion, the interaction with the operating organisation in particular has been good and this collaboration would be beneficial in every project design phase. Operating comments on project

Commissioning advisers in design 173

materials have also allowed energy consultants to receive support from the operating organisation and strengthen their own arguments against the architect in the design process. For example, it was often pointed out that the technical rooms were too small. This became the architect's decision, but they had trouble defining the floor area needed to meet the objectives for the building. This example shows how comments on project materials and the dimensions of technical rooms changed design practice by giving higher priority to technical rooms than other spaces, thus changing the boundary practice between architect and energy consultant.

Viewed from the perspective of the operating organisation, entering construction projects introduces a new responsibility. The move from sitting as an outsider and criticising the final result, 'the physical building', to active participation in the design process and being responsible for the choices made was an issue for the technical staff in the operating organisation. At the same time, they mentioned that they have gained a better understanding of the construction project, the choices made and the priorities established. This also means that the requirements for operations must be substantiated and not just based on personal opinion. The operational log was therefore sent to the in-house technical specialists in Copenhagen Property to ensure that the choices had a strong foundation. This has been a time-consuming learning process. In this case, the ongoing dialogue about the project material has redefined the operating organisation's understanding of the design process and led to changes in required practices, from personally based requirements to organisationally based ones.

A second version of the operating log was made within the framework of BtO based on all projects in which it is used. The adjustment is about reformulating the requirements so that they can be better understood by consultants: reformulating simplifications so that there are no contradictions, eliminating requirements because they are covered by legislation and adding new conditions. It is a dynamic tool which evolved from being a communication tool into one which represents functional requirements of the operating organisation. The function of the log today is to ensure that Copenhagen Property is heard in construction projects and can be seen as a functional programme which may form the basis for the preparation of a building programme. In this sense, the operational log has integrated knowledge from both the operating organisation and the design team. Further to this, it has integrated the client's need for definitive requirements. The operational log can be viewed as a boundary object that has the potential to support communication and coordination in a construction project across the various actors. The log represents a new type of information integration tool in a construction project, thus transcending traditional professional boundaries.

Dilemmas that the use of commissioning causes for the various organisations

Commissioning is seen as a new process in which new objects are introduced and therefore influences boundary practices in the design phase. Table 10.2 points out the different dilemmas that the various parties face in the design phase when new

174   Forman

Table 10.2 Overview of dilemmas that the use of commissioning causes for the various parties

Actor	Dilemmas that the use of commissioning causes for the various organisations
Client	**Benefits:** • increased process documentation which in turn increases the possibility for controlling the project material; • increased possibility for articulating operational requirements for the construction project which in turn optimises the coordination task in relation to the operating organisation. **Drawbacks:** • increased costs for the new consultancy service; • increased contract uncertainty; difficult to maintain roles which in turn introduces a new financial risk in relation to design responsibility.
Operating organisation	**Benefits:** • possible access to expert assessment solutions; • transfer of tasks to commissioning adviser; • systematic meeting structure gives rise to deep insight into the construction project process, which in turn may contribute to generalisation of the operational log; this may strengthen the operating organisation's role in the design phase. **Drawbacks:** • transfer of tasks to commissioning adviser which may hamper organisational development of the operating organisation; • the technical system is separated from other operational conditions, which may increase the risk of a low priority of other operational condition; this may lead to poor solutions in other areas; • uncertainty in roles; the energy consultant company has also focused on operation and withdrawal of the operating organisation as a new user.
Energy consultant	**Benefits:** • priority of operational conditions and visibility of the 'new user' can identify the client's requirements for the building; this makes it easier to deliver a building that lives up to expectations; • systematic meeting structure may increase the possibility of knowledge creation across boundaries, which in turn may shape new priorities in the design process in favour of operational conditions. **Drawbacks:** • uncertainty in roles since the commissioning adviser is not accountable; • use of scrutiny detracts from the design process as it takes time to fill in answers. Competing engineering paradigms may keep both design methods and solutions in traditional design mode. In the worst case scenario, it may take focus away from adaptation and development of new design methods and solutions based on experience from operating buildings.

boundary objects influence communications and coordination across organisational boundaries.

Strategies for new information integration structures

The increased focus on and requirements for more efficient energy performance of buildings challenges both traditional project-organised companies and large operating organisations. It has also intensified the need to ensure consistency between energy requirements, the design process, construction process and the operation of completed construction projects. This has brought the need for information integration across construction projects and large operating organisations into the limelight.

In Denmark, commissioning is one of the emerging methods for handling this problem. Commissioning is a quality-assurance system that is designed as a project delivery strategy. The case study presented here has however shown that transitioning the construction sector towards buildings with high energy performance is not only about integrating information across life-cycle phases based on existing knowledge. Rather, there is a need to develop and adapt the practices of involved stakeholders through a mutual learning process.

This highlights the importance of critically evaluating the development needs of the various parties to ensure that the energy performance-based construction will mature in a way that suits the various organisations as well as the common demand for enhanced energy performance in buildings. This requires determining where and how knowledge is to be obtained for new practices by project-based companies (design teams) or operating organisations, or through outsourcing the commissioning advisory role as a new project delivery activity within each project. At the same time, it is necessary to assess which boundary objects may function as information integration structures in construction projects and support communication and coordination across life-cycle phases. The case study highlighted three development areas that are central to the interaction between project-based companies and large operating organisations in transitioning the building and construction industry towards high energy performance.

Firstly, there has not been a tradition of operations and maintenance employees being involved in the design of new buildings. Although it is common to develop construction programmes that account for end-user general requirements, they seldom take into consideration operational staff's needs. Involvement of operational staff has been more ad hoc and on a personal level. At the same time, the operations division of large building portfolios is mostly based on historic factors, developed before the technique became an integral part of the building as a result of increased energy-performance requirements. This brings two challenges for large operating organisations. First, with increased use of technology in buildings there is a need to adapt and educate the operating organisations so they match the new competence requirements. Second, there is also a need to develop more generalised operating requirements for buildings on which future building programmes can be established or, in other words, an information integration

176 *Forman*

structure that can support communication and coordination across the operating phase and design phase in a construction project.

Secondly, design of technical systems is based on the expected use of buildings such as the number of users, number of occupant hours and type of activities. Consultants use simulation models based on expected use of a building, meaning that a specific use of the building is inscribed in the building's integrated technique, which then largely determines the use of the building. Due to changes in needs, such as changes in organisations, users commonly use buildings in ways that are different to those originally planned. This can very easily cause indoor pollution and reduce the building's energy performance. These issues raise the need for consultants to develop new methods for designing buildings that can support flexible use. Lessons from the case study indicate that combining design and engineering processes can provide solutions that help achieve this.

Thirdly, integrating various user-needs into the design phase of new buildings must be ensured through new project practices. Aspects of commissioning can contribute to this, but this solution should be assessed based on the relevance and functionality of each element. Furthermore, it is relevant to consider where the commissioning activity should be placed. Although perhaps a costlier option, outsourcing the commissioning advisory role can provide clients with a higher degree of control over the design and hand-over process. However, this approach can also promote competing engineering priorities, reducing input from new users and hindering design innovation. When the commissioning advisory role is awarded to the consulting engineers, it can support the development of new design methods and become a natural progression of their design delivery processes. This is also a less costly solution because only one engineering consulting firm is required to understand the technical system in detail. However, it also implies a lower level of control for the client. If the commissioning advisory role is carried out by the operating organisation, this function can strengthen their capabilities in building technologies generally and support their organisational development. It may, however, require specialised knowledge that can be difficult to locate in operating organisations.

Concluding remarks

Integrating information across new actors will most likely remain a priority area in the built environment industry for the foreseeable future. It may seem that 'concepts' such as quality-assurance processes and standards are sometimes prioritised ahead of 'ongoing learning processes' in the interstices between different knowledge boundaries. There is a tendency for 'concepts' and not 'ongoing learning processes' to remain after a project ends. It seems then that a strategic solution including the development of standards and new consulting services can lead to increased fragmentation across construction projects and large operating organisations rather than an increased integration of information across life-cycle phases.

The case study presented earlier suggests that, in future, it will be important to develop an understanding about socio-technical aspects of new forms of information integration in construction projects. Understanding the importance of project mechanisms for the realisation of large properties' long-term strategies is also crucial. It is recommended that new information integration infrastructures fulfil construction project needs as well as the long-term needs of the operating organisation. Commissioning is a project activity where problems related to the building's energy performance are attributed to a lack of direction and control of the project in terms of the building's technical system and initial design. This type of project activity may counteract experiential learning across heterogeneous organisations and multiple projects with a focus on the development of new types of information integration structures that can guide the organisations' collaboration across different practices.

References

Ágústsson, R. and Jensen, P., 2012. Building commissioning: What can Denmark learn from the U.S. experience? *Journal of Performance of Constructed Facilities*, 26(3), 271–278.

Bougrain, F., Forman, M., Gottlieb, S. C. and Haugbølle, K., 2014. *Complex Performance in Construction: Danish Building Research Institute*, Copenhagen: SBI Forlag.

Carlile, P. R., 2002. A pragmatic view of knowledge and boundaries: Boundary objects in new product development. *Organization Science*, 13(4), 442–455.

Danish Standards, 2014. *The Commissioning Process in Buildings – Installation Services in New Buildings and Major Renovations*, DS 3090, 1st edn, Copenhagen: Danish Standards.

Gal, U, Lyytinen, K. and Yoo, Y., 2008. The dynamics of IT boundary objects, information infrastructures, and organisational identities: The introduction of 3D modelling technologies into the architecture engineering and construction industry. *European Journal of Information Systems*, 17, 290–304.

Star, S. L., 1999. The ethnography of infrastructure. *American Behavioral Scientist*, 43(3), 377–391.

Star, S. L., 2010. This is not a boundary object: Reflections on the origin of a concept. *Science, Technology, and Human Values*, 35(5), 601–617.

Star, S. L. and Griesemer, J. R., 1989. Institutional ecology, ʹtranslationʹ and boundary objects: Amateurs and professionals in Berkeley's museum of vertebrate zoology, 1907–39. *Social Studies of Science*, 19(3), 387–420.

Værdiskabende Byggeproces (Value Creating Construction Process), 2013. *Commissioning-Processen (Commissioning Process)*, Copenhagen: Værdiskabende Byggeproces. Available at: www.vaerdibyg.dk/index.php?option=com_docman&task=doc_view&gid=156&Itemid= [Accessed 14 March 2017].

11 Turning energy data into actionable information

The case of energy performance contracting

Frédéric Bougrain

Introduction

Reducing energy consumption in buildings is one of the biggest challenges facing the built environment operations industry. Tackling this issue requires gathering energy consumption data and other details about building characteristics. Before cutting waste and making the appropriate investment decisions, it is necessary to know more about the energy use, the characteristics of the buildings and the equipment, the occupancy rate and the behaviour of the occupants. Having reliable information is also necessary to enforce contracts. For example, in energy performance contracts, when the information is not reliable, the risks taken by private partners are bigger. In addition to this, there is often information asymmetry due to the occupants' tendency to hide some of their behaviours. All of this results in the public paying higher prices for the contract.

On the one hand, the need to lower prices has to be balanced against the cost of information gathering. On the other hand, while data is necessary, the main issue is being able to turn this data into 'actionable information'. For example, smart sensors produce data about energy consumption and are also seen as a solution to monitoring energy consumption, but they can also be costly to implement and maintain. Additionally, according to Schneider Electric, one of the major companies operating in this field, most facility managers use only a limited amount of the potential functionality of their building management systems (Moore, 2014). This may be due to the specific competence required to use complex facility management tools. Energy performance contracting (EPC) may help address some of these issues by enabling more efficient coordination and information integration.

This chapter will first define EPC and present the issue of information integration within the context of EPC. One of the key questions asked is how to identify how much data is required to understand building energy performance. Finding the right balance between investment and the data required for effective building management will help maximise the return on metering investment. Another important issue relates to the interpretation of the information in order to make it actionable. Two case studies will be presented to illustrate this point. The chapter will end with recommendations for public and private actors who intend to benefit from energy saving performance contracts.

Energy performance contracting: definition and issues

Performance-based models are becoming more and more important in construction. This change of approach results from the pressures of demanding clients who need better value from their projects and spur the actors of the built environment supply chain to improve their quality standards and to provide better services. Consequently, traditional design and build contracts based on input specifications are increasingly being replaced by service-led contracts where the output to be delivered is specified (Hoezen et al., 2010). Under this scheme a comprehensive performance measurement system containing key performance indicators often becomes the backbone of operational management.

The use of EPC is growing among performance-based contracted projects, with a number of public actors having already implemented this model for retrofitting existing buildings (Lee et al., 2015).

> Energy performance contracting means a contractual arrangement between the beneficiary and the provider of an energy efficiency improvement measure, verified and monitored during the whole term of the contract, where investments (work, supply or service) in that measure are paid for in relation to a contractually agreed level of energy efficiency improvement or other agreed energy performance criterion, such as financial savings.
>
> (European Parliament, 2012)

Beneficiaries are either public or private organisations, and the provider of an energy efficiency improvement measure is referred to as the energy service company (ESCO). ESCO is:

> a natural or legal person that delivers energy services and/or other energy efficiency improvements measures in a user's facility or premises, and accepts some degree of financial risk in doing so. The payment for the service delivered is based (either wholly or in part) on the achievement of energy efficiency improvements and on the meeting of the other agreed performance criteria.
>
> (European Parliament, 2006)

The European ESCO markets vary widely in terms of development and size (Bertoldi et al., 2014). A steady growth of this sector has been observed since 2010 due to an expansion of interest from public clients who look forward to alternative solutions for financing energy renovations. This was also due to the implementation of dedicated ESCO legislation and measures. This was done both at a national (e.g. in Italy and Croatia with a new legislative framework) and European Union (EU) level with the energy efficiency obligation schemes which became mandatory in EU States via the Energy Efficiency Directive. Moreover, public tenders started to integrate issues such as life-cycle costs and energy efficiency. Several standardised documents and guides were also published

180 *Bougrain*

in many European countries to accommodate actors interested in ESCO projects. The aim was also to decrease transaction costs. Several factors have, however, been perceived as barriers to the development of the ESCO market and the promotion of EPC. The lack of certification schemes, the contradicting interpretation of legislation, the procurement process, the lack of proper measurement and verification practices and the absence of partnerships between ESCOs and sub-contractors were seen as the main problems. Compared with other European markets, the French market was considered as one of the most developed, benefiting from strong growth between 2010 and 2013.

The French ESCO market

According to the ESCO Market Report 2013 (Bertoldi et al., 2014), the market volume for ESCO-type projects in France was worth EUR3.2 billion (USD3.6 billion).[1] Within this, the market for EPC projects, 'which incorporates the cost of complete contracts, including audit, measure implementation and measurement and verification (M&V) for public sector projects', was estimated between EUR75-100 million per year in 2013.[2] In France, there are three types of contracts which are usually identified (Bertoldi et al., 2014):

1   EPC of type A covers operation, maintenance and energy purchase.
2   EPC of type B includes investments on equipment, energy monitoring, energy performance guarantee and services, and may also incorporate the services of type A.
3   EPS of type C includes building renovation (mainly insulation works) and part or total services of types A and B.

Financing problems, low awareness about successful projects and high transaction costs are among the main barriers to ESCO projects in France. Transaction costs in energy service contracts are mainly due to the preparation, negotiation, establishment, execution, monitoring and enforcement of the contract (Sorrell, 2007). These costs increase with the lack of data which makes projects more risky. The availability of energy consumption data and details on buildings characteristics are the cornerstone of the contractual arrangement between parties. Once the ESCO is in charge of operating the building, the issue is also how to integrate information about occupants' behaviour, equipment inventory and conditions in order to reach the energy target and to complement existing evaluation parameters with new data.

Information integration in EPC

Writing an EPC that ties compensation to performance is a complex issue because there is information asymmetry between parties involved in the relationship.[3] Having reliable data contributes to reducing this asymmetry and the risks associated with the project. It also helps improve the quality of the relationship between the partners before and after the signature of EPC in the following ways:

- during the call for tenders, good knowledge of the status, occupation level, equipment inventory and conditions, and energy consumptions of buildings helps ensure a quicker and better informed bidding process. All these elements contribute to defining the baseline of the contract and to the speed of the candidate selection. It also reduces information asymmetry and the risks for each candidate. Given that it is necessary to verify that the baseline has been properly defined, there is a negotiation around this issue once the preferred bidder has been selected. Finally, collecting complete data about energy consumption per building covered by the contracts reduces the costs of procurement for bidders and contributes to the project success. It helps potential candidates to propose solutions and to identify the payback period;
- during the life of the contract, baseline information is used to integrate any changes that may occur and modify the equilibrium of the contract. It also contributes to enforcing the contract and monitoring the ESCO by measuring its performance. 'There are two conditions for measures to be effective: observability and verifiability. The first is the possibility for the principal to observe the performance of the agent. The second is the capacity of the principal to verify observations and supply evidence by measures' (Aubert et al., 1996).

Although gathering reliable data is paramount, as mentioned earlier, the key issue is to turn these data into actionable information. The data–information–knowledge–wisdom (DIKW) hierarchy appears useful to explain how basic data are transmitted and integrated to become strategic elements.

According to theoreticians in management information systems, data are symbols and information is data that have been processed/structured to become meaningful (Rowley, 2007). Knowledge is based on information and it allows individuals to make decisions. Going from information to knowledge requires understanding, which is based on experience and skills (Bellinger et al., 2004). This idea was already expressed by Nonaka (1994): 'information is a flow of messages, while knowledge is created and organised by the very flow of information, anchored on the commitment and beliefs of its holder'. Nonaka distinguishes explicit knowledge which is codified and close to information and tacit knowledge which 'is deeply rooted in action, commitment, and involvement in a specific context' (Nonaka, 1994). He also considers that there are four modes of knowledge creation (Figure 11.1): '(1) from tacit knowledge to tacit knowledge, (2) from explicit knowledge to explicit knowledge, (3) from tacit knowledge to explicit knowledge, and (4) from explicit knowledge to tacit knowledge'.

In the first mode called 'socialisation', 'the key to acquiring tacit knowledge is experience'. Computerised communication networks are frequently used for the second mode of creation which is referred to as 'combination'. Tools such as

182 *Bougrain*

	Tacit knowledge	Explicit knowledge
Tacit knowledge	Socialisation	Externalisation
Explicit knowledge	Internalisation	Combination

Figure 11.1 Modes of knowledge creation (Nonaka, 1994).

Building Information Modelling (BIM) and a centralised control station can contribute to the reconfiguration of explicit knowledge and lead to new knowledge. The third mode of knowledge conversion is called 'externalisation' and requires dialogue, trust and coordination between team members to be activated. When redundancy of information exists within the team, dialogue and trust are easier to build. The fourth mode is referred to as 'internalisation' and is very similar to learning. 'These conversion modes capture the idea that tacit and explicit knowledge are complementary and can expand over time through a process of mutual interaction.'

This framework indicates that turning data into actionable information in EPC can go through several modes of knowledge creation.

Case studies

According to Tellis (1997) and Eisenhardt (1989), the selection of the case is one of the most important issues when using a case study approach. Among EPC, two different and very significant projects were selected. The first was awarded in July 2010. It concerned the renovation, maintenance and operation of 18 high school buildings for 15 years. The goal was to reduce energy consumption by 42 per cent and greenhouse gas emissions by 58 per cent. The total costs of the project reached EUR80 million and the annual unitary payment is EUR5.2 million. It was one of the first EPCs with a primary focus on wall insulation. It was possible to obtain data and feedback from the schools in operation thanks to the years of service. The second case study was awarded in December 2015. It concerned the renovation, maintenance and operation of 14,000 dwellings for 15 years. The goal of the landlord, Habitat 76, was to reduce energy consumption by 40 per cent. The total costs of the project reached approximately EUR160 million. It was the fourth EPC launched by Habitat 76. The size of the project was quite unusual and the project benefited from the digitalisation of the whole housing stock by the company.

Both projects were procured through a competitive dialogue procedure. This process allows public organisations to describe the output/outcome needed instead of prescribing the input wanted from the bidders. Thus, technical specifications and prices are defined during the dialogue rather than being predetermined. In the first and second case, the procurement was respectively organised in three and four discussion rounds between the public body and the candidates.

Regarding data collection, the case study concerning the high schools was based on documents from the projects, face-to-face interviews with key actors and several meetings with the stakeholders of the projects.[4] The interviews focused on the organisation of the projects, their origins and goals (mainly energy and environmental issues), the characteristics and impacts of main innovative solutions on the operating costs, the competencies of the different stakeholders of the projects, the nature of the contractual agreements, the responsibilities in case of poor performance, the performance of the building in operation, and users' involvement during design, construction and operation. Data concerning the second case study was gathered through public documents, phone interviews with three people working for the housing company, one construction company and an architect.

Energy performance contracting in 18 high schools

The public organisation is a regional political administration that owns high schools and is in charge of operating them. As the owner of 106 high schools (95 per cent of its assets), it aims at reducing greenhouse gas emissions, energy consumption of its buildings and its energy bill. In 2009, it considered that EPC was the best solution to reach its objectives. Two different public procurement schemes were considered: public-private partnership (PPP) and traditional public procurement. Finally, after assessing that it offered better value for money, the PPP tender proposal was chosen for the financing, renovation, maintenance and operation of 18 high schools.

In 2010, after three rounds of competitive dialogue, a private consortium led by one of the largest construction companies in France was selected. Operating and optimising energy systems were new activities for this enterprise. Thus, it did not have any experience with EPC. This project was a way to develop new competencies and to expand its business towards facility management and energy efficiency services. In the typology of ESCO proposed by Duplessis et al. (2012), the enterprise can be classified in the category of 'public works companies strong enough to give saving guarantees. They start with new construction but will have to take in charge important refurbishments of building envelops'.

The use of energy data and building characteristics during the call for tenders

An audit was launched in 2008 by the head of the Energy Department of this regional public authority to find out more about the energy consumption of each of its high school assets, the efficiency of heating equipment and the quality of building façades. The classification of these datasets helped the Energy Department to select 18 schools based on criteria such as high energy consumption, poor building quality and no ongoing investment to improve energy performance.

This information with operational use was integrated in the tender documents. It helped bidders prepare for site visits during the tender stage and develop proposals. It also contributed to reducing operational risks. Indeed, dynamic

184 Bougrain

thermal modelling and simulation is a complex activity, which is illustrated by the frequent gap between predicted energy performance of buildings and measured energy use once buildings are operational (de Wilde, 2014). During the competitive dialogue bidders were able to get further details about current building conditions.

During the negotiation that followed the award of the contract and before the signing of the contract, one of the main issues was around the establishment of an accurate energy-use baseline. The ESCO considered that it was necessary to organise further site visits under the supervision of an engineer working for a thermal design office member of the winning consortium. They found that many important areas were not visible after the initial site visits. This engineer, who had to validate the energy conservation measures for the ESCO, integrated the supplementary information in his simulation model and considered that some solutions were not appropriate to reach the guaranteed energy savings. Thus, new technical solutions detailing the required works and guaranteed energy were proposed by the consortium. Works combined both building insulation and energy systems refurbishment.

The diffusion of information concerning operation and energy consumption

The ESCO must ensure the energy savings and the service quality, such as a minimum level of temperature in classrooms, during the 15 years of the contract. Operation and maintenance are critical to ensuring the project's performance. To avoid any downward spiral concerning energy savings, the ESCO invested in meters and sensors in order to:

- check the temperature of the high schools in use (classrooms, dormitories, offices, houses, laboratories and gymnasium);
- enable the monitoring of the energy consumption;
- report the performance of the energy conservation measures that have been installed in line with the measurement and verification plan.

A centralised control station was set up in every high school and completed this investment. With the help desk, this centralised control station is the backbone of the EPC. It allows the ESCO to monitor and optimise equipment performance and to verify, on a day-to-day basis, that cost saving goals are met. If consumption is above the target, the ESCO can implement rectification measures.

Information coming from the centralised control station and concerning all high schools are only seen by people working at the regional headquarter of the ESCO. This information is shared with the employees of the Energy Department of the public body who led the project to completion. However, they have no possibility to modify the parameters. The administrative agents of the high schools can also see most of the information concerning their buildings but are frequently not able to interpret it.

There is a technical agent in every high school who monitors that the ESCO fulfils its contractual agreement. The ESCO provided these agents with basic

Energy performance contracting   185

training to help them execute their duties. The aim was also to rely on representatives who are always in the schools. However, most of them did not have the required skills to understand and absorb the information delivered during the training sessions.

Occupants such as teachers and students are informed about energy savings and reduction of CO_2 emissions through a screen installed in the main lobby. Every year, the ESCO organises activities for the students to make them aware of climate change and the importance of reducing energy consumption. The administrative employees and the teachers can use an online platform when temperatures are too low and equipment is failing. Then, the ESCO is in charge of finding a solution to fulfil the agreed service level. The records and history of all calls and work orders addressed to the platform are contained in an information system that provides service performance statistics. When the ESCO does not respect the service level agreement then penalties are levied.

The performance of the ESCO

For the last three years, savings achieved were below the target, and the operator was penalised for not responding in due time to the needs of some high schools. However, every year the ESCO is getting closer to the initial guaranteed energy savings (Table 11.1). According to the ESCO, this inability to reach the target is due to not being able to integrate the actual water-heating energy usage into the calculations used to establish the energy consumption baseline, which in turn only included estimated values. This imprecision explains part of the difficulties in reaching the expected energy target.

However, this inability of the ESCO to reach the project's target performance is also due to a lack of experience. Firstly, there was a high internal turnover among technicians working in the ESCO. On average, one technician was in charge of three high schools located in the same region. The managers of the ESCO first thought that operation and maintenance would not require specific knowledge. Since information is available in the centralised control station, it is easy to use it. Indeed, with the centralised control station, information is available to any newcomer and is not attached to one person. However, there is a need to know about teaching activities, building characteristics, equipment in use and occupants' behaviour to operate each high school. Thus, each time a technician was promoted, transferred to another site or resigned, there was a loss of corporate knowledge. There was no planning ahead for this transition to ensure that properly trained staff would be available. Every new employee had to learn everything from scratch. This situation contributed to worsening the relationship

Table 11.1 Results of the EPC project concerning 18 high schools

Energy savings compared with baseline	2011	2012	2013	2014
Guaranteed savings (in %)	−10	−25	−42	−42
Real savings (in %)	−14	−20	−27	−36

186 *Bougrain*

between the ESCO and the administrative agents of the high school. To circumvent this problem, the ESCO decided to employ two people for every three high schools. One of them was not full-time but at least was able to share information with their colleague that was not always available in the centralised control station.

Further to this, more sensors were placed in the schools in order to improve the precision of the information. A centralised control station was also developed in order to expand existing functionalities. While the control over the energy consumption was initially done on a monthly basis, the ESCO was progressively able to know whether it reached the guaranteed energy savings on a weekly and then daily basis. Technicians working for the public body in every high school also received further training on the information delivered by the meters.

On the other side, the public body underestimated the cost of the monitoring process. It has to double check the information provided by the ESCO. Expenditures linked to the operation and maintenance plan were carefully examined by the financial department of the public body. Additionally, people working for the high schools, such as teachers and administrators, do not always signal faults on the online platform, letting, for example, too low temperatures and failing equipment go unnoticed. They have difficulties taking ownership of this new tool and frequently prefer directly contacting the agents working for the ESCO.

Energy performance contracting in social housing

The social housing company is located in Normandy, France. It employs about 820 people, and its turnover is around EUR128 million. It owns and manages 32,836 dwellings with approximately 65,000 tenants. In 2013, it launched its first EPC for five years on 119 dwellings. In 2014, it undertook two new contracts covering 330 and 700 dwellings for eight years. The total project costs reached, respectively, EUR4.8 million, EUR11.4 million and EUR19.9 million. The first project was very ambitious. Energy consumption was supposed to decrease by 60 per cent after replacing equipment and insulating dwellings. Works also concerned the interior of the dwellings. However, to reach its target the ESCO tended to decrease the comfort temperature and the renting department of the housing company received several complaints from tenants. Thus, the housing company preferred to be less ambitious for future projects in order to reduce the pressure on the ESCO. The targets were to reduce energy consumption by between 35 and 55 per cent for the second and third contract.

In 2010, the housing company started the digitalisation of its housing stock which concluded at the end of 2014. As a client who owns and operates dwellings, the aim was to improve its strategic asset management. Before this digitalisation, most of the information concerning the dwellings was in the hands (and the head) of the estate managers. They used common and traditional 2D plans. The digitalisation offers the housing company a 3D representation of its housing stock and a shared knowledge resource for its facility management and operations department. By gathering information concerning the surfaces of buildings' façades, dwellings and common areas in one digital database, the aim was to promote information sharing, monitoring and transfer and to integrate 3D modelling with facility management and operations processes. The EUR2 million representing the total cost of the digitalisation could be split into three parts:

- the salaries of four employees (the project team) who had to handle all aspects of the process;
- the cost of the software and the engineer delivering the training, EUR300,000;
- a design office had to redesign the drawings, to digitise them and to integrate them in the software used for asset management at a cost of about EUR32 per dwelling.

The impacts of this digitalisation were numerous and immediate:

- information concerning buildings was integrated in the call for tenders. It allowed the housing company to provide candidates with more details. In the case of calling for tenders concerning the restoration of buildings' façades, for example, it was easier to evaluate the quantity of painting required and to monitor painting trades workers;
- cleaning contracts were re-examined so they would represent the exact surface areas;
- building contractors saved time and money when they had to evaluate project costs;
- all employees are able to get more information from the desk of their office. Managers also know that the information is more reliable. Thus, strategic asset-management decisions are taken in an environment where uncertainty has decreased.

From BIM in design and construction to BIM in operation

The social housing company is also among the sector's pioneers for using BIM in design and construction. In December 2012, it launched its first project based on BIM. A design-build delivery system was chosen for the construction of 53 dwellings. This project and the ones that followed were successful. Buildings were available for use on time with no increase in costs. Field conflicts were also reduced since most issues that traditionally appeared at the building site were addressed during the design. Moreover, virtual dwelling visits were organised by using the 3D visualisation aspects of BIM. However, the integration of the information in the digital database was and still is problematic. The information linked to a new project is more detailed but some of the information, such as data related to the foundations, is not used during operation. Thus, some information is lost during the transfer from BIM in design and construction to BIM in operation.

188   *Bougrain*

Energy performance contracting for 14,000 dwellings

The digitalisation of the housing stock and the knowledge and experience accumulated after the first three EPC led the housing company to undertake an EPC project for 15 years on behalf of its housing stock in April 2014. The call for tenders was divided into seven 'batches'. This division was based on the heating process of the different estates. A traditional public procurement scheme was chosen and the housing company financed the project by borrowing money. This was made possible because in 2012 the national legal framework changed. Design, build and operate (DBO) became possible with the traditional procurement scheme, while before 2012 PPP was considered as the only option. PPP also integrated finance.

Every consortium (the ESCO) consisted of four companies:

- a general contractor who was the authorised representative of the consortium;
- an architect in charge of the design and the administrative procedures;
- a thermal design office who did the dynamic thermal modelling and simulation;
- a facility manager/heating and cooling operator.

Every general contractor had in-house competencies in facility management. However, their experience with social housing and tenants was limited, and they considered that they would not be able to develop a proposal offering the best value for money. Consequently, they preferred to be associated with large facility managers who had experience with 'chauffage (heating) contracts' (combined operation and maintenance of heating, ventilation and air-conditioning (HVAC) systems) in the social housing sector.

While in the past it was necessary to consult a building surveyor to make precise plans, information concerning buildings' façades was now immediately available thanks to the digitalisation of the housing stock. The avoided cost linked to this task was estimated at EUR300,000. Candidates received detailed information about the energy consumption of the buildings. This information was available because, in 2009, the housing company established incentives contracts with facility managers in charge of the operation and maintenance of HVAC systems of its housing stock. The performance attached to these contracts requires a perfect knowledge of the installations (Duplessis et al., 2012) and a detailed follow-up of the energy consumption. General contractors who led every bidding consortium considered that the information concerning the façades and the energy consumption was a central issue; a majority of energy conservative measures to be installed concerned building insulation. In some cases, it could represent up to 50 per cent of the project's costs (Table 11.2). However, the housing company considered that this information was not required for the contract due to discrepancies between real data and data coming from the digitalisation that can occur on rare occasions. All bidders checked the data on a limited number of buildings and took the risk by considering this information as accurate.

Energy performance contracting 189

Table 11.2 Characteristics of the call for tenders concerning the EPC of 14,000 dwellings

Batches	Heating process	Number of buildings	Number of dwellings	Dwellings with work on the envelop	Building with work on the envelop	Number of bidders	Total value of the market (EUR 1,000)
Batch 1	Heating network	64	2,425	976	20	3	29,761
Batch 2		38	1,069	499	10	3	13,525
Batch 3		18	588	550	6	4	18,022
Batch 4	Individual heating	41	1,230	784	31	Unsuccessful	
Batch 5	All types of heating except individual and network	84	1,441	912	54	2	32,367
Batch 6		107	2,657	812	54	4	43,228
Batch 7		43	1,199	501	21	4	23,362

Source: BOAMP (2016).

In some cases, the transfer of 3D information was problematic and required supplementary drawings because the link between software was inadequate. Similarly, the social housing company mentioned that some interoperability issues between BIM packages are expected once the renovation is finalised. Importing data leads to the overwriting of the initial database.

The evaluation of the proposals was based on cost and pricing (55 per cent of the criteria) and quality of the technical approach; this is the quality of the architecture, percentage of refurbished building envelops and energy savings. During the four rounds of the competitive dialogue, there was no exchange of information between bidders. General contractors and facility managers developed proposals for different 'batches'. Depending on the batch, facility managers frequently cooperated with different general contractors. Nevertheless, even within the same company, information exchange was limited. During the competitive dialogue, the housing company decided to limit the works done inside the dwellings in order to reduce the costs of the renovations. While the first round of competitive dialogue was organised at the beginning of 2015, the EPC was signed in December that year, for 15 years; renovation works are supposed to last 14 months.

Lessons learned

After defining EPC and presenting why information integration in EPC is a central issue, this chapter described two case studies. The two EPC projects followed the same goal, reducing energy consumption of renovated buildings by about 40 per cent, and have the same duration: 15 years. In both cases, works combined building insulation and energy systems refurbishment. The first EPC project concerning the high schools was in operation for five years while the

190 *Bougrain*

second EPC linked to the refurbishment of social dwellings is still in its infancy since renovation is yet to start. Works are still ongoing and it is not possible to know whether the expected energy target is reached. However, some lessons concerning information integration can already be drawn from the two cases.

In the first case, the public body was familiar with the issue of energy performance. However, it had never practised EPC before the call for tenders. Similarly, the laureate, a general contractor, had no experience. Information delivered to the candidates was limited and mainly based on energy audits. Then, once the EPC was in operation, the ESCO implemented a centralised control station in every high school. This equipment reconfigured energy data and explicit knowledge into a new form of explicit knowledge, a process called 'combination' according to Nonaka (1994). It is the backbone of the EPC since it allows the ESCO to monitor energy performance, compile information on energy performance and communicate with the public body who owns the high schools. However, it took time before the data gathered through the centralised control station could be used for improving the energy performance.

The limited experience of the ESCO was a barrier for understanding and interpreting information. Similarly, people working for the public body in the high schools had limited competences and were not always able to effectively interact with the ESCO. However, year by year, the capacity of the centralised control station was improved by expanding its functionalities and monitoring energy consumption on a daily basis. This evolution was partly done through an iterative process of trial and error. According to Nonaka (1994), this process of learning by doing is the source of conversion of explicit knowledge into tacit knowledge. This is also why the ESCO decided to employ two people for every three high schools. Redundancy was needed to share experience and limit the development of individual tacit knowledge. Otherwise, this would have jeopardised the performance of the contract in the long run when key employees would have left the company or changed position. Without this organisational change, energy consumption would have probably been even further from the expected target.

The length of the contract enabled the ESCO to make specific investments. It also provides partners with opportunities to learn from one another. It contributes to the establishment of common codes of information between the public body and the ESCO. The information coming from the centralised control station provides the ESCO and the public body with an identical set of references. It favours communication, contributes to the stability needed for exchange and enhances the efficiency of coordination and information integration.

The second case concerned experienced stakeholders. Information integration was already a very sensitive issue for the social housing company who decided to digitise all of its housing stock to favour information sharing and limit actions based on tacit knowledge. The investment in 3D modelling contributed to the conversion of tacit and hidden knowledge into explicit knowledge.

The housing company spent time and money gathering reliable information to digitise its housing stock. Without having done this, the housing company would have never been able to launch such an ambitious EPC on half of its housing stock. Due to its past experience with EPCs, the housing company knew what to expect from the candidates of the call for tenders. Moreover, it was able to deliver precise and abundant information to the candidates who were able to assimilate and use this information because of their experience with EPC. Members of the winning consortium already worked for the social housing company. This was very helpful during the competitive dialogue since they were able to develop common codes of language. All these elements contributed to a successful call for tenders and bred trust among the future partners.

The main limiting factor mentioned by the actors concerned the software linked to BIM. Apparently, the software suppliers have not been able to develop tools that allow a perfect transfer of information from renovation to operation. As a pioneer with BIM for social housing, the company has been a victim of the underdeveloped nature of the market for this sector.[5] Similarly, bidders had to check some of the data and took risks by considering the information coming from the digitalisation as accurate.

Conclusions

The two case studies have implications for future research and decision-makers.

At the research level, it appears that the framework proposed by Nonaka (1994) was quite apt to interpret how information was integrated and converted. Different modes of knowledge conversion were involved. In the case of the high schools, the centralised control station contributed to the development of new information and reconfiguring of explicit knowledge to explicit knowledge aimed at improving energy efficiency. In the case of the housing company, BIM was seen as a way to transfer tacit knowledge into explicit knowledge and to promote information sharing. Indeed, before digitalisation, most of the information concerning the housing estates was in the head of the estate managers. There was no real information sharing. When key employees retired, information and knowledge were lost.

The impacts of technical tools on knowledge conversion and on information integration changed according to the context. The research is limited to two cases, and further work should be carried out to understand the process of knowledge creation, transmission and coordination when tools such as BIM or centralised control stations are involved. The use of these tools will likely increase with the need to reduce energy consumption and to improve the productivity of the built environment industry over the life-cycle of a building.

Both cases also show to practitioners that turning energy data into actionable information is not a straightforward process and requires contextual awareness. Tools such as BIM and centralised control stations have strong organisational impact; this is also partly due to the conversion of knowledge that takes place. They contribute to providing actors involved in a refurbishment project leading to EPC with an identical set of references. They enhance the efficiency of coordination tasks and information integration. However, there are both

192   *Bougrain*

technical and human barriers to this process. Information tools may not be apt to deal with the complexity of renovation projects leading to EPC. In the case of the social housing company, the transfer of 3D information was problematic because existing software suffers from a lack of interoperability between renovation and operational stages. Conversely, tools can be well developed but human resources or management may not follow. The energy service company in charge of reducing energy consumption in the high schools did not fully realise at the beginning of the contract that internal turnover led to a loss of knowledge and expertise. To counterbalance this situation, it modified its organisation and created redundancy by having several employees in charge of operating high schools. Thus, actors have to be aware that information integration cannot only be based on technology. There is a need to integrate and balance technical tools and human resources.

Finally, the impact of tools such as BIM and centralised control stations on information integration has to be evaluated over the long term since it takes time to appropriate them. During a process of trial and error, tools will be developed until their efficiency is proved. Moreover, time contributes to the establishment of common codes of information between actors, which brings trust. For example, the information coming from the centralised control station did not favour communication between the ESCO and the public body during the first years of the contract. But there was a learning curve and after a certain point the tool genuinely enhanced the efficiency of coordination tasks and information integration.

Acknowledgements

This chapter acknowledges and expresses thanks for the support provided by ADEME, the funding agency that sponsored part of this research; ADEME is a French national agency active in the implementation of public policy in the areas of the environment, energy and sustainable development. The deepest gratitude is also extended to Yann Baduel, Fabrice Laversanne, Julien Martin, Philippe Souchal and Henri Turbelin who provided valuable information concerning both cases presented in this chapter.

Notes

1 EUR1 = USD1.11 (average exchange rate for 2015).
2 The market potential has been estimated at around EUR250-500 million.
3 Within the framework of the agency theory, clients are commonly referred as to 'principal' and ESCO as to 'agent'.
4 This EPC benefited from a three-year follow-up by the CSTB, Centre Scientifique et Technique du Bâtiment.
5 Software issues are frequently considered as a major negative effect for BIM projects. Software unable to handle large amounts of data, inability of packages to exchange data and a lack of knowledge and experience of software programming were among the main limits identified in a survey covering 35 case studies found in academic journals (Bryde et al., 2013).

References

Aubert, B. A., Rivard, S. and Patry, M., 1996. A transaction cost approach to outsourcing behaviour: Some empirical evidence, *Information and Management*, 30 (2), 51–64.

Bellinger, G., Castro, D. and Mills, A., 2004. *Data, Information, Knowledge, and Wisdom*, The Way of Systems. Available at: www.systems-thinking.org/dikw/dikw.htm [Accessed 10 August 2016].

Bertoldi, P., Boza-Kiss, B., Panev, S. and Labanca, N., 2014. *ESCO Market Report 2013*, Joint Research Centre, European Union, Report EUR 26691. Available at: http://iet.jrc.ec.europa.eu/energyefficiency/system/tdf/jrc_89550_the_european_esco_market_report_2013_online.pdf?file=1&type=node&id=8869 [Accessed 10 August 2016].

BOAMP, 2016. *Conception-réalisation exploitation maintenance pour l'amélioration de la performance energétique de 14 000 logements chauffés collectivement sur le département de la Seine-Maritime à Rouen* (Design, build and operate to improve the energy performance of 14,000 collectively heated dwellings in Rouen and its surroundings in the French department of Seine-Maritime). Available at: www.boamp.fr/avis/detail/16-13399/officiel [Accessed 10 August 2016].

Bryde, D., Broquetas, M. and Volm, J. M., 2013. The project benefits of Building Information Modeling (BIM), *International Journal of Project Management*, 31 (7), 971–980.

De Wilde, P., 2014. The gap between predicted and measured energy performance of buildings: A framework for investigation. *Automation in Construction*, 41, 40–49.

Duplessis, B., Adnot, J., Dupont, M. and Racapé, F., 2012. An empirical typology of energy services based on a well-developed market: France. *Energy Policy*, 45, 268–276.

Eisenhardt, K. M., 1989, Building theories from case study research. *Academy of Management Review*, 14(4), 532–550.

European Parliament, 2012. Directive 2012/27/EC of the European Parliament and of the Council of 25 October 2012 on Energy Efficiency, Amending Directives 2009/125/EC and 2010/30/EU and Repealing Directives 2004/8/EC and 2006/32/EC. *Official Journal of the European Union*, 14 November 2012. Available at: http://eur-lex.europa.eu/legal-content/EN/TXT/PDF/?uri=CELEX:32012L0027&from=en [Accessed 10 August 2016].

European Parliament, 2006. Directive 2006/32/EC of the European Parliament and of the Council of 5 April 2006 on Energy End-Use Efficiency and Energy Services and Repealing Council Directive 93/76/EEC. *Official Journal of the European Union*, 27 April 2006. Available at: http://eur-lex.europa.eu/legal-content/EN/TXT/PDF/?uri=CELEX:32006L0032&from=FR [Accessed, 10 August 2016].

Hoezen, M., Van Rutten, J., Voordijk, H. and Dewulf, G., 2010. Towards better customized service-led contracts through the competitive dialogue procedure, *Construction Management and Economics*, 28(11), 1177–1186.

Lee, P., Lam, P. T. I. and Lee, W. L., 2015. Risks in energy performance contracting (EPC) projects. *Energy and Buildings*, 92, 116–127.

Moore, B., 2014. *Optimizing Buildings Using Analytics and Engineering Expertise*. Schneider Electric White Paper.

Nonaka, I., 1994. A dynamic theory of organizational knowledge creation, *Organization Science*, 5(1), 14–37.

Rowley, J., 2007. The wisdom hierarchy: Representations of the DIKW hierarchy, *Journal of Information Science*, 33(2), 163–180.

Sorrell, S., 2007, The economics of energy service contracts, *Energy Policy*, 35(1), 507–521.

Tellis, W. M. (1997). Application of a case study methodology. *The Qualitative Report*, 3(3), 1–19.

12 Stakeholder perspectives and information exchange in AEC projects

Torill Meistad, Marit Støre-Valen, Vegard Knotten, Ali Hosseini, Ole Jonny Klakegg, Øystein Mejlænder-Larsen, Eilif Hjelseth, Fredrik Svalestuen, Ola Lædre, Geir K. Hansen and Jardar Lohne

Introduction

Management of construction projects is mainly about managing people, materials and information. Architecture, engineering and construction (AEC) projects bring together a large number of people with various professions and organisational affiliations as well as numerous building components and considerations to attend to. Together, this creates the complexity typically characteristic of AEC projects. Adding to this complexity, the multiple phases of the construction project life-cycle introduce further challenges due to the number of stakeholders involved at each stage that have significantly different individual interests and needs.

The AEC industry is under constant pressure to improve performance and increase productivity. This requires handling and analysing large amounts of information to enable the best possible decisions to be made at all stages during the life-cycle of the building. Handling information, most notably by integrating various information sources and information carriers, is a decisive factor to meet the need for improvements.

Numerous stakeholders are involved in AEC projects. This includes the owners, the executive parties and the users. In addition, there are external stakeholders including regulating authorities and market actors. A stakeholder perspective on project (and asset) management requires an understanding of the objectives of the various stakeholders. The traditional task-based approach of AEC project management is insufficient within this context of high degrees of external and/or internal complexity (Aarseth, 2012). Therefore, relationship management has developed as a new branch of project management. This relates to increased awareness of the value of relational competence in project-based industries, especially in innovating projects. Partnering is emerging as a new model that relies heavily on relational competencies (Lampel, 2001). Early involvement of end-users is gaining interest to improve value and usability of buildings in use (Baharuddin et al., 2013) through, for example, modern, energy-efficient office buildings (Meistad et al., 2013). This is also the experience of the respondents to the survey conducted by the OSCAR project[1] which reported increased value for users and owners when involved in the early planning phase (Støre-Valen et al., 2016; Spiten et al., 2016).

Project execution models are seen as a way to systematically deal with the various stakeholders and their perspectives on a project. Execution models can also be considered as a platform to integrate information and engage people throughout all phases of a construction project.

The challenges regarding information exchange increase with the number of stakeholders involved and with demands for early involvement of end-users. However, tools for digital information handling provide opportunities for better integration of various stakeholder objectives and for improving value for owners, operators and end-users. Existing execution models are being challenged by digital developments and the AEC industry has yet to fully explore their potential. A stakeholder perspective can be a guide for further exploring digital information handling within these models for the benefit of public and private assets.

This chapter will first explore the concepts of information integration, execution models and stakeholder perspectives. Four case projects are then used to showcase attempts to increase the level of information integration and strengthen project execution in construction projects. The following sections then explore opportunities and challenges of systematically managing construction projects. These sections have a special focus on how the various stakeholder perspectives can be integrated and how digital information carriers can contribute to decision points (stage gates) throughout the project phases (life-cycle). Finally, the chapter concludes by providing suggestions for future research and investment priority areas.

Information integration and execution models

The word 'integration' can have different meanings when related to building and construction industries (Gielingh and Tolman, 1991):

- integration of building and construction processes;
- integration of construction technologies;
- integration of information technology components; and
- integration of data or information.

Information integration across project phases and across organisational and professional borders is one of the core issues of project management, possibly the most important one. Within this context, execution models are systems that meet the first meaning of 'integration' on Gielingh and Tolman's list, as they are used to coordinate information and to ensure control over the product quality and the schedule. Integration of data or information, however, forms the basis for all the other types of integration, including integration of building and construction processes (Gielingh and Tolman, 1991), and is one of the building blocks of execution models.

Integrated design and delivery solutions (IDDS) have been used in manufacturing and service industries to improve the quality of production and to deliver complex new products and services (Owen et al., 2009; 2013). The construction industry has developed a variety of such approaches; namely the so-called 'integrated design process' (IDP). This process was introduced mainly to meet the challenges of designing sustainable buildings. IDP is described as a collaborative process that focuses on the design, construction, operation and occupation of a building over its complete life-cycle (Larsson, 2002). The process includes the client and other stakeholders, and allows the development and realisation of functional, environmental and economic goals and objectives. IDDS and IDP cover all the meanings of integration suggested by Gielingh and Tolman.

Integrated energy design (IED) is a type of IDP that focuses on environmental sustainability and energy efficiency. The methodology emphasises the importance of integrating information and engaging all relevant stakeholders during the concept and design phases. This focus on the early phases of a project is based on the understanding that the costs associated with changes increase as the project progresses, while the ability to affect the outcomes decreases in later life-cycle phases (Figure 12.1).

The methodology focuses just as much on the social process for integrating knowledge and expectations among the parties as on optimising energy efficiency, architecture and cost efficiency of the building concept (Hestnes et al., 2009). In these early planning phases, the focus is on what the clients want. Dialogue

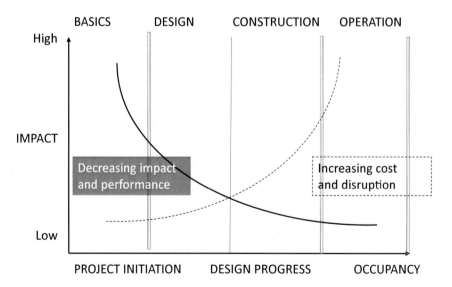

Figure 12.1 Integration in early phases offers opportunities for large impact on performance (based on work by Andresen et al., 2005).

Stakeholder perspectives 197

between design teams and the owners, operators and users is therefore essential to the process. Andresen et al. (2005) summarise the process in the following nine steps:

1 Select a multidisciplinary design team from day one, which are skilled in energy/environmental issues and are motivated for close cooperation and openness.
2 Analyse the boundary conditions of the project and the client's needs and demands and formulate a set of specific goals for the project.
3 Develop a Quality Assurance Programme and a Quality Control Plan that is implemented throughout the project.
4 Arrange a kick-off workshop to make sure that all team members have a common understanding of the design task.
5 Facilitate close cooperation between the architect, engineers and relevant experts through co-location or through a series of workshops during concept design phase.
6 Update the Quality Control Plan and document the energy performance at critical points (milestones) during the design.
7 Develop and implement contracts that encourage integrated design and construction.
8 Motivate and educate construction workers and apply appropriate quality tests.
9 Make a user manual for operation and maintenance of the building.

(Andresen et al., 2005)

By contrast, Samset (2010) suggests monitoring projects at two levels: tactical and strategic. The tactical level deals with cost, time and quality as success indicators for a project. The strategic level looks at effect, relevance and sustainability indicators when considering project success. Different stakeholders can be interested in the strategic or tactical level, depending on their role in the project. The owner is typically most interested in the strategic performance of the project, while the executing parties tend to limit their interest to the tactical performance.

Execution models are tools used in project management to coordinate information and to ensure control over the product quality and the schedule. Traditionally execution models are understood as consisting of 1) enterprise form, 2) contract form, and 3) procurement form (Lædre et al., 2006; Løkkeberg, 2015). In simpler words, project execution models can be defined as 'the way we work and deliver projects'. These models normally present projects as a series of phases, each of which has a clear purpose and defined roles for the stakeholders involved. The handover from one phase to the next is critical to the flow of information in the project. The work of each phase leads to a point of decision whether to proceed or not, sometimes referred to as a stage gate (Knotten et al., 2016b). Each shift of phase is also a handover of information to the parties/ enterprises responsible for the next phase. Ensuring the relevant information is

198 *Meistad et al.*

handed over at the appropriate time is therefore a key issue for project management professionals. The constant drive to improve information flows, quality and productivity leads to project execution models being continuously developed to adapt to changes in regulations and market expectations.

Traditionally, due to competition and procurement rules, there have been adversarial interactions between execution parties and projects have been negotiated as a zero-sum game. Information integration is an alternative approach for project management that highlights the overall purpose of the project. By letting professionals get together and analyse various information sources, the process allows them to develop innovative solutions that would not have emerged from an adversarial process (Lampel, 2001).

Stakeholders as information integrators and their perspective

There are two basic approaches to exploring the issues related to integrating information during project execution, depending on the type of information carrier that is of interest: *the people* handling the information or the *physical carrier* of the data. The interest around information integration has been driven by technical developments, especially the ongoing revolution of digital information handling. IDP and IDDS register and exploit a large amount of data in the search for optimal logistics and project solutions. Less explored, however, is the guidance that IDP and IDDS provide for facilitating team-building and high-performance attitudes.

People are information integrators, independently of the physical information carriers. This includes the client, other project parties and the users, all of whom are often labelled 'stakeholders'. This chapter will explore the role of the various stakeholders during project execution and the challenges associated with integrating information across the various perspectives of the many stakeholders involved in the different life-cycle phases of a built asset.

Three major groups of stakeholders are involved during the various phases of a construction project, each bringing their own point of view. Samset (2010) refers to these points of view as perspectives and lists them as owner perspective, user perspective and executing perspective. The owner is the initiating and financing party; owners normally have a long-term interest in the investment that the project represents. The user is the party who is going to utilise the end result (e.g. the building) to operate their business. The executing party, or parties, is formed by the architects, engineers and contractors who will be executing the project on behalf of the owner.

Four research projects: learning across trades, parallelism, scan-to-BIM and Next Step

Project management requires the above-mentioned perspectives to be identified and managed. Knotten et al. (2016a) suggest identifying them at an early stage to understand the various focal points and to coordinate or possibly change the

Stakeholder perspectives 199

attitudes regarding purpose and success of the project. Mejlænder-Larsen (2016) further suggests identifying drivers to secure alignment to common goals in the project team. The following case studies will show different approaches to dealing with the challenges brought by these differing perspectives.

The first case study compares the AEC industry with the shipbuilding and offshore construction industries. In this case, varying perspectives are handled by implementing a higher degree of standardised designs, predefined interconnections between parts of systems and use of in-house design teams. The second case study explores how general contractors are able to handle parallelism between these two phases and thereby shorten the timeframe of construction projects. The third case study, based on Hjelseth et al. (2016), focuses on the facility management perspective and explores the use of Building Information Modelling (BIM) for multiple purposes by different stakeholders in the in-use phase of buildings. This case discusses topics such as information integration for existing buildings, decision processes and usefulness of scan-to-BIM for the purpose of asset management. The final case study provides a new systematic approach to plan and execute AEC projects, clarifying phases and roles throughout the life-cycle of a building construction project (Knotten et al., 2016a). All four cases are based on research carried out in Norway and represent front-end research on current challenges for the Norwegian construction industry.

Improving the design phase of AEC industry projects: learning across trades

The design phase is crucial for value creation in a project. A major question for the AEC industry today is whether it can improve this phase by using insights and practices from other industries. This question was explored by Knotten et al. (2016a) in a comparative study of design management in shipbuilding, offshore construction and the AEC industry. The study explored the characteristics of some of the key processes in these three industries.

Shipbuilding (SB), offshore construction (OC) and AEC are project-based industries. Unique products are designed and manufactured for different customers, and there is a high level of complexity. These similarities make the comparison of these three industries possible and useful in understanding the lessons that can be learned across sectors. The comparison revealed that design processes vary between the industries, especially regarding reciprocal and sequential processes. Figure 12.2 shows a comparison of the design process in the three industries, highlighting individual characteristics that lead to differences in the design process and management.

The design process is more standardised in SB than in AEC. This allows SB designers to often use previous designs as a starting point, adjusting it to the client's requirements. The engineering process is often parallel between design and production, narrowing the options of change as the parts are finished. The engineering team consists of in-house personnel, though they can be located in multiple offices. The planning of engineering is based on delivering drawings to the production office. This is monitored by a computerised planning system

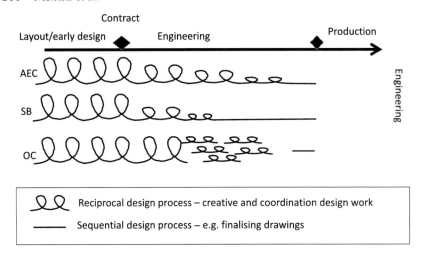

Figure 12.2 Design process in different trades (Knotten et al., 2016a).

linking working hours to drawings. However, this process does not monitor the value-creation processes.

OC companies deliver parts of larger production systems. Therefore, there are a lot of predefined interfaces in space, weight and technical requirements. OC companies have their own design teams and their members remain the same throughout the whole design process. This ensures that the knowledge gathered in the early design phase is carried through the whole design process. Moreover, to ensure that the knowledge from construction is brought into design, key members of the construction team are engaged during the early design phase. The in-house team also shares the same organisational culture. This ensures that their work is aligned with a single set of organisational and project goals and is based on pre-existing trust between the team members.

In OC, the design phase follows a stage-gate model with clear deliverables at each stage. Key members of the design team agree on the maturity level needed to proceed to the next stage. The designers use BIM as a main tool for design. When the correct maturity level is reached in an area of the model, this area is 'frozen' and no further interdisciplinary changes are possible. The finished model has to be approved by the client before detailed drawings are produced in 2D.

The three industries are similar in that the design and engineering services are required to transform the needs of clients and users into a finished product. However, the complexity of the products, processes and context varies between them. While the contractors in the AEC industry to a large extent do not use their own design team, this is more common in OC and SB. The AEC industry also does not often operate from a common production site and prefabrication is limited. OC and SB, on the other hand, use a common production site and prefabrication is the norm.

While all three industries have reciprocal design processes after contract, the OC industry process is divided into smaller concrete tasks and finalising the drawings is a sequential process. The AEC industry can learn from this by implementing planning and execution methods used by OC. Of special interest is the new way of planning and executing engineering that the OC industry has implemented, thus exploiting more of the benefits of BIM. 'By producing production drawings at the last responsible moment, they let the coordination process last longer, leaving time for the design to evolve and mature' (Knotten et al., 2016a).

Parallelism between phases

Parallelism, or concurrent engineering and construction, is gaining popularity due to the increased demand for shorter project timeframes. However, it presents greater challenges for contractors to control the work and the extent of the challenge depends on the client's requirements. The client sets the scene in terms of how complex the process becomes, in part by setting the timeframe from the signing of the contract to the delivery date. While longer timeframes allow for more predictability between phases, shorter timeframes allow a higher degree of parallelism between the phases. The more parallelism there is in a project, the greater the demands put on the participants.

In OC, engineering, procurement and construction (EPC) contracts are common. In EPC contracts, construction is often pushed in parallel with engineering services. This parallelism is a challenge for information integration due to the need to ensure that the drawings and materials are available when they are needed.

Mejlænder-Larsen (2016) presents a case study about a Norwegian EPC contractor. This company developed a building sequence where the engineering phase is divided into stages with corresponding milestones (see M2A-C in Figure 12.3). At the third milestone, the engineering sub-contractor reaches a defined quality level to start issuing drawings from BIM so that construction can start. Engineering influences all project phases and is developed to a quality level where the design and all interfaces between disciplines are frozen. 'Engineering for procurement' is then developed during the procurement phase and 'engineering for fabrication' is developed in the construction phase (Kvaerner, 2013; 2014).

The ambition of the EPC contractor is to get the work 'right the first time'. For this purpose, a project execution model (PEM) was developed, where progress and quality requirements were aligned at the relevant milestones. The first requirement for being able to govern construction projects according to this model, is that the PEM uses a standard methodology which is well known to the team. Secondly, common incentives and drivers are required, including the possibility to use a joint venture between the engineering sub-contractor and the EPC contractor. Thirdly, it requires the utilisation of technology, including a 3D design environment such as BIM, to support a desired build sequence for the EPC contractor. Accurate lead times for equipment and timely availability of

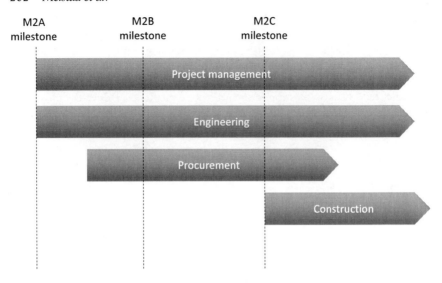

Figure 12.3 Parallelism between engineering, procurement and construction (as published by Mejlænder-Larsen, 2016).

correction vendor information are critical to efficient fabrication assemblies. The EPC contractor's building sequence is an example of how defined milestones can be used to better integrate information between engineering and construction teams. There is, however, a need to better understand this process, the people involved and the technology in use before this execution model can be leveraged to its full potential by the built environment industry.

Scan-to-BIM for the in-use phase

This section presents a case study based on a scan-to-BIM test project. The study carried out by Hjelseth et al. (2016) explored alternative technologies for 3D scanning and modelling, and collaboration between project stakeholders to establish a BIM model for the purpose of facility management of an old apartment building. The execution model suggested for management of the scan-to-BIM process is based on a step-by-step process framework for ordering scan-to-BIM services. The framework combines relevant technology, processes and human resources.

Hjelseth et al. (2016) present an ordering guide as a starting point for buyers ordering a scan-to-BIM service and seeking to receive the best cost–benefit ratio possible. This includes three steps:

1 Establishing a development plan that focuses on collaboration. This plan outlines those stakeholders who have vested interests in the project and those who should have access to the resulting model.

2   Developing an overview of the different challenges faced by the collaboration team and how BIM can be used as a tool to solve these issues. One of the questions this addresses is what level of accuracy is required for the data capture.
3   Integrating the measurements into a BIM model. This can be a simple volume model, a volume model with standard objects added, or added attributes and relations to get a full BIM model. The full BIM model version is more useful but also more expensive.

Further, Hjelseth et al. (2016) focus on the process from the decision to capture geometrical data about the building to establishing a BIM model for one or more purposes. This process includes three main steps, namely scanning, BM-ing and BIM-ing, as illustrated in Figure 12.4.

There are various instruments for capturing geometric data (scanning in Figure 12.4). Technology not only includes the physical data-collection devices but also the software required for processing and enriching of scanned data (this is referred to as BIM-ing in Figure 12.4). Most software packages are plug-ins into typical architectural design authoring software.

The result from the test project was good. The board of directors of the managing organisation wanted to use BIM as a tool to link the maintenance history, day-to-day status and scheduled maintenance directly to building objects in the model. The existing drawings were old and inaccurate. The BIM model can be used for area calculation and was shown to be a good starting point for establishing a maintenance plan (Direktoratet for byggkvalitet, 2015).

This test project reveals that there are a number of technologies available, both hardware and software, and that digital information integration provides better decision support for existing buildings. The 3D presentation is good for communicating information to the various stakeholders and is supportive for shared decision-making on operation and maintenance. The study reveals the challenge of selecting the most suitable level of data accuracy and provides a guide to find the level of details (and costs) matching the purpose of the scan-to-BIM.

'Next Step': a new systematic execution model

How can one ensure that decisions are made at the right time and at the right organisational level? This is a basic challenge for project management. Norwegian

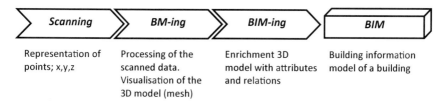

Figure 12.4 Main stages in the scan-to-BIM process (Hjelseth et al., 2016).

204   Meistad et al.

construction industry and researchers have been working together over the last couple of years to develop a systematic approach to overall project management. The new framework, called Next Step, identifies the key steps and tasks in a project life-cycle from the definition to the termination of the building. The framework focuses on project execution, critical decisions at a corporate level, involvement of the proper stakeholders' perspectives and sustainable development of the AEC industry.

The eight steps of the project are indicated at the top of Table 12.1. Each step has a clear purpose and together they cover all the different phases of a project. Termination can refer to the termination of ownership, where the owner sells the property, or the demolition of the building in order to utilise the site in a different way.

Next Step is inspired by the RIBA Plan of Work (Royal Institute of British Architects, 2013). The framework is based on a systems-thinking approach that includes input, process and output logic as well as creating decision gates after each step. An output can become the input of the next step or lead to the termination of the project. The process includes the actual tasks that need to be completed in order to advance the project (Klakegg et al., 2010).

Next Step is also based on the principles of 'project governance' (Klakegg et al., 2009; Müller, 2009) and advocates a structure of clearly defined phases in construction project processes. It has a special focus on decision gates at the end of each phase before handing over information and tasks to the team(s) in the next phase(s). The gateway is a key element of an adequate implementation strategy. The purpose of a decision gate, as seen from a project owner's perspective, is to make sure the formal decision-making supports the success of the organisation, business corporation or public entity. Ensuring the right decisions are made requires decision-makers to be as informed as possible. However, the right information is a question of what is available or known at the time, versus the cost of obtaining more/better information and the risk associated with making

Table 12.1 Outline of the Next Step framework

Step	1 Strategic definition	2 Brief development	3 Concept development	4 Detailed designing	5 Production	6 Handover	7 In use	8 Termination
Core process		**Owner** perspective						
		User perspective						
		Supplier perspective						
		Public perspective						
Management process		Planning						
		Procurement						
		Communication						
		Sustainability - economics						
		Sustainability - environment						
		Sustainability - social						

Source: Knotten et al. (2016a).

the decision based on less-than-perfect data. Decision gates are often characterised by having defined procedures for assessments, control and decision-making; defined roles and responsibilities; criteria for acceptance and a gatekeeper (owner of the gateway process) who decides whether the project is allowed to (enter) pass the gateway and continue to the next step or not.

The framework divides the processes into two major categories: core and management processes. The core processes deal with the perspective of four stakeholders: the owner, the user, the executor and the public. All these perspectives have to be dealt with to ensure that the different mindsets are integrated in both the input (requirements) and output (deliverables) of the project. The management processes include planning, procurement, communication, and the economic, environmental and social aspects of sustainability.

The purpose of the model is to facilitate stronger governance of construction projects, and its focus is on clarifying the phases and the stakeholders related to each phase. The decision gates are a key concept in the model. Control of documents and assumptions must be made before making a decision to accept a project or to close one phase and enter the next.

Each decision gate is seen from different perspectives by the various stakeholders. For owners, a decision point is a time for them to look forward and focus on how the project can support the success of the business corporation, public entity or user organisation. From the construction team's perspective, predicting or proving whether the project is a financial success may be a milestone.

The intention of Next Step is to give the industry a common language and a collective reference for AEC projects. Next Step can help achieve success for owners and users by defining the necessary steps for going from a problem to a solution. The framework forces the parties to consider the long-term issues and to holistically assess sustainability of alternative concepts. The right choices are expected to become the natural outcome from such a process. The new framework/general standard is expected to improve planning and control of project executions by providing a step-by-step system that eliminates non-conformance and miscommunication compared to a model where each company uses their own execution model. However, it may take time for the many actors involved in the AEC industry to approve the model and change their practices.

Concluding remarks

Although expectations are high, the process of developing and deploying systems for information integration within the construction industry suffers from a series of problems or challenges, including the following (Shen et al., 2010):

- it is difficult to access accurate data, information and knowledge in a timely manner in every phase of the construction project life-cycle;
- conventional programme plans and designs are optimised for a limited set of parameters in a limited domain; the capability to support 'total best value' decisions does not exist;

206 *Meistad et al.*

- life-cycle issues are not well understood and therefore modelling and planning do not effectively take into account all life-cycle aspects such as operations, maintenance and environmental impact.

The first three case studies presented in this chapter highlighted two dominating approaches to information integration:

1 Refine and improve existing project management systems using information technology as a tool to support the overall process.
2 Explore the potential gains from using information technology for various purposes in the construction industry in order to improve performance and quality.

In both approaches, the use of technology for information integration is a helpful tool for decision-making. Hjelseth et al. (2016) expect BIM for existing buildings to become more relevant with the increasing number of solutions available to combine technology, processes and human resources. They suggest breaking down the overall process into small work packages or services that can act as options for further processing and enabling of new purposes. An outcome of this approach can be add-on services that enable the reuse of previous work. This can result in extended use of BIM for multiple purposes in asset management.

Followed by facility managers, clients have the most to gain financially from implementing BIM during the operation phase. A study of BIM users published in 2013 demonstrated that collaboration aspects have the highest positive impact on the success of the implementation effort, followed by process aspects and, finally, software aspects (Eadie et al., 2013). In 2004, a US study concluded that inadequate interoperability across all life phases of facilities resulted in efficiency losses for the whole industry and especially for owners and operators (Gallaher et al., 2004). The ongoing digital transformation provides tools to manage and communicate product and project data between stakeholders and throughout life-cycle phases, and thereby to improve interoperability. 3D design and just-in-time management tools provide benefits not only for cost, time efficiency and quality (the tactical level), but also for responsiveness, further investments and other strategic issues (Sanchez et al., 2016).

The design phase is crucial for value creation in the project. Typically, the AEC industry is characterised by a strong sequential mindset which influences design management (Knotten et al., 2014). Here, it has been suggested that using reciprocal design processes will help to overcome the barriers between phases and professionals, and to improve the potential value of the project. Such reciprocal processes are well recognised in theory but only implemented in the AEC industry to a limited degree. This relates to risks and challenges which are difficult to plan and manage (Hansen and Olsson, 2011).

Further research is needed to understand how information integration can be used as a tool for improving the overall value of the project, especially the strategic level, and dealing with effect, relevance and sustainability. The social aspects of information integration also need to be investigated; in particular, the dialogue between stakeholders, exchange of perspectives and expectations, and the potential to explore options that maximise performance and reduce costs simultaneously. Finally, research is also needed about how information integration technologies can be used as tools for innovation and whether standard information systems support or hamper creativity and learning for industrial development.

Note

1 OSCAR is a research and development project funded by the Norwegian Research Council to develop knowledge about and methodology for increasing the value for the owners and users of buildings. Multiconsult is leading the project with the involvement of a broad range of industry partners, public owners and managers, and research and education institutions: www.oscarvalue.no.

References

Aarseth, W., 2012. An empirical study of organizational cooperation in large traditional and global project execution. PhD thesis. Trondheim: Norwegian University of Science and Technology.

Andresen, I., Hestnes, A. G., Kamper, S., Jørgensen, P. F., Bramslev, K., Hammer, E., Førland-Larsen, A., Lehrskov, A., Rynska, D., Holanek, N., Synnefa, A., Wilson, M., Solomon, J. and Sander, K., 2005. *Integrated Energy Design IED. A Guide to Integrated Energy Design*. Available at: www.integrateddesign.eu/downloads/Guide-1-rev11.pdf [Accessed 8 March 2017].

Baharuddin, H. E. A., Wilkinson, S. and Costello, S. B., 2013. *Evaluating Early Stakeholder Engagement (ESE) as a Process for Innovation*, paper presented at CIB World Building Congress, Brisbane, Australia, 5–9 May.

Direktoratet for byggkvalitet, 2015. *Bruk av BIM i Borettslag og Sameier* (Use of BIM in cooperatives and condominiums), Oslo: Direktoratet for byggkvalitet (DIBK, The Norwegian Building Authorities).

Eadie, R., Browne, M., Odeyinka, H., McKeown, C. and McNiff, S., 2013. BIM implementation throughout the UK construction project lifecycle: An analysis. *Automation in Construction*, 36, 145–151.

Gallaher, M. P., O'Connor, A. C. and Dettbarn, J. L., 2004. *Cost Analysis of Inadequate Interoperability in the US Capital Facilities Industry*. Gaithersburg, MD: National Institute of Standards and Technology (NIST).

Gielingh, W. F. and Tolman, F. P., 1991. Information integration in the building and construction industries. *Computer-Aided Civil and Infrastructure Engineering*, 6(4), 329–334.

Hansen, G. K. and Olsson, N. O. E., 2011. Layered project – Layered process: Lean thinking and flexible solutions. *Architectural Engineering and Design Management*, 7(2), 70–84.

Hestnes, A. G., Andresen, I., Kamper, S., Jørgensen, P. F., Bramslev, K., Hammer, E., Førland-Larsen, A., Lehrskov, A., Rynska, D., Holanek, N., Synnefa, A., Santamouris, M., Wilson, M., Solomon, J. and Sander, K., 2009. *Integrated Energy Design IED. Some Principles of Low Energy Building Design*, Indesign. Available at: www.integrateddesign.

208   *Meistad et al.*

eu/downloads/Some_principles_revised_NormalQuality.pdf [Accessed 01 November 2016].

Hjelseth, E., Oveland, I. and Maalen-Johansen, J., 2016. *Framework for Enabling Scan to BIM Services for Multiple Purposes – Purpose BIM*, paper presented at CIB World Building Congress, Tampere, Finland, 30 May–3 June.

Klakegg, O. J., Williams, T. and Magnussen, O., 2009. *Governance Framework for Public Project Development and Estimation*. Philadelphia: Project Management Institute.

Klakegg, O. J., Williams, T., Walker, D., Andersen, B. and Magnussen, O. M., 2010. *Early Warning Signs in Complex Projects*. Philadelphia: Project Management Institute.

Knotten, V., Svalestuen, F., Aslesen, S. and Dammerud, H., 2014. *Integrated Methodology for Design Mangement – A Research Project to Improve Design Management for the AEC Industry in Norway*. Trondheim: Akademia forlag.

Knotten, V., Hosseini, A. and Klakegg, O. J., 2016a. *'Next Step' – A New Systematic Approach to Plan and Execute AEC Projects*, paper presented at 20th CIB World Building Congress, Tampere, Finland, 30 May–3 June.

Knotten, V., Svalestuen, F., Lohne, J., Lædre, O. and Hansen, G., 2016b. *Design Management – Learning across Trades*, paper presented at 20th CIB World Building Congress, Tampere, Finland, 30 May–3 June.

Kvaerner, 2013. *Kvaerner Jackets*. Available at: www.slideshare.net/Kvaerner-slides/kvaerner-jackets-brochure [Accessed 1 November 2016].

Kvaerner, 2014. *Project Execution Model*. Available at: www.kvaerner.com/PageFiles/103/Konsernbrosjyre_070814.pdf [Accessed 1 November 2016.

Lædre, O., Austeng, K., Haugen, T. and Klakegg, O. J., 2006. Procurement routes in public building and construction projects. *Journal of Construction Engineering and Management*, 132(7), 689–696.

Lampel, J., 2001. The core competencies of effective project execution: The challenge of diversity. *International Journal of Project Management*, 19(8), 471–483.

Larsson, N., 2002. *The Integrated Design Process*. Report on a National Workshop held in Toronto in October 2001, Toronto: Buildings Group, CETC, Natural Resources Canada, Canada Mortgage and Housing Corporation, Enbridge Consumers Gas.

Løkkeberg, T., 2015. *Gjennomføringsmodell. Betyr disse Noe for Verdi for Byggeier?* (Implementation model. How does it affect value for building owners?), Oslo: OSCAR value.

Meistad, T., Støre-Valen, M. and Lohne, J., 2013. *Use of Collaborative Working in Projects with High Energy Ambitions*, paper presented at the 7th Nordic Conference on Construction Economics and Organization, Tampere, Finland, 12–14 June.

Mejlænder-Larsen, Ø., 2016. *Improving Transition from Engineering to Construction Using a Project Execution Model and Building Information Model*, paper presented at 20th CIB World Building Congress, Tampere, Finland, 30 May–3 June.

Müller, R., 2009. *Project Governance*. Aldershot, UK: Gower Publishing.

Owen, R. L., Palmer, M. E. and Dickinson, J., 2009. *CIB White Paper on IDDS Integrated Design & Delivery Solutions*, CIB Publication 328. Rotterdam: International Council for Research and Innovation Building and Construction.

Owen, R., Amor, R., Dickinson, J., Prins, M. and Kiviniemi, A., 2013. *CIB Research Roadmap Report Integrated Design & Delivery Solutions (IDDS)*, CIB Publication 370. Rotterdam: International Council for Research and Innovation Building and Construction.

Royal Institute of British Architects, 2013. *Plan of Work*. Newcastle, UK: RIBA.

Samset, K., 2010. *Early Project Appraisal: Making the Initial Choices*. New York: Palgrave Macmillan.

Sanchez, A. X., Hampson, K. D. and Vaux, S., 2016. *Delivering Value with BIM: A Whole-of-life Approach*. London: Routledge.

Shen, W., Hao, Q., Mak, H., Neclamkavil, J., Xie, H., Dickinson, J., Thomas, R., Pardasani, A. and Xue, H., 2010. Systems integration and collaboration in architecture, engineering, construction, and facilities management: A review. *Advanced Engineering Informatics*, 24(2), 196–207.

Spiten, T. K., Haddadi, A., Støre-Valen, M. and Lohne, J., 2016. *Enhancing Value for End Users: A Case Study of End-User Involvement*, paper presented at 24th Annual Conference of the International Group for Lean Construction, Boston, USA, 20–22 July.

Støre-Valen, M., Boge, K. and Foss, M., 2016. *Contradictions of Interests in Early Phase of Real Estate Projects – What Adds Value for Owners and Users?*, paper presented at 20th CIB World Building Congress, Tampere, Finland, 30 May–3 June.

Part 3
Added Value

13 The concept of value of buildings in use

Marit Støre-Valen, Torill Meistad, Knut Boge, Margrethe Foss, Leif D Houck and Jardar Lohne

Introduction

'Value' is one of the concepts dominating contemporary literature on construction management. However, it often remains an abstract idea and is not taken forward into concrete project actions. The concept has many facets: financial value, relationship value, user or exchange/market value, or social, individual and collective value. Several researchers have reviewed the literature to find a common definition used within the context of construction projects; see for example Haddadi et al. (2015) and Drevland and Lohne (2015). The latter also discuss the implications of value-related concepts within the construction phase of the project, especially value for 'whom' and 'perceived value'.

For the contractors and design team, the largest portion of value is typically created during the design and construction phases. For the owner, value is first captured when the built asset is handed over and used to produce income. For the end-user, the value creation taking place in a workplace is both individual and collective. Collectively, it relates to what is important for the organisation and its work environment to successfully run their business. Individually, it relates to whether their individual values fit with the core values of the organisation and allow the person to thrive, grow and create value (Støre-Valen et al., 2016; Frow et al., 2015).

Additionally, according to Drevland and Lohne, value can be seen as a relationship between 'benefit' and 'cost' that results from a value judgement. They argue that the complexity of the concept can be simplified by answering the question of 'What does the customer want?' However, during the operations phase, the changes in use and the demands that such changes put on built assets heavily influence its value. This topic has been gaining special interest over the last few years within the debate on facilities management (FM) (Støre-Valen et al., 2014; Baharum and Pitt, 2009). This attention has particularly focused on the subject of so-called adaptive reuse of buildings, which is typically based on a building's ability to adapt and handle changes in practice and subsequent changes of workspace.

This chapter first discusses various aspects of what creates value in real estate projects and how information integration can contribute to increasing value for

214 Støre-Valen et al.

users and owners in the operational phase. Of particular concern here are the questions of so-called value-driven leadership and added value in the context of FM. The chapter then discusses some recent approaches to the value concept that are considered relevant for information integration in buildings during the in-use phase. It further reflects on the challenges of optimising a construction project for end-users and owners by focusing on the value aspects of the stakeholders (Støre-Valen et al., 2016; Frow et al., 2015). These suggested paths of analysis can be summarised in the following questions:

- how is value for the end-users and owners realised?
- are there any conflicts of interest between the owner and user?
- does the information gathered in the early planning phase have a real impact on the decision-making process?
- how important is user involvement in the early concept and design phase?
- how can information integration technology be a key enabler for innovation and value creation for building in use?
- how can organisational and individual values be translated into innovative design and solutions?

This chapter also presents some of the main research findings of studies conducted recently about Norwegian hospital buildings, university and campus buildings as well as office buildings.

Value-driven leadership

The concept of value-driven leadership stems from the notion that an organisation governs its business towards collective values specific to that organisation. A main idea underlying this concept is that when this value drive is clearly perceived by society, the organisation attracts people with individual values that resonate with those of the organisation. Such organisations with value-driven leadership do in fact exist. For example, one of the largest construction companies in Norway has evolved from being a local contractor to being a nation-wide, well-recognised contractor. Its success stems largely from clear value-driven leadership through long-term development of competencies within the organisation. Their business philosophy, clearly stated in their strategy document, is to create value through partnership with stakeholders in order to create a built environment that has a positive impact on and is largely enjoyed by society. This collaborative way of working is based on close relationships with the project owner and relevant stakeholders. The contractor has completed several projects with great success in achieving economic profits, new innovative ways of working, social outcomes and competence development by finding new and innovative technical solutions. These outcomes seem to be mainly the result of achieving close interaction and integration between the design team and the project owner.

Despite success stories like this, the challenges of implementing value-driven leadership within the Norwegian context are significant. More than 80 per cent

The concept of value of buildings in use   215

of the country's building and construction contractors are small and medium-sized enterprises (SMEs) dealing with small-scale projects. This characteristic is also shared by many other countries across the world such as Australia and the United States (US). This represents a challenge for the implementation of information and communication technologies (ICT), in particular because small-scale projects are mainly handled using traditional project governance and execution models. Hampson et al. (2014) point out that the construction industry is largely a site-based industry with high levels of specialist knowledge. A report by the National Research Council (NRC) in the US also summarises additional challenges for the US construction industry: low innovation levels, low productivity, the need for higher efficiency and improved quality of construction processes and outcomes, better time–cost effectiveness and more sustainable construction projects (NRC, 2009).

The report concludes with five key actions, titled 'Opportunities for Breakthrough Improvements', based on input from over 50 experts from the US construction industry. Two that are particularly relevant for this chapter are:

1   'Widespread deployment and use of interoperable technology applications, also called Building Information Modeling (BIM)'.
2   'Improved job-site efficiency through more effective interfacing of people, processes, materials, equipment and information' (NRC, 2009).

The other three opportunities relate to the industrialisation of construction elements and off-site fabrication processes, innovative and widespread demonstration of installations, and effective performance measurements to drive efficiency and support innovation. Aspects related to finding effective performance measures are discussed further in Chapter 12 (Meistad et al.).

The NRC report (2009) says that these actions are interrelated and implementation of one will enable others. The report also points out that this requires a strategic, collaborative approach led by project owners that are interested in benefits such as lower-cost, high building quality and sustainable buildings (NRC, 2009). In Norway, this is beginning to be put into action, particularly in large companies, among public owners and SMEs with a genuine interest in applying new technology.

The requirements stated in the NRC report are in line with Meistad's study on exemplary projects in Norway (Meistad, 2015). All the projects studied had high ambitions for finding sustainable solutions and low energy performances. Furthermore, Meistad found that the key to success lies in developing a sustainable building practice that incorporates a collaborative approach towards the involvement of users, owners and other stakeholders to satisfy functional and social needs, and user and owner objectives. The study also highlights the importance of involving the right competencies corresponding to the expertise needed during the early planning phase. However, there is still a long way to go for SMEs to achieve this as they have small profit margins and need to be connected to a knowledge platform or suitable network to develop such knowledge.

216 *Støre-Valen et al.*

Whether enterprises apply smart technologies and BIM as tools is also a question of leadership and strategies to overcome barriers when implementing them (Valen et al., 2010). Today, only major actors and big enterprises within the Norwegian context use BIM and other ICT platforms. In addition, even though the practice is spreading, handheld devices are still rarely used for project delivery among SMEs. The development of handheld devices and apps that communicate with the BIM model is also an area that needs improvement and maturity among SMEs, both in the construction and the operations phase.

SMEs lagging behind their larger competitors in their use of technologies constitutes a main challenge for the industry; although SMEs' profit margins are small, such technologies have great potential to help them create new competencies among their workers, thus increasing profit margins and creating new markets (Sanchez et al., 2016). SMEs will also benefit from a project owner that takes a strategic collaborative approach involving expertise at an early stage of the project, including dialogue with the SMEs before the procurement process begins or the contract is signed. If this is put into practice, this will help ensure that sustainability is at the heart of the industry and will be clear evidence that we are heading towards value-driven leadership.

Value creation in FM

FM typically assesses the building as a physical unit that facilitates the core business to reach their organisational goals. Ideally, the building ought to be considered a strategic means to reach the owner's objective. There is an acute need to translate the requirements and values of the owners and their customers into a design that includes suitable flexible spaces. The understanding of these needs is in fact a novel insight, requiring a cultural shift among actors involved in the planning and design of buildings. It also requires knowledge about how to facilitate such processes and value management. These are both essential to finding sustainable and innovative solutions (Støre-Valen et al., 2016).

This insight is, however, not new. In Norway, user involvement in the early phase is common, with a user coordinator representing different user groups being involved to provide input for the early phase team. Nevertheless, other processes take place when it comes to the decision phase. Several researchers have documented the need to involve the FM personnel during the briefing and design phase but this is not commonly practised (see for example Larssen, 2011; Støre-Valen et al., 2014; Frow et al., 2015). Thus, the decisions made do not necessarily result in solutions that give optimal value to the end-user or that ease the operations phase of the building.

Several studies of the OSCAR project[1] have investigated value-enhancing elements for users of hospital offices and university/campus buildings (Hareide et al.; 2015; Ravik et al., 2016; Hulbak, 2016; Spiten et al., 2016). This will be described in more detail in the section titled 'Case studies'. However, more research is still needed to fully understand this issue more broadly. There is also a need to explore the potential benefits of using BIM and other smart digital

The concept of value of buildings in use   217

platforms in terms of information integration from the early phases, throughout the design and construction phases, and into the operations phase. This requires strategies concerning how to handle information throughout the life-cycle of projects, including the implementation of digital devices. It also requires a clear vision of the use of the building, which emerges as an important applied research topic for the future.

The Internet of Things (IoT) also offers huge potential when applied to buildings. The devices in different buildings are connected, thus enabling a suite of buildings to communicate through a platform and gather information that can be used to improve and develop better practices for operations and FM services (See Chapter 4, Abanda and Tah). Davies (2015) says that IoT applied to a building makes it a 'smart building'. These buildings would then be fitted with sensors and monitors that measure activity through several devices connected to a communication platform that enables FM personnel to monitor the building and its use.

Building automation then has significant potential for IoT applications.

> For the FM staff this will ease the operation of the building, the maintenance planning, analysing the use of space as well as individual regulation of lighting and temperature control. This technology is already applied when optimising the energy performance of the building.
>
> (McHale, 2015)

Further to this, when IoT is applied to buildings to form BIoT,

> All aspects of the building's technical performance, together with improving the performance of the business enterprise within many of the verticals, can now be brought together. This has been made possible by the Internet of Things (IoT), which allows one common IP [Internet Provider] platform to link all the sensors and devices together to interchange information and through analytical software, commonly called Big Data, optimize the controls automatically. The terminology applied to buildings to describe this is called the Building Internet of Things (BIoT), and the process of morphing all the BAS services [Building Automation Systems] into one whole system is now underway.
>
> (McHale, 2015)

IoT and BIoT have huge potential to change the construction industry. Not only in the construction process but especially in the operational phase. For FM suppliers this means a significant change. By moving from BAS to BIoT systems they can optimise the operations of the building as well as integrate the information about usability and FM service performance across building portfolios. The BIoT and exchange of information, or the gathering of big data, gives access to information that can help FM personnel and owners to develop new knowledge about how the facilities are being used and how well the spaces are functioning.

218   *Støre-Valen et al.*

Information integration in construction

The literature describes several challenges and barriers to improving the construction industry. Of particular importance is a need for more innovative and research-based practice rather than keeping the traditional focus on low cost. Eadie et al. (2013) describe collaborative and integrated approaches as a co-creation process that involves the customers in the design of the building. Frow et al. (2015) developed a co-creation design framework based on the expertise of eight senior executives in leading enterprises. The framework can be used to facilitate a process that involves the customer in order to create innovative products. This framework addresses several dimensions such as co-creative motives, co-creation form, engaging factors, engagement platform, level of agreement and duration of agreements. The eight executives reported improved insight about the customer's needs as well as experiencing improvement and innovation. A similar framework can be applied to the design process of a construction project to improve the built quality of the building. Other researchers report on several studies that state that the success of a collaborative relationship to co-create value leads to successful value delivery to all stakeholders (Haddadi et al., 2015; Jensen et al., 2012; Sarasoja and Aaltonen, 2012).

So, what does it take to engage the customer in the early phase? What are the barriers to doing so? BIM has been hailed as a key enabler to resolve some of the challenges typically found in projects that require high levels of integration. As early as 1992, Gielingh and Tolman described these challenges as fragmentation, the large number of sub-contractors, lack of a common understanding of the project objectives and high transaction costs. They pointed towards the solution of 'integration' as a way to handle these challenges.

The 1990s saw great expectations about the digitisation of information bringing the solution to many problems in the construction industry. There was hope that using digital platforms and ICT could increase productivity and reduce the fragmented nature of the industry. Some researchers also had visions and ideas that BIM could be a good management tool for the involved actors in the construction process (NIST, 2004; Construction Cost Programme, 2008; Valen et al., 2010). When BIM was introduced to the construction industry in the mid-2000s, it was primarily embraced by designers and planners as it was well suited for early adoption of technology for the conceptual and design phases. However, it was less well received by other professional groups. Many contractors and suppliers are characterised as practitioners and craftspeople, a fact that is often found to be a barrier when implementing new technologies (Valen et al., 2010).

Of particular interest within this context is the manner in which BIM permits the integration of information that enables value creation for FM staff and end-users of buildings. A major challenge, from both a research and a practical perspective, is to understand how these needs can be managed in the early planning phase as well as throughout the construction process and the life-cycle of the building.

The concept of value of buildings in use   219

Shen et al. (2010) refer to the US construction industry and highlight interoperability as one of the major problems this industry faces. In their review, they conclude that the integration of information and development of collaboration platforms are key to enabling the construction industry to improve its productivity and efficiency. Of the solutions on offer, they suggest that integration solutions loosely coupled with intelligent web platforms prove to be the most promising for the time being. Such platforms have the potential to revolutionise the construction industry both for small and large enterprises.

Eadie et al. (2013) investigated the effect of BIM use throughout the project life-cycle. This research confirmed that such technologies are mostly used during the early stages. They concluded that the collaboration aspects produce the highest positive impact and that the process aspects are more important than the technology. The survey showed that the facilities manager benefited the most from a financial perspective. However, facilities managers are not commonly recognised as having substantial expertise in using BIM. Competence and maturity within the FM organisation therefore has significant potential to be developed.

The role of information integration and the social dimension of sustainability

The relationship between owner and user is a crucial element in achieving the intended use of the building. For the owner, the financial and physical aspects of a building are equally some of the most important elements to take into consideration. However, the question of integrating the user perspective within the context of value-enhancing elements is not straightforward. The following sections examine the influence of users on their work environment by exploring the social dimension of sustainability.

Haddadi et al. (2015) explain that value creation in a construction project depends on three main roles: owners, suppliers and users. The relationship between the main roles during the project and their requirements contribute to value creation, as illustrated in Figure 13.1.

The academic literature on the topic suggests that the involvement of the user and other relevant stakeholders during early phases of the planning processes provides benefits and ultimately added value to the end-user during operations. These benefits include greater ownership of the building and flexibility of the space, better understanding of use and innovative better-fitted solutions (Thyssen et al., 2010; Pemsel et al., 2010; Frow et al., 2015; Gemser and Perks, 2015; Spiten et al., 2016). Within this context, how can information integration increase the understanding of the users' perspective and help ensure that the value-enhancing elements are promoted?

In order to ensure high design quality of a building, the Construction Industry Council (CIC) (Spencer and Winch, 2002) identified three quality fields: functionality, build quality and impact. These fields overlap and their interplay determines the level of design quality. CIC developed a short questionnaire based

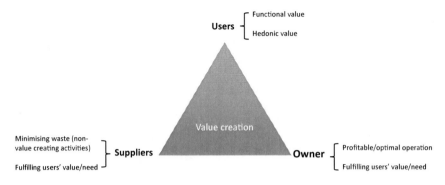

Figure 13.1 Construction project main roles and their relationship when creating values (Haddadi et al., 2015).

on these three indicators. The respondent is asked to score the design quality of a building based on these three fields by giving their opinion about a range of subsets. This helps the respondents define the level of quality of the building. Spencer and Winch (2002) say that this is a simple method to raise the quality of a building but it also enables the clients/owners to understand the value of good design. The method allows the participant to compare the completed building against their specified intended use and to evaluate the functionality of the building in use, look at the build quality and look at what impact the building has on society and its surroundings.

Figure 13.2 illustrates the complexity of the construction process and the different views held by various actors about what the concept of 'value' actually means.

Decision-making is typically carried out according to the stakeholders' power and influence, and they illustrate their viewpoints throughout the design and construction phase. This influences the client's decision-making process. Naturally, the value systems and view points of the stakeholders will influence the decisions and choices of the client that affect the quality of building. However, the process itself (time, cost and resources available) also influences the value system. The contractor/client relationship and the economics of the project regulate the power and the decisions made. CIC suggested using a 'structured role' that is loosely coupled with the other stakeholders involved in the design and construction phase (see Figure 13.2). The purpose of the role is to balance power between stakeholders and network groups and to advocate the user's needs at the decision table. In the case of the College School of Bergen (HiB), this role was key to its success. For HiB, this involved a user–coordinator role and their technical competencies were essential to attaining a high level of quality for that building (Spiten et al., 2016). In this case, information integration was handled systematically. In Norway, however, there is no standard procedure to ensure such involvement nor its use throughout the design and construction phases.

A survey carried out by the OSCAR project in 2015 found that the FM role was poorly represented and understood during the early planning phase and the

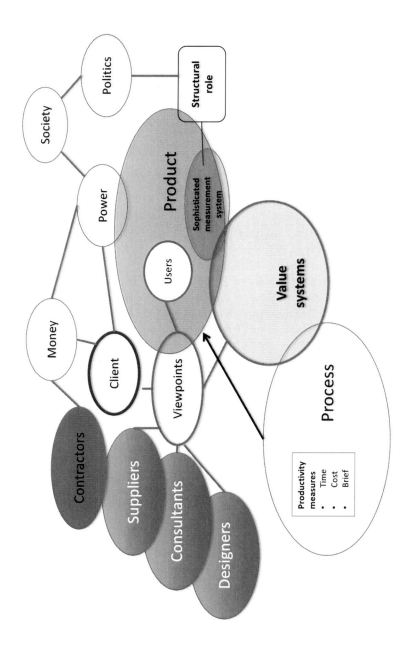

Figure 13.2 Complicity of stakeholders involved in the construction process (adapted from CRISP Design Group, 2000, cited in Spencer and Winch, 2002).

222 *Støre-Valen et al.*

conceptual phase. Even though users are represented and involved in the early phase, ensuring that important information is handed over and not lost during the transfer from the design to the construction phase is typically challenging (Støre-Valen et al., 2016). Although BIM solutions may be useful within this context, the maturity and competence of FM personnel in the use of digital platforms and BIM remains a key issue for this stakeholder group. Eadie et al. (2013) studied the use of BIM throughout the project life-cycle in several projects in the United Kingdom (UK). They confirmed that BIM is mostly used during the early phase stages and less often during the operations stage.

The potential for FM suppliers lies in the utilisation of IoT technologies applied to the building, including BIM technology. Using IoT employs sensors to measure the use of the space and track the users' working patterns. This can reveal important trends about the usage potential and inefficient working hours.

Case studies

The following sections will discuss a series of case studies to illustrate what gives value to the involved stakeholders. The findings illustrate current practices in hospitals and the university sector in Norway and point towards the need for a value management process in the construction phase that ensures quality for the end-user. In general, information integration is essential to all these sectors. The following topics are presented:

- the OSCAR project: increased value for owner and user (Støre-Valen et al., 2016);
- cases of hospital buildings (Hareide et al., 2015);
- cases of university and educational buildings (Spiten et al., 2016);
- choice of façade materials (Houck, 2016).

Buildings in use – What adds value from a stakeholder's perspective?

The OSCAR research and development (R&D) project (2014–2017), *Better Values for End-Users and Owners of Buildings,* is an industry involvement project partly funded by the Norwegian Research Council. The Norwegian industry partners include architects, consultants, contractors, public owners and research and education institutions. The main objective of the project is to enable optimisation of the building design and contribute to value creation for the owner and users throughout the asset's life.

The underlying idea of the project is to develop new knowledge, methods and tools that are tested, developed and implemented in real-life construction projects. Contractors, consultants and architects work together regardless of their competition when bidding on projects. This collaborative work, through the sharing of knowledge and practices, creates the opportunity to develop new insights, both by finding new ways to involve stakeholders in the early planning phase and by developing models that stimulate value thinking.

The project is organised in three parts:

1 Development of new knowledge in early planning phases.
2 Execution models and processes.
3 Methods and assessment tools.

During the first two years, the OSCAR project organised workshops and conferences and conducted several surveys among a wide range of stakeholders across the Norwegian building industry. The surveys investigated questions about what gives value to users and owners in buildings, and which stakeholders are involved during the early planning phase. The respondents ranked a set of statements that described four aspects of sustainability: physical, social, economic and environmental. Støre-Valen et al. (2016) analysed the social and economic dimensions of the survey and looked at what elements gave the most significant difference in value for owners and users. They found that there is a significant difference between users' and owners' views on economic value. The FM roles were also rarely involved in early phases. Respondents from a user perspective seemed to be more concerned about the financial aspects than cost-efficient operations. By contrast, respondents from an owner perspective seemed to be more concerned with cost-efficient operations than with other issues. As for the social dimension, owners and users shared their most important values: that the workplaces facilitate social interaction and are flexible spaces.

Previous studies have revealed the complexity of the user role in building projects. Users may not clearly understand what their needs are or they may find it difficult to communicate their needs and understand the value of a good building design. User involvement in the early phases creates ownership of the decisions made and higher experienced value of the final building.

Findings from the literature review and survey confirm that there is a need for increased competence and understanding of value management through the building process. The literature and examples from practice point towards co-creative collaborative working models as a way to improve the understanding of what creates value for the user (Frow et al., 2015; Gemser and Perks, 2015). A co-creation process requires involvement of relevant stakeholders together with user coordinators, FM and owner representatives to find the optimal design and innovative solutions. Such a process can deepen the understanding of customer needs and create better solutions. The process also helps the users identify what their real needs are. This is in agreement with the findings of Spiten et al. (2016) who studied the process for HiB.

Value-enhancing elements for end-users of university buildings and hospitals

Only half of Norwegian hospital buildings are reported to be adequate for today's procedures. There is a backlog of maintenance, development and operational services due to insufficient prioritisation of FM functions in hospital budgets. This has led to an estimated upgrading cost of NOK40 billion (USD4.9 billion).

224 Støre-Valen et al.

As an initiative to improve hospital buildings, the Ministry of Health and Care Services introduced the Sykehusbygg HF trust (SBHF) in late 2014. SBHF's purpose is to aid health authorities during planning and construction of new hospital buildings, as well as the development of FM services for existing hospital buildings.

Hareide et al. (2015) investigated strategies for optimising the value of Norwegian hospital buildings. They researched the following questions:

1 What is value and how can buildings add value?
2 What creates value within hospital buildings?
3 Which strategies ought to be present for future development of hospital buildings?

These questions were researched through a literature review and a series of case studies involving four Norwegian hospital buildings and SBHF. Document studies and semi-structured interviews were conducted and constitute the main sources of information. The research revealed that a valuable hospital building is one that creates optimal conditions for effective delivery of healthcare services. Value management is a recommended approach for value creation. The process of identifying, classifying, evaluating and optimising needs and requirements at an early stage of a project provides the criteria and specifications to create a valuable building. From the literature review, adaptability and life-cycle costs (LCC) are recommended focus areas when preparing for future developments. The findings of the case studies show that these concepts are well known. Most of the newest hospitals also have a high degree of flexibility and elasticity. Pre-design documents additionally describe LCC as an important assessment deliverable throughout the projects. However, interviews and document analysis indicate that both the adaptability and LCC are inadequately utilised. Reported usability conditions and the backlog of maintenance, development and operational services support this.

Strategic involvement of FM is a goal for future planning of hospital buildings. Currently, there seems to be ample room for improving FM services. There is however an increasing awareness of the need for involvement of FM practitioners in Norwegian hospitals. There are high expectations for SBHF; the interviewees expect SBHF to make a difference in knowledge transfer and sharing of hospital planning. They also said that they expect an increased focus on achieving a high degree of adaptability and FM involvement in the planning phase of future projects. SBHF aims to contribute to large hospital building projects and develop competence and knowledge transfer routines for the whole sector. Continuous benchmarking and measuring of experience data will contribute to increasing competence and knowledge in the sector.

Strategies for optimisation of value in pre-design contributes to enhanced future planning of Norwegian hospital buildings. In brief, three strategies were identified:

1 Focus on adaptability.

2 Ensuring LCC analysis is completed.
3 Strategic involvement of facility managers.

SBHF aims to ensure that the sector utilises the strategies for future development of hospital buildings (Hareide et al., 2015).

Value-enhanced elements of users in educational buildings and offices

Spiten et al. (2016) did research on defining the value-enhancing elements among staff and students of university buildings in Norway. They also looked at which end-user involvement strategies are required during the early planning phase to achieve end-user value. They did a broad survey among students, FM staff and educational personnel to investigate the most important factors for the social and environmental aspects of the users of university buildings as well as several case studies of fairly new university school buildings. They found that informal meeting places such as coffee shops and a good, spacious indoor learning environment were significant factors for these groups.

Ravik et al. (2016) investigated the value-enhancing elements for office workers using a survey and one case study of an office building in Oslo. They found that indoor climate and environment is the most important factor to create value for the users of office buildings. The office workers also valued the possibility of having individual adjustable blinds, temperature and ventilation to a greater extent than professional consultants.

What makes a building become an asset that not only facilitates an enterprise's operations but also adds value to their business? Researchers agree on the importance of having a clear vision and objective at the strategic level. It is also essential to engage at the tactical and operative level during the early planning and design phases in order to translate this into innovative solutions (Pemsel et al., 2010; Gemser and Perks, 2015; Frow et al., 2015). The literature also points towards the need for adapting a systematic approach to value management processes during the concept phase (Kelly et al., 2015). They indicated that BIM models are useful to communicate the value-enhanced elements through the visualisation of the suggested solutions and keeping track of the early concept ideas and innovative solutions that seem to get lost along the way in the project. This indicates that an approach that promotes greater integration can help better understand what constitutes value.

The choice of façade material – values and beauty

In recent years, there has been an increased focus on environmentally friendly buildings in Norway and across the globe. Official policies, legislation and support from different organisations advocating sustainable buildings aim to affect the level of sustainability in the building industry. With the need for more environmentally friendly or 'green' buildings in mind, the objective of this case study was to investigate what factors influence the choice of façade materials in contemporary Norwegian building projects. This section presents the results of the analysis of seven public construction projects.

It was found that in some projects, the choice of façade material was influenced by the general guidelines originating from public authorities, the user, the client or special building programmes. Methods such as LCC calculations, greenhouse gas accounting and life-cycle assessments were used in all the projects that were examined. For some projects, these analyses and calculations were only used to fulfil documentation requirements and thus had no major impact on the decision-making process. For other projects, these tools were used to evaluate alternative product choices (Houck, 2016).

The findings in this investigation show that sustainability, aesthetics and cost are considered to be the most important factors influencing the choice of façade material (Figure 13.3). Participation in different sustainability programmes seems to have influenced the evaluation and choice of façade material, both through developed processes and the required use of assessment tools such as environmental performance declaration analysis, LCC calculations, greenhouse gas accounting and life-cycle analysis.

Although the different projects used the same calculation tools, the project processes and goals were still different. Therefore, each tool could be used in different ways and the results were in some cases more advisory than prescriptive. In other projects, quantitative goals were set based on these results. In one case, however, the zoning and planning requirements overruled conclusions based on sustainability criteria. Different building functions may also lead to different

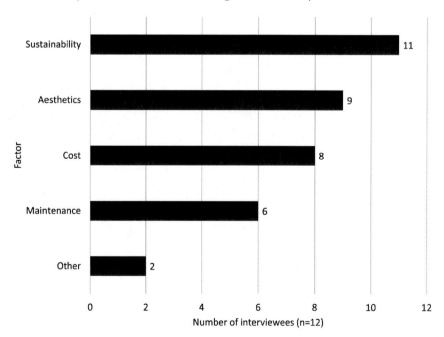

Figure 13.3 Factors emphasising choice of façade materials (Houck, 2016).

The concept of value of buildings in use 227

choices of materials. In this investigation, some clients and architects were sceptical about using timber façades for school buildings.

To conclude, despite environmental ambitions in all of the investigated projects, the decision-makers chose materials such as different timber solutions and bricks. The explanation for this is partly found in the projects' different functions, how the different project teams executed and evaluated the environmental analyses, and the differences in project goals. Governmental incentive programmes like 'The WOOD Program' were applied to increase the use of wood as a building material and particularly as a façade material.

Conclusions and priority areas

This chapter examined various aspects and challenges related to the concept of value of buildings in use. How can these challenges be overcome by using an information integration approach?

The owner's main concerns in the decision-making processes are naturally related to the financial and physical aspects. The motivation for investing extra time and money in the early design phase and demanding user processes will normally be found in a building that generates some income by housing a business that creates value in the future. When the owner sees how this economic benefit pays off in the long run, they are willing to put time and money into innovative and adaptable solutions, architectural design and sustainable materials and products, as well as looking at the technical aspects of the operation of the building using technology to better interact and understand the users' needs (Houck, 2016; Støre-Valen et al., 2016; Hareide et al., 2015). However, the owner must still acquire this knowledge and involve the necessary expertise. This field is yet to be fully explored.

The FM role and its potential for increasing productivity and adding value to the core business has been discussed. According to the literature in this field, there is no doubt about the owner benefiting when FM suppliers are involved in the early planning phase. The literature also suggests that FM suppliers can benefit financially from using BIM but that this opportunity is not yet fully exploited (Støre-Valen et al., 2016; Eadie et al., 2013; Pemsel et al., 2010; Jensen et al., 2012). Experience from practice shows that FM suppliers are not overly present in the early planning phase and that the tactical or strategic level could significantly benefit from their involvement (Støre-Valen et al., 2016). Another benefit highlighted in the chapter was easing the workload of FM personnel by developing IoT and smart development of handheld devices and digital platforms that can easily improve workflow efficiency and information flow between spaces and facilities (Davies, 2015; McHale, 2015).

The chapter also discussed how a greater understanding of value management processes increases added value benefits and how co-creation and collaboration can integrate these processes. One example of such an enquiry is whether value management processes can constitute an integral part of the construction process. This requires new skills and competences and at least awareness of the necessity to do so.

228 *Støre-Valen et al.*

The users' concern for what creates value in a building naturally varies according to the purpose of the building. This chapter discussed surveys from users at universities, hospitals and office spaces. The survey results highlighted the need for a representative of the users to find creative and innovative solutions that can adapt to their future needs. The survey results also indicated that office workers and students value social functions, such as coffee shops, in addition to the normal working functions found in office spaces and teaching facilities. The literature reviewed recommends a co-creation framework to help find the solutions that add value to the operation phase by involving stakeholders and expertise in the early planning and design phase (Støre-Valen et al., 2016; Pemsel et al., 2010; Gemser and Perks, 2015).

Based on the areas highlighted in this chapter, there is a need for further exploration of:

- better ways to ensure that the scope of the project fits with the needs of the user;
- use of a coordinator role and interactive guidelines that can challenge traditional practices;
- increased focus on co-creation and collaborative work schemes in early planning phases, adapted as a way to find innovative solutions and designs that increase the building quality, add value to the end-user and become a sustainable practice;
- involvement of the FM organisation in the early planning phase and design phase of construction projects;
- development of handheld digital devices and apps that ease the workload of FM personnel.

Information integration is essential to ensure smooth information flows and critical interfaces between the early project phases, design and construction phases, and the handover and operational phases of the building. Developing standards and procedures to handle the information exchange as well as developing tools and competences that strengthen the utilisation of the information to make the right decisions among stakeholders is, however, challenging. Nevertheless, stakeholders who use technology and develop the potential for innovation and new ways of working will have a competitive advantage in the market.

Note

1 OSCAR is a research and development project funded by the Norwegian Research Council to develop knowledge about and methodology for increasing the value for the owners and users of buildings. Multiconsult is leading the project with the involvement of a broad range of industry partners, public owners and managers, and research and education institutions. For more information visit: www.oscarvalue.no.

References

Baharum, M. R. and Pitt, M., 2009. Determining a conceptual framework for green FM intellectual capital. *Journal of Facilities Management*, 7(4), 267–282.

Construction Cost Programme, 2008. *Construction Cost Programme – Steering Document.* Available at: www.byggekostnader.no [Accessed 1 September 2016].

Davies, N., 2015. *How the Internet of Things will Enable 'Smart Buildings'.* Available at: http://search.proquest.com/docview/1696184683?accountid=12870 [Accessed 7 September 2016].

Drevland, F. and Lohne, J., 2015. *Nine Tenets on the Nature of Value*, paper presented at IGLC 23: Global Problems – Global Solutions, Perth, Australia, 29–31 July, pp. 475–485.

Eadie, R., Browne, M., Odeyinka, H., McKeown, C. and McNiff, S., 2013. BIM implementation throughout the UK construction project lifecycle: An analysis. *Automation in Construction*, 36, 145–151.

Frow, P., Nenonen, S., Payne, A. and Storbacka, K., 2015. Managing co-creation design: A strategic approach to innovation. *British Journal of Management*, 26(3), 463–483.

Gemser, G. and Perks, H., 2015. Co-creation with customers: An evolving innovation research field. *Journal of Product Innovation Management*, 32(5), 660–665.

Gielingh, W. F. and Tolman, F. P., 1992. Information integration in the building and construction industries. *Microcomputer in Civil Engineering*, 6(4), 329–334.

Haddadi, A., Temeljotov-Salaj, A., Foss, M. and Klakegg, O. J., 2015. *The Concept of Value for Owners and Users of Buildings – A Literature Study of Value in Different Contexts*, paper presented at 29th International Project Management World Congress: The way to project management in multicultural context, Panama City, Panama, 28–30 September.

Hampson, K. D., Kraatz, J. A. and Sanchez, A. X., 2014. *R&D Investment and Impact in the Global Construction Industry*. London: Routledge.

Hareide, P. J., Bjørberg, S., Støre-Valen, M., Haddadi, A. and Lohne, J., 2015. *Strategies for Optimization of Value in Hospital Buildings*, paper presented at 29th International Project Management World Congress: The way to project management in multicultural context, Panama City, Panama, 28–30 September.

Houck, L. D., 2016. The choice of façade material – Values and beauty, in M. Prins, H. Wamelink and B. Giddings (eds), Volume II – Environmental Opportunities and Challenges. Construction and Commitment and Acknowledging Human Experiences, proceedings of the CIB World Building Congress 2016, Tampere Finland, 1–3 June, pp. 141–152.

Hulbak, M. B., 2016. Verdi for Brukere av Universitets – og Høgskolebygg. Effekten av Brukermedvirkning på Brukerverdien (Value for users of universy and college buildings. The effect of user involvement on the user value). Master's thesis in Norwegian, Norwegian University of Science and Technology, Trondheim: Tapir Press.

Jensen, P. A., van der Voordt, T. and Coenen, C., 2012. *The Added Value of Facilities Management. Concepts, Findings and Perspectives.* Kongens Lyngby: Polyteknisk Forlag.

Kelly, J., Male, S. and Graham D., 2015. *Value Management of Construction Projects.* Chichester, UK: Wiley Blackwell.

Larssen, A. K., 2011. Bygg og Eiendoms Betydning for Effektiv Sykehusdrift (Buildings' impact on hospital effectiveness). Doctoral thesis, Norwegian University of Science and Technology, Trondheim: Tapir Press.

230 Støre-Valen et al.

McHale, A., 2015. *Why the Building Internet of Things Will Enable Smart Buildings*. Available at: www.facilitiesnet.com/buildingautomation/article/Why-the-Building-Internet-of-Things-Will-Enable-Smart-Buildings-Facilities-Management-Building-Automation-Feature--16039 [Accessed 7 September 2015].

Meistad, T. R., 2015. Sustainable building – from role model projects to industrial transformation. Doctoral thesis, Norwegian University of Science and Technology, Trondheim: Tapir Press.

NIST, 2004. *Cost Analysis of Inadequate Interoperability in the U.S. Capital Facilities Industry*, NIST GCR 04-867. Available at: http://nvlpubs.nist.gov/nistpubs/gcr/2004/NIST.GCR.04-867.pdf [Accessed 7 September 2016].

NRC, 2009. *Advancing the Competitiveness and Efficiency of the U.S. Construction Industry*. Washington, DC: The National Academies Press.

Pemsel, S., Windén, K. and Hansson, B., 2010. Managing the needs of end-users in the design and delivery of construction projects. *Facilities*, 28(1/2), 17–30.

Ravik, K. M., Haddadi, A., Bjørberg, S., Foss, M. and Lohne, J., 2016. *Characteristics That Enhance Value for Users of Offices—Focus on Buildings and Stakeholders*, paper presented at 24th Annual Conference of the International Group for Lean Construction. Boston, US, 20–22 July.

Sanchez, A. X., Hampson, K. D. and Vaux, S., 2016. *Delivering Value with BIM: A Whole-of-life Approach*. London: Routledge.

Sarasoja, A. L. and Aaltonen, A., 2012. Green FM as a way to create added value, in P. A. Jensen, T. van der Voordt and C. Coenen (eds), *The Added Value of Facilities Management. Concept, Findings and Perspectives*, Kongens Lyngby: Polyteknisk Forlag, pp. 195–203.

Shen, W., Hao, Q., Mak, H., Neelamkavil, J., Xie, H., Dickinson, J. K., Thomas, J. R., Pardasani, A. and Xue, H., 2010. Systems integration and collaboration in architecture, engineering, construction and facilities management: A review. *Advanced Engineering Informatics*, 24(2), 196–207.

Spencer, N. C. and Winch, G. (eds), 2002. *How Buildings Add Value for Clients*. Construction Industry Council, London: Thomas Telford Publishing.

Spiten, T. K., Haddadi, A., Støre-Valen, M. and Lohne, J., 2016. *Enhancing Value for End Users – a Case Study of End-User Involvement*, paper presented at 24th Annual Conference of the International Group for Lean Construction, Boston, US, 20–22 July.

Støre-Valen, M., Larssen, A. K. and Bjørberg, S., 2014. Buildings' impact on effective hospital services – The means of the property management role in Norwegian hospitals. *Journal of Health Organization & Management*, 28(3), 386–404.

Støre-Valen, M., Boge, K. and Foss, M., 2016. Contradictions of interests in early phase of real estate projects – What adds value for owners and users? in K. Kähkönen and M. Keinänen (eds) Volume I – Creating Built Environments of New Opportunities, proceedings of the CIB World Building Congress 2016, Tampere, Finland, 1–3 June, pp. 285–296.

Thyssen, M. H., Emmitt, S., Bonke, S. and Kirsk-Christoffersen, A., 2010. Facilitating client value creation in the conceptual design phase of construction projects: A workshop approach. *Architectural Engineering and Design Management*, 6(1), 18–30.

Valen, M. S., Klakegg, O. J. and Hustad, S., 2010. *Barriers and Bridges in Construction Processes*, paper presented at the IPMA Conference, Istanbul, Turkey, 1–3 November.

14 Information integration and public procurement

The role of monitoring, benchmarking and client leadership

Adriana X. Sanchez, Jessica Brooks and Keith D. Hampson

Introduction

In 2013, just the site-based activities of the construction sector contributed between three and ten per cent of gross domestic product (GDP) globally. When the broader supply network is included, this contribution increases to 10 to 30 per cent of a country's GDP (Hampson et al., 2014). Up to 80 per cent of the whole-of-life cost associated with built environment assets is incurred during the operational phase (buildingSMART, 2010). This chapter takes the whole-of-life asset management view of procurement proposed by Sanchez and Hampson (2016). Based on this approach, asset procurement and management starts at the strategic planning phase and ends with the operations and decommissioning phase.

Procurement plays a strong role throughout the life-cycle of built environment assets and in driving progress towards many higher social goals such as sustainability and equity (Hardy, 2013; Dovers, 2005). At a local, state and national level, government procurement in particular often represents a significant share of national GDP, accounting for up to 20 per cent for some countries (Garcia-Alonso and Levine, 2008). Therefore, in sectors where government entities commonly constitute the largest client, government procurement practices have a significant impact on the industry (Sanchez et al., 2014b; Brown et al., 2006). In Australia for example, a report issued by Engineers Australia suggested that changes to public procurement models can have a significant impact on the national budget. This report estimated that, in 2012, a 1 per cent improvement in the efficiency of national procurement organisation, processes, technology and performance management could generate over AUD600 million in savings (Yates, 2012).

Integrating information across life-cycle phases, asset delivery and management systems, and stakeholder organisational boundaries can significantly reduce efficiency losses. A study of the United States (US) capital facilities management

232 *Sanchez et al.*

industry in 2002 showed that, based on conservative estimates, the sector loses almost USD16 billion annually due to inadequate interoperability between asset delivery and management systems, with owners and operators losing over USD10 billion (Gallaher et al., 2004).

Advanced information and communication technologies (ICT) and socio-technical systems such as Building Information Modelling (BIM)[1] are a core dimension of knowledge-based economies (Trewin, 2002). These can improve procurement processes such as: 'approval, design, specification and documentation, as well as the tendering, appointment and contract management stages of a project by increasing data integration and information sharing, as well as reducing design and documentation shortcomings' (Allen Consulting Group, 2010). However, these represent new ways of carrying out traditional tasks, bringing a paradigm shift in the way the industry and organisations deliver projects and manage assets. This requires changes in practice standards, legal arrangements and industry norms (Kraatz and Sanchez, 2016). Effective client leadership is needed to overcome the challenges brought by this change (Dossick and Neff, 2008; 2010; Brown et al., 2006). This chapter will explore some of the roles that public clients can play in achieving information integration across software domains and organisational boundaries as well as the successful implementation of whole-of-life industry benchmarking systems. It will do this by drawing from Australian examples to provide valuable insight into these issues.

The role of public clients

Public clients can reduce tensions, empower particular players and lower the transaction costs associated with the uptake of new technologies, which puts them in an empowered position to promote change (Sørensen and Torfing, 2009; Hovik and Vabo, 2005). Furthermore, as long as the central and regional objectives are broad enough to permit local adjustments and amendments, they can facilitate and structure policy interactions (Sanchez et al., 2014a).

Public clients can promote access to different skills and enable actors to learn new ones. They can do this through collaborative networks, directly promoting new technologies and providing conditions necessary for organisational change to occur (Damgaard and Torfing, 2010; Keast and Hampson, 2007). They are engines of change that can create flows of knowledge, resources and motivation across public and private networks (Duyshart et al., 2003). While doing this, public clients have the benefit of maintaining the 'ability to influence the scope, process and outcomes of networked policy-making' (Sørensen and Torfing, 2009). They can therefore spread and accelerate change beyond their organisational boundaries through the choices they make while procuring services (Bonham, 2013). The following sections provide examples of three types of clients taking a leadership role to drive whole-of-life information integration.

Informed/ expert client

The Sydney Opera House (SOH) is an iconic Australian building that has been estimated to have a replacement value of AUD4.6 billion (USD3.5 billion)[2] and contribute AUD775 million (USD583 million) annually to the Australian economy (Deloitte, 2013). The SOH has a long-standing history of innovative information management. This history started with a challenging design and construction process, which prompted what could be the first field-to-finish system for surveyors in Australia, creating great efficiency gains. It now continues with the implementation of what is expected to be a fully integrated BIM for asset management interface (Linning et al., 2016).

The SOH team has taken an informed and expert client approach to the development of their BIM guidelines and requirements. They have carried out extensive research into international and national practices, significant stakeholder engagement and collaboration, as well as building and maintaining close ties with industry research groups. Within the industry, this role is widely recognised and acknowledged by frequently inviting SOH representatives to speak at conferences about BIM for asset and facility management. They are also often at the centre of BIM-related publications as an exemplar case for BIM implementation (CRC for Construction Innovation, 2007a; CRC for Construction Innovation, 2007b; Sanchez et al., 2015a; Schevers et al., 2007; Arayici et al., 2012).

The SOH has taken this proactive approach for several reasons:

- the SOH embarked on their BIM journey in the early 2000s when the industry was still discovering the potential behind BIM. At the time, they started developing their BIM for facilities management (BIM4FM) strategy in collaboration with the Australian Cooperative Research Centre (CRC) for Construction Innovation and found that the industry was lagging behind in providing BIM products for the operations phase;
- as a well-visited heritage-listed performance venue and architecturally complex iconic facility, the SOH considered that an off-the-shelf application would not be suitable to meet their needs as client–operator;
- due to its heritage status, the SOH required the development of a strategy that would mean the system could use legacy databases as well as withstand the test of time so the data can still be accessed in 100 years and beyond. They considered that the only way to ensure this level of interoperability and integration into the future and past would require a well-informed in-house developed strategy.

Visionary client

The Perth Children's Hospital (PCH) is an AUD1.2 billion project carried out under a two-stage managing contract model between the Government of Western Australia and constructor John Holland. The project used BIM for the design and

234 *Sanchez et al.*

construction of the hospital and required a facilities management BIM model as a key deliverable. This hospital is due to open in late 2016 and will become Western Australia's dedicated children's hospital providing best possible clinical care and paediatric research. It forms one of the cornerstones of the Western Australian state government strategy to deliver major social infrastructure for future generations. Though this project has been the subject of some political and media criticism, particularly regarding the relationship between the contractor and sub-contractors and problems with some suppliers' quality assurance, these criticisms have been unrelated to the BIM tools and processes implemented.

Although the Western Australian Government (WAG) did not have a high level of experience in the use of BIM at the start of the project, the team endeavoured in carrying out extensive consultation to develop their strategy and help drive broader industry development. Here the public client aimed to set an example to other Western Australia organisations and the industry itself; to demand more of its project delivery practices by demonstrating the benefits that can be realised by embracing modern ICT technology. WAG's key driver was the belief that owners are the ultimate beneficiaries of such an embrace, and that it would help achieve efficiency gains across the life-cycle of their new hospital asset. WAG decided they had to remain involved in the development process as well as lead and facilitate change until it becomes self-sustaining. This would allow them to leverage a wider range of benefits during all life-cycle phases and across other assets in their portfolio (Sanchez et al., 2015b).

Performance-driven client

The New Generation Rollingstock (NGR) project is an AUD4.4 billion project that will increase South East Queensland's passenger rail fleet by 30 per cent. By the expected commencement date for full operations in December 2018, at least 50 per cent of all operational trains in the region will be NGR trains. The purpose-built train maintenance depot is located at Wulkuraka, west of Brisbane, Queensland. The Queensland Department of Transport and Main Roads (QTMR) awarded the contract to an international consortium comprising Bombardier Transportation, John Laing, ITOCHU Corporation and Uberior. Laing O'Rourke was the design and construct contractor for the Rollingstock Maintenance Depot using a range of BIM tools and related processes.

QTMR is taking an incremental approach to learning about the impact of using BIM for their procurement process through a series of pilot projects of increasing complexity. The BIM implementation strategy of the NGR, however, was initiated by the project delivery team in an effort to satisfy performance-driven objectives set by the client and as a cornerstone to its own group's ongoing operation and maintenance business requirements. Bombardier will act as concessionaire for part of the operational life of the asset through a 32-year exclusive servicing agreement and is therefore also invested in QTMR's performance-driven goals. These goals relate to cost savings in construction and operation, risk sharing and project schedule requirements. Implementing BIM was also used to help document progress payments and meeting design obligations early enough so that full payments were made based on timely task achievements (Utiome et al., 2015).

The role of benchmarks in public procurement

Benchmarks are the result of a systematic process using established performance and value metrics[3] for measuring and comparing individuals, teams, organisations and industries across business activities. This process allows lessons learned from others and challenges set from within to be used to establish improvement targets and to promote changes across organisations and the industry as a whole (Costa et al., 2006). Within the BIM context, benchmarks often combine quantitative and qualitative measures 'of the "hard" and "soft" aspects of BIM' (Sebastian and van Berlo, 2010).

The key benefit of benchmarking systems is that they help answer the question 'has a change actually made any difference, and what sort of difference has been made?' and leverages off the concept of 'what gets measured, gets done'. This process allows lessons learned from others to be used to establish improvement targets and to promote changes within the organisation and industry (Costa et al., 2006). Benchmarks can be internal when relating to a single organisation or external when referencing other organisations. They can be used at a project level within or across life-cycle phases, at a regional or even at a global level (CURT, 2005).

In the coming decades, countries will face ongoing challenges as a result of economic and political volatility, changing climate and technological changes. Throughout the life-cycle of community assets, new technologies such as BIM can provide benefits that are not yet being taken full advantage of and concurrently address challenges associated with managing assets in these circumstances. This is especially the case for constructed community assets where key performance indicators (KPIs) relevant to benefits from using BIM could address long-term impact factors which are transferable throughout the asset life-cycle and across asset portfolios (Sanchez and Joske, 2016).

Although public domain performance data are important, the industry's reluctance to make these public is a barrier to achieving a more economically and environmentally sustainable built environment (Tuohy and Murphy, 2015). In the construction industry, there have been many efforts to establish productivity benchmarks. For example, in 2008 the American National Institute of Standards and Technology (NIST) launched 'a multiyear, collaborative research effort that aims to supply the measurement science needed to bring major gains in construction productivity' (Suermann, 2009). The Construction Industry Institute (CII) also has the CII Performance Assessment programme and associated system (Construction Industry Institute, 2016). Brazil, Chile, Hong Kong, the United Kingdom (UK) and other countries have also sought to develop construction benchmarks (Du et al., 2014). However, benchmarking information continues to be disappointingly limited, a deficiency which the Australian

236 *Sanchez et al.*

Productivity Commission said 'must be addressed in a future structure for infrastructure decision making as a whole' (Australian Government Productivity Commission, 2014).

Privately, individual organisations develop their own internal benchmarks to measure their performance through metrics such as returns on investment. However, organisations may also need to extend their evaluation beyond their legal boundaries and therefore require benchmarking strategies that are aligned with their value-adding networks (Lockamy and McCormack, 2004). Having access to standard metrics that may form the basis of future internal benchmarks as well as industry benchmarks would help resolve this issue. The existence of public benchmarks may also help the industry as a whole to understand the gaps and target specific processes to increase performance in a more direct and effective manner (Tuohy and Murphy, 2015).

BIM benefits benchmarking – an Australian case study

The challenges associated with measuring benefits from investment in ICT systems are not unique to the construction industry, but are a global issue experienced in all types of business sectors and organisations (Becerik and Pollalis, 2006). These are however additionally complicated by the fact that each construction project is different in terms of its parameters and characteristics, such as financing and delivery method, site characteristics, inter-organisational relationships and end-user requirements. The construction industry also lacks established cross-organisational benchmarks to build on or measure against. Its fragmented supply chain and undercapitalisation further complicate the development and implementation of meaningful metrics that can be used for industry benchmarks (Becerik and Pollalis, 2006; Becerik-Gerber and Rice, 2009).

Confidentiality in contractual requirements also remain a large impediment to data collection and therefore to validating developed metrics (Becerik and Pollalis, 2006) as well as to making the data public. While this issue is also not unique to the construction industry, it does represent a significant barrier to establishing industry benchmarks. Tuohy and Murphy (2015) for example discuss the building industry's reluctance to publish actual building performance data. They further advocate for political leadership in making the publication of such data mandatory as it is in the motor and electronics industry.

Additionally, the involvement of multiple stakeholders can increase the complexity of the data-gathering exercise and resources required to measure it (Furneaux et al., 2010). It may introduce errors due to different interpretation of metrics and levels of data accuracy. It is also difficult to establish benchmarks that are simple enough, useful to all stakeholders and that encompass all the relevant dimensions required to quantify the issue being benchmarked (Morris et al., 2006).

To address this challenge, a national group of researchers in Australia is developing the online open access tool 'BIM Value Benchmark'. This effort, through the Sustainable Built Environment National Research Centre (SBEnrc), successor to the Australian CRC for Construction Innovation, builds on industry research carried out to develop an interactive BIM Value information delivery tool (SBEnrc and NATSPEC BIM, 2016) and draws from crowdsourcing principles to overcome the challenges outlined earlier in this chapter. The following sections provide some background to this research and what is planned to be delivered in 2017.

Crowdsourcing industry benchmarks

Crowdsourcing is an increasingly popular and powerful technique used to complete complex tasks and collect and manage large amounts of data (Pierce and Fung, 2013; Amsterdamer and Milo, 2014). The term has traditionally been linked to sourcing labour for technological advance of information-based industries provided by the public either free or for a lower rate than paying a traditional employee (Howe, 2006). The concept has, however, continued to evolve as new ways to engage the public in creating new services have been found. Doan et al. (2011) for example define a system that uses crowdsourcing as one that 'enlists a crowd of users to explicitly collaborate to build a long-lasting artefact that is beneficial to the whole community'. This definition views crowdsourcing as a general-purpose problem-solving method.

These systems face similar challenges to those hindering the development of benchmarking systems for BIM benefits in the built environment industry. These include how to recruit and retain users, what contributions users can make, how to combine user contributions, and how to evaluate users and their contributions (Doan et al., 2011).

There are many types of crowdsourcing systems, all varying in objective and approach to sourcing data. Some systems are based on explicit forms of collaboration while others are based on implicit collaboration; some are stand-alone while others are linked to other systems; some are 'open world' and others are 'closed world', which has to do with the amount of data available and how queries are sourced; and in some cases, crowdsourcing focuses on humans capturing or generating new data, while in others it focuses on ranking and enhancing existing data (Franklin et al., 2011; Parameswaran et al., 2012; Doan et al., 2011). A stand-alone implicit system may for example provide a service that allows users to collaborate indirectly to solve a problem when using it (Doan et al., 2011). Some well-known examples include websites such as Wikipedia and OpenStreetMap (OSM) as well as software such as Linux.

Users can be motivated by different types of needs or drivers. This may be an economic need that is met by payment, a social need met through achieving recognition, and in more specific cases, it may be the opportunity to develop a skill or use the service for some other reason (Estellés-Arolas and González-Ladrón-de-Guevara, 2012). Perhaps one of the best-known services that utilises the data of their users to answer intricate questions is Facebook. In this sense, Facebook provides a platform for a billion users globally that actively and daily share their statuses, post photos, exchange messages, chat in real time, check-in to physical locales, play games and shop. In return, the organisation has access to

238 *Sanchez et al.*

an unprecedented wealth of data that, when aggregated and mined, provides them with valuable insight into the intricacies of social and consumer behaviour (Russell, 2013).

The idea behind the Australian-developed BIM Value Benchmark tool is to provide a service to industry by developing an internal benefit benchmarking tool that is free and open access. The tool will include functionalities that allow the user to select from a range of metrics that can be associated to specific benefits and BIM uses and processes. These metrics will then be visualised in different ways and produce reports that can help organisations piloting new BIM tools and processes to build a business case for wider use. It will also allow comparing performance across a portfolio of projects or assets and including contextual information that may help better understand why some projects perform better than others. The data introduced will be saved by the server in an aggregated and unidentifiable format. Once a critical mass of users has been achieved, this data can be used to publish valuable industry benchmarks to compare against. The tool also will allow users to provide different levels of access to a number of stakeholders to view, edit and add data.

Potential for procurement

A concept that is common among Australian clients is 'value for money'. In the transport sector, for example, it is referred to as 'contribution to the advancement of Government Priorities and cost related factors including whole-of-life costs' (QTMR, 2009); as any feature that provides a benefit to the government and the community (MRWA, 2012); and as satisfying assessment criteria as well as providing additional details at the lowest price and 'in the spirit of cooperative contracting' (NSW RMS, 2011). A barrier for risk-adverse clients to driving the use of BIM and a more whole-of-life integrated approach to information management is the lack of industry-wide shared data to prove this value for money.

The BIM Value Benchmark tool would allow clients to have a single platform where all the data related to benefits of new investments on different BIM tools and processes can be centrally stored and analysed across time and the complete portfolio. In the long term this would mean having both an internal benchmarking system and access to industry benchmarks encompassing a large set of performance metrics. These can then be used as a basis for future investment decisions and lessons learned. This tool will allow the user to have a continuous record of specific metrics across the life-cycle of a built environment asset and the impact of introducing different BIM-related tools and processes. In the short-term, it means having an open access tool that allows ongoing monitoring of progress towards specific organisational performance goals and their relation to different tools and processes being implemented. This could be used, for example, for project review meetings, allowing their teams to make adjustments to the implementation strategy as required.

Additionally, the research used to develop the tool highlighted that, to allow the system to evolve as technology changes and maintain user participation, it was necessary for users to be able to introduce their own performance metrics. These could then be proposed to the hosting organisation to be standardised into the public list of metrics offered by the tool. This means that although the system is being created for the purpose of monitoring and benchmarking benefits from implementing BIM, it could potentially be expanded to monitor and benchmark the introduction of virtually any change to the standard practice. Philp and Thompson (2014) proposed in their 'Built Environment 2050' industry development strategy document an evolutionary path through digitising the built environment industry and integrating information across assets. This report published by the UK Construction Industry Council predicted innovations, such as industrial 3D printing, self-procuring and robotics, which promise great productivity gains for the industry. Having access to such long-lasting performance benchmarks may be very useful in assessing the impact of these innovations to come.

Role of public clients

Although the knowledge about BIM-related benchmarking and crowdsourcing systems' success is limited, there are some indications about what factors may affect it. Recruiting participants is perhaps one of the most challenging factors to the success of systems that depend on the input from crowds. One of the main strategies that has been suggested to drive usage is an authority requiring participation, such as public clients requiring their service providers to use a certain system or provide information for it (Doan et al., 2011).

Another way in which public clients can help promote such a tool in order to create a benchmark that would benefit the industry as a whole is by endorsing it to some extent. Trust in the system is an important factor contributing to usage. Users tend to provide more accurate data if they trust the system (Zheng et al., 2011). This trust can 'play a role in overcoming the potential barriers related to sharing knowledge openly with predominantly unfamiliar people' (Kosonen et al., 2013). However, perhaps one of the most crucial challenges that organisations face in crowdsourcing is 'creating the environment of mutual trust between the crowd and the organisation itself' (Jain, 2010). This trust can be achieved by association to a sponsoring agency who is considered as trustworthy by the target community or by knowing other users who are well respected within the community (Zheng et al., 2011).

Trust in the system is also reliant on perceived properties of the system itself or the institution that hosts it (Kosonen et al., 2013). BIM Value Benchmark will be hosted by NATSPEC, an Australian not-for-profit organisation that aims to improve the construction quality and productivity of the built environment through leadership of information. NATSPEC already hosts a previous SBEnrc-developed online tool: BIM Value. The new tool is also being developed in collaboration with some of Australia's largest public clients in transport infrastructure and buildings. The challenge for the industry, government and research team will be to promote trust in the system based on this support and the

240 *Sanchez et al.*

system functionalities to achieve the critical mass of users required to produce meaningful industry benchmarks.

Limitations of the tool

The manual effort required for contributors to use digital platforms depends on the level of automation of the system and can be a significant barrier to adoption of such crowdsourced benchmarking systems (Doan et al., 2011; Uchoa et al., 2013). The lack of a well-established comprehensive set of standard performance evaluation metrics that can be used for benchmarking purposes means that much of the data required for the metrics proposed in BIM Value Benchmark is not currently being gathered. Although the tool's original design would allow users to directly import and export to basic standard formats from a variety of project and asset management tools, if the data is not created to begin with, then, regardless of its interoperability, users will have to gather and input the data manually. This has the potential to limit its use unless standard procedures are put in place so the required data can be readily collected and imported into the tool.

One of the options for future development of the tools is to follow a similar path to that of the BIM Cloud Score (BIMCS) tool proposed by Du et al. (2014), created to benchmark an organisation's BIM implementation progress. In this case, a BIM plugin that automatically logs the required data from common BIM tools was developed to work in conjunction with the BIMCS tool (Du et al., 2014). Creating a plugin to automatically collect data from common BIM software would save users time, reduce errors and importantly, reduce manual effort.

Conclusions and future research priorities

The UK Construction Industry Council has predicted that the adoption of digital built environments is just the beginning of a complete shift in the way that societies and industries operate (Philp and Thompson, 2014). These new technologies have the potential to help countries deliver more resilient and sustainable infrastructure and better manage existing community assets. To be prepared for this, it is important that clients understand the potential consequences of adopting these new technologies and include these considerations into future implementation and procurement strategies.

This chapter has introduced different roles that public clients can play as well as their motivation for enabling information integration across organisational boundaries through the use of BIM.[4] Later sections explored the potential role of internal and industry benchmarks of benefits achieved from BIM in public procurement and presented a case study from Australia. This collaborative industry research example has explored the use of crowdsourced systems to overcome traditional barriers to the creation of comprehensive performance benchmarks in the built environment industry.

Further research and industry collaboration is required into the validation processes for crowdsourced data that can be used for BIM benchmarking systems that allow organisations to share performance data across organisational boundaries, enabling industry benchmarks to be published. Additional research is required into how ICT technologies and processes already in place can be used to automate the collection of this data.

Notes

1 Building Information Modelling (BIM) is 'a digital process that encompasses all aspects, disciplines and systems of built assets within a single virtual model' (Sanchez et al., 2016).
2 AUD1 = USD0.7521.
3 BIM value metrics are those that can be used to measure organisational performance improvement across a number of fields to support investment decisions, compare and rank areas of return on investment, and provide targets for success and benchmarking. These can be used across life-cycle phases to measure benefits of implementing BIM over the whole-of-life of built environment assets (Sanchez and Hampson, 2016).
4 BIM in this publication includes other associated terms such as Digital Engineering and virtual design and construction (VDC).

References

Allen Consulting Group, 2010. *Productivity in the Buildings Network: Assessing the Impacts of Building Information Models*, Sydney: BEIIC.

Amsterdamer, Y. and Milo, T., 2014. Foundations of crowd data sourcing. *ACM SIGMOD Record*, 43(4), 5–14.

Arayici, Y., Onyenobi, T. C. and Egbu, C. O., 2012. Building Information Modelling (BIM) for facilities management (FM): The MediaCity case study approach. *International Journal of 3D Information Modelling*, 1(1), 55–73.

Australian Government Productivity Commission, 2014. *Public Infrastructure. Productivity Commission Inquiry Report. Volume 2*, Canberra: Australian Government.

Becerik, B. and Pollalis, S. N., 2006. *Computer Aided Collaboration in Managing Construction*, Design and Technollogy Report Series 2006-2, Cambridge, MA: Harvard University Graduate School of Design.

Becerik-Gerber, B. and Rice, S., 2009. The value of Building Information Modeling: Can we measure the ROI of BIM? *AECbytes*, August 31, 47.

Bonham, M. B., 2013. Leading by example: New professionalism and the government client. *Building Research & Information*, 41(1), 77–94.

Brown, K., Hampson, K. and Brandon, P., 2006. *Clients Driving Construction Innovation: Moving Ideas to Practice*. Brisbane: Icon.Net Pty Ltd.

buildingSMART, 2010. *Investors Report: Building Information Modelling (BIM)*, London: UK Department of Business Innovation and Skills and buildingSMART.

Construction Industry Institute, 2016. *CII Performance Assessment System*. Available at: www.construction-institute.org/nextgen/index.cfm [Accessed 18 January 2016].

Costa, D. B., Formoso, C. T., Kagioglou, M. and Alarcón, L. F., 2006. Benchmarking initiatives in the construction industry: Lessons learned and improvement opportunities. *Journal of Management in Engineering*, 22(4), 158–167.

242 Sanchez et al.

CRC for Construction Innovation, 2007a. *Adopting BIM for Facilities Management: Solutions for Managing the Sydney Opera House*, Brisbane: Cooperative Research Centre for Construction Innovation.

CRC for Construction Innovation, 2007b. *FM as a Business Enabler*, Brisbane: Cooperative Research Centre for Construction Innovation.

CURT, 2005. *Construction Measures: Key Performance Indicators*, Cincinnati, OH: The Construction Users Roundtable.

Damgaard, B. and Torfing, J., 2010. Network governance of active employment policy: The Danish experience. *Journal of European Social Policy*, 20(3), 248–262.

Deloitte, 2013. *How Do you Value an Icon? The Sydney Opera House: Economic, Cultural and Digital Value*, Sydney: Deloitte.

Doan, A., Ramakrishnan, R. and Halevy, A. Y., 2011. Crowdsourcing systems on the world-wide web. *Communications of the ACM*, 54(4), 86–96.

Dossick, C. and Neff, G., 2008. *How Leadership Overcomes Organizational Divisions in BIM-enabled Commercial Construction*, paper presented at LEAD – The Engineering Project Organisation Society (EPOS) Lake Tahoe, US, 16–19 October.

Dossick, C. S. and Neff, G., 2010. Organizational divisions in BIM-enabled commercial construction. *Journal of Construction Engineering and Management*, 136(4), 459–467.

Dovers, S., 2005. *Environment and Sustainability Policy: Creation, Implementation, Evaluation*. Leichhardt, Australia: The Federation Press.

Du, J., Liu, R. and Issa, R. R., 2014. BIM cloud score: Benchmarking BIM performance. *Journal of Construction Engineering and Management*, 140(11), 1–13.

Duyshart, B., Walker, D. H., Mohamed, S. and Hampson, K. D., 2003. An example of developing a business model for information and communication technologies (ICT) adoption on construction projects – the National Museum of Australia project. *Engineering Construction and Architectural Management*, 10(3), 179–192.

Estellés-Arolas, E. and González-Ladrón-de-Guevara, F., 2012. Towards an integrated crowdsourcing definition. *Journal of Information Science*, 38(2), 189–200.

Franklin, M. J., Kossmann, D., Kraska, T., Ramesh, S. and Xin, R., 2011. *CrowdDB: Answering queries with crowdsourcing, Proceedings of ACM SIGMOD/POD Conference*, Athens, Greece, 12–16 June, pp. 61–72.

Furneaux, C. W., Hampson, K. D., Scuderi, P. and Kajewski, S. L., 2010. *Australian Construction Industry KPIs*, paper presented at CIB World Congress – Building a Better World, Salford Quays, United Kingdom, 10–13 May, pp. 1–12.

Gallaher, M. P., O'Connor, A. C., Dettebarn, J. L. and Gilday, L. T., 2004. *Cost Analysis of Inadequate Interoperability in the U.S. Capital Facilities Industry, NIST GCR 04-867*. Available at: www.nist.gov/manuscript-publication-search.cfm?pub_id=101287 [Accessed 8 November 2012].

Garcia-Alonso, M. D. C. and Levine, P., 2008. Strategic procurement, openness and market structure. *International Journal of Industrial Organization*, 26, 1180–1190.

Hampson, K. D., Kraatz, J. A. and Sanchez, A. X., 2014. The global construction industry and R&D. In: *R&D Investment and Impact in the Global Construction Industry*. London: Routledge, pp. 4–23.

Hardy, R., 2013. *The Role of Procurement in Embedding Sustainability Along the Life Cycle of a Construction Project*, paper presented at the CIB World Building Congress, Brisbane, Australia, 5–9 May.

Hovik, S. and Vabo, S. I., 2005. Norwegian local councils as democratic meta-governors? A study of networks established to manage cross-border natural resources. *Scandinavian Political Studies*, 28(3), 257–275.

Howe, J., 2006. The rise of crowdsourcing. *Wired Magazine*, 14(6), 1–4.

Jain, R., 2010. Investigation of governance mechanisms for crowdsourcing initiatives. *Proceedings of the Sixteenth Americas Conference on Information Systems*, Lima: AIS Electronic Library (AISeL), pp. 1–8 .

Keast, R. L. and Hampson, K. D., 2007. Building constructive innovation networks: Role of relationship management. *Journal of Construction Engineering and Management*, 33(5), 364–373.

Kosonen, M., Gan, C., Olander, H. and Blomqvist, K., 2013. My idea is our idea! Supporting user-driven innovation activities in crowdsourcing communities. *International Journal of Innovation Management*, 17(3), 1–18.

Kraatz, J. A. and Sanchez, A. X., 2016. Leadership in implementation. In: A. X. Sanchez, K. D. Hampson and S. Vaux (eds), *Delivering Value with BIM: A Whole-of-life Approach*. London: Routledge, pp. 23–45.

Linning, C., Sanchez, A. X. and Hampson, K. D., 2016. Implementation tips with hindsight. In: A. X. Sanchez, K. D. Hampson and S. Vaux (eds), *Delivering Value with BIM: A Whole-of-life Approach*. London: Routledge, pp. 81–101.

Lockamy, A. and McCormack, K., 2004. The development of a supply chain management process maturity model using the concepts of business process orientation. *Supply Chain Management: An International Journal*, 9(4), 272–278.

Morris, J. et al., 2006. *Sydney Opera House – FM Exemplar Project*, Brisbane: Cooperative Research Centre for Construction Innovation.

Main Roads Western Australia (MRWA), 2012. *Tender Submission Document*, Perth: Main Roads Western Australia.

NSW RMS, 2011. *Engineering Contract Manual*, Sydney: New South Wales Roads and Maritime Services.

Parameswaran, A. G., Park, H., Garcia-Molina, H., Polyzotis, N. and Widom, J., 2012. Deco: Declarative crowdsourcing, *Proceedings of the 21st ACM International Conference on Information and Knowledge Management*, Maui, USA, 29 October–2 November, pp. 1203–1212.

Philp, D. and Thompson, N., 2014. *Built Environment 2050: A Report on our Digital Future*, London: Construction Industry Council.

Pierce, C. and Fung, N., 2013. *Crowd Sourcing Data Collection through Amazon Mechanical Turk*, Adelphi, MD: Army Research Lab, Computational and Information Science Directorate.

QTMR, 2009. *Main Roads Project delivery System. Volume 2: Tendering for Major Works*, Brisbane: Queensland Government. Department of Main Roads. Engineering and Technology.

Russell, M. A., 2013. *Mining the Social Web: Data Mining Facebook, Twitter, LinkedIn, Google+, GitHub, and More*. Sebastopol: O'Reilly Media.

Sanchez, A. X. and Hampson, K. D., 2016. BIM, asset management and metrics. In: A. X. Sanchez, K. D. Hampson and S. Vaux (eds) *Delivering Value with BIM: A Whole-of-life Approach*. London: Routledge, pp. 3–22.

Sanchez, A. X. and Joske, W., 2016. Benefits dictionary. In: A. X. Sanchez, K. D. Hampson and S. Vaux (eds), *Delivering Value with BIM: A Whole-of-life Approach*. London: Routledge.

Sanchez, A. X., Kraatz, J. A. and Hampson, K. D., 2014a. *Research Report 1 – Towards a National Strategy*, Perth: Sustainable Built Environment National Research Centre (SBEnrc).

244 *Sanchez et al.*

Sanchez, A. X., Lehtiranta, L. M., Hampson, K. D. and Kenley, R., 2014b. Evaluation framework for green procurement in road construction. *Smart and Sustainable Built Environment*, 3(2), 153–169.

Sanchez, A. X., Hampson, K. D. and Mohamed, S., 2015a. *Sydney Opera House: Case Study Report*, Perth: Sustainable Built Environment National Research Centre.

Sanchez, A. X., Hampson, K. D. and Mohamed, S., 2015b. *Perth Children's Hospital: Case Study Report*, Perth: Sustainable Built Environment National Research Centre.

Sanchez, A. X., Hampson, K. D. and Mohamed, S., 2016. Delivering value with BIM: A framework for built environment practitioners, paper presented at CIB World Building Congress, Tampere, Finland, 30 May–2 June.

SBEnrc and NATSPEC BIM, 2016. *BIM Value*. Available at: http://bimvaluetool.natspec.org/ [Accessed 29 February 2016].

Schevers, H., Mitchell, J., Akhurst, P., Marchant, D., Bull, S., McDonald, K. and Drogemuller, R., 2007. Towards digital facility modelling for Sydney Opera House using IFC and semantic web technology. *ITcon*, 12, 347–362.

Sebastian, R. and van Berlo, L., 2010. Tool for benchmarking BIM performance of design, engineering and construction firms in the Netherlands. *Architechtural Engineering and Design Management*, 6(4), 254–263.

Sørensen, E. and Torfing, J., 2009. Making governance networks effective and democratic through metagovernance. *Public Administration*, 87(2), 234–258.

Suermann, P. C., 2009. Evaluating the impact of Building Information Modelling (BIM) on construction – Doctoral Thesis, Gainesville: University of Florida.

Trewin, D., 2002. *Measuring a Knowledge-based Economy and Society: An Australian Framework*, Canberra: Australian Bureau of Statistics.

Tuohy, P. G. and Murphy, G. B., 2015. Closing the gap in building performance: Learning from BIM benchmark industries. *Architectural Science Review*, 58(1), 47–56.

Uchoa, A. P., Esteves, M. G. P. and de Souza, J. M., 2013. *Mix4Crowds – Toward a Framework to Design Crowd Collaboration with Science*, paper presented at 2013 IEEE 17th International Conference on Computer Supported Cooperative Work in Design (CSCWD), Whistler, Canada, IEEE, pp. 61–66.

Utiome, E., Mohamed, S., Sanchez, A. X. and Hampson, K. D., 2015. *New Generation Rollingstock Depot: Case Study Report*, Perth: Sustainable Built Environment National Research Centre.

Yates, A., 2012. *Government as an Informed Buyer: How the Public Sector Can Most Effectively Procure Engineering-intensive Products and Services*, Sydney: Engineers Australia.

Zheng, H., Li, D. and Hou, W., 2011. Task design, motivation, and participation in crowdsourcing contests. *International Journal of Electronic Commerce*, 15(4), 57–88.

15 Four metaphors on knowledge and change in construction

Kim Haugbølle

Introduction: Danish context, yet a global problem

Creating a more effective and reflective industry and delivering benefits to public and private asset owners ranks high on the agenda of many policy-makers, industry professionals and academics within the real estate, housing, building and infrastructure design, construction and asset management sectors. This ambition points to the need for a better understanding and management of the creation, accumulation and application of knowledge across the various actors and domains of the built environment as well as across the life-cycle of constructed assets.

This challenge or 'problem of knowledge' is particularly manifest and intriguing with regard to refurbishment of existing buildings and facilities. In most industrialised countries, the construction markets are increasingly turning from new buildings towards refurbishment of existing buildings. Refurbishment entails a somewhat different set of challenges compared to new buildings, which calls for new strategies, knowledge and development of practices among construction professionals, policy-makers and knowledge institutions. Hence, a range of new initiatives and research projects are being launched worldwide to address these challenges such as the European Union's extensive schemes for energy renovation of buildings (European Commission, 2011; IEA, 2013; Buildings Performance Institute Europe, 2014). Despite all of these efforts, the problem of knowledge seems to be a 'sticky' one that does not disappear easily.

Denmark is no exception to this general line of development. Thus, two large private philanthropic foundations recently initiated and funded a think tank to develop a new comprehensive refurbishment strategy for the built environment. The think tank was composed of some 30 members representing all major actors of the built environment. Over the course of a year, this think tank held a set of consultations with experts, professionals and policy-makers to identify relevant challenges and formulate corresponding initiatives to address them. Eventually, this group published its strategy with seven initiatives, among others, related to improving refurbishment statistics and knowledge-sharing, accelerating innovation and strengthening education (Tænketank for bygningsrenovering, 2013). One of the ensuing initiatives encompassed a more substantial analysis of the problem of knowledge. This chapter will report on the results of this analysis.

246 *Haugbølle*

Based on a case study of the Danish knowledge system in construction, this chapter will take a closer look at where, why and how problems occur with regard to knowledge. Although the focus will be on a specific national context, it is the contention that the lessons learned will be relevant to an international audience either because the lessons learned mirror similar lessons in other national contexts or because the lessons learned may function as a kaleidoscope that provokes new fruitful insights. It will take as its starting point that knowledge is not simply about data (e.g. facts or numbers) or information about something (e.g. case studies) or even integration of information (e.g. in databases). It will also assume that the problem of knowledge is not 'just' about knowledge and learning, but rather about putting knowledge into action in order to transform practices. This chapter will identify four different metaphors for knowledge and discuss how these understandings and metaphors articulate knowledge differently as a problem, thus controlling the corresponding solutions. Consequently, this chapter will purposefully transcend the perspective of information integration and focus on the issue of actionable knowledge.

Theoretical framework: SCOT theory

This chapter builds on a background report in Danish (Haugbølle and Storgaard, 2014) and a conference paper by Haugbølle (2016). It adopts a constructivist perspective emphasising that knowledge and human agency are governed by the intimate linkages between technical objects and social relations (see for example Bijker et al., 1987; Bijker and Law, 1992; Oudshoorn and Pinch, 2003). The concept of 'technological frame' was introduced by Bijker (1995) as part of his social construction of technology (SCOT) theory to analyse interactions involving knowledge and technology within and between different relevant social groups. To describe a technological frame, Bijker (1995) proposed a tentative list of elements including goals, key problems, problem-solving strategies, requirements to be met by problem solutions, current theories, tacit knowledge, testing procedures, design methods and criteria, users' practice, perceived substitution function and exemplary artefacts. This list is tentative, as some elements may be irrelevant for some social groups and other elements may need to be added for other social groups. Since actors generally belong to more than one relevant social group, they will be simultaneously involved in different technological frames with different degrees of inclusion (Bijker, 1995).

This chapter will apply the concept of technological frame to better understand the various framings of 'the knowledge problem' in construction and the possible implications thereof. For practical reasons, the analysis will only elaborate on some of the tentative elements of a technological frame including goals, key problems, problem-solving strategies and current theories.

Research design: case study

This chapter builds on a selection of international studies on knowledge to establish an analytical framework. The selection is not a complete or comprehensive literature review, but it does cover a range of the most prominent studies on knowledge that can be considered seminal. The analysis of the national development is based on a comprehensive coverage of written Danish sources combined with qualitative research interviews with seven key representatives of major actors of the construction industry. These include representatives from vocational training, research institutions, information offices, contractors, trade unions, wholesalers, and funders of research and development (R&D).

Flyvbjerg (2006) states that paradigmatic cases develop 'a metaphor or establish a school for the domain that the case concerns'. Identifying a case as paradigmatic is particular challenging as, by their very nature, they transcend any sort of rule-based criteria (Flyvbjerg, 2006). The analysis of the Danish knowledge system on construction and refurbishment may be considered paradigmatic in the sense that it shares a number of characteristics and similarities with other industrialised countries. Denmark in particular shares characteristics with many other countries in the European Union due to the European internal market. These include the distribution of construction output as new build versus refurbishment, the application of the same European rules of procurement and the use of a range of similar construction products. However, there are a number of differences in national building codes and industry structure, among others. These make it less likely to use the Danish knowledge system as an exemplar for construction knowledge systems in general.

Denmark, however, also has elements of a critical case, which makes it possible to draw conclusions or generalisations of the kind that 'if it is valid for this case, it is valid for all (or many) cases' (Flyvbjerg, 2006). These arguments relate to characteristics of the Danish construction industry like Danish construction professionals being considered as highly skilled and knowledgeable, a well-paid workforce with social status, high degree of digital literacy and extensive access to information and communication technology (ICT) solutions. Hence, if problems of knowledge exist in the Danish construction industry, the same kind of problems, and maybe even in an accentuated form, are likely to be found in other national construction industries as well.

Case: the problem of knowledge in Danish construction

This section presents four different metaphors on the problem of knowledge in Danish construction. Each guiding metaphor is briefly introduced followed by a short theoretical overview with reference to some of the seminal works within the perspective, a closer look at relevant activities within Danish construction and a brief summary describing the technological frame guiding interaction within this perspective.

248 *Haugbølle*

Metaphor 1: knowledge as flows

The first metaphor is about knowledge as a resource, where knowledge flows like a river between different actors ('knowledge flows'). Based on such a metaphor solutions tend to look for removing obstacles to ensure an unimpeded flow of data and information.

Proponents of this perspective tend to consider knowledge as isolated islands which explain why knowledge is shared or not. Armed with such a metaphor, solutions tend to map islands of knowledge with the aim of strengthening linkages between them. Hence, isolated islands of knowledge can be bridged through establishing strong or weak ties (see for example Pryke, 2004). Gann and Salter (2000) have adopted a more systemic perspective on construction, where innovation must be understood within the context of both the technological infrastructure for knowledge and the regulatory and institutional environment. The individual elements are connected through knowledge flows. Haugbølle et al. (2012) build on this systemic perspective to also explicitly include the building's users. They suggest that the links between the individual elements should be predominantly understood as policy processes, learning processes and business processes between actors in three different types of markets related to building products, construction services and property.

Considering the problem of knowledge as a matter of removing obstacles to ensure unimpeded flows has been a dominating topic in Danish construction policy. Knowledge-sharing or rather the lack thereof has frequently been suggested to be a core problem for improving the overall performance, productivity and sustainability of construction. This has led to a range of policy analyses along with development initiatives and research studies in Denmark. The policy analyses include, among others:

- the use of technological services in construction (Bang, 1997);
- a survey on learning and knowledge-sharing (Alsted Research, 2003);
- the characteristics of the communication landscape in construction as formulated by the development programme Project House (Christoffersen, 2000);
- the repeated call for improved dissemination of building knowledge (see for example Carlsen et al., 2005).

A number of other policy studies have focused on activities related to R&D in construction. These include, among others, an analysis by the Danish Building Development Council on production, use and dissemination of technical building knowledge (Dræbye, 1997) and several mappings of construction-related R&D activities (see for example Christoffersen and Bertelsen, 1990; Boligministeriet, 1993; Det Offentlige Forskningsudvalg for Byer og Byggeri, 2000; Haugbølle and Clausen, 2002). These were later followed by a public action plan on R&D activities by the Committee on Construction Research in Denmark (Udvalget vedrørende byggeforskning i Danmark, 2002) and a proposal for strengthening research and learning in construction issued jointly by industry, government and knowledge institutions (Koordinations- og initiativgruppen for viden i byggeriet, 2009).

The majority of these policy studies on the knowledge problem in Danish construction have focused on how new research-based knowledge can come to be used in project-based companies in construction. Most of these policy studies criticise in particular the universities and similar knowledge institutions for not disseminating new knowledge appropriately to the project-based firms of construction, notably contractors and consultants. Thus, the preferred solution to the knowledge problem is to improve dissemination from the universities and the like through, for example, mandatory dissemination plans for R&D projects and joint information services as a sort of one-stop-shop solution (Figure 15.1).

Such a perspective is, however, too narrow and restrictive. As observed by Haugbølle (2015), the vast majority of R&D activities are not carried out in public research institutes but rather in manufacturing industries.

> Second, the level of R&D expenditures at the core public institutions in 2001 is estimated at around DKK175 million (USD 30 million) plus DKK100–150 million (USD17–26 million) at other public research institutions, totalling approximately DKK275–325 million (USD48–56 million). The R&D expenditures in private companies within the building/ housing resource area were estimated at DKK1.2 billion (USD207 million) corresponding to the R&D efforts in primary and manufacturing industries, and half of the calculated R&D efforts in support and service industries. This estimate is marked by significant uncertainty due to the methodology applied in collating these statistics. Although the exact figures may be disputable, private R&D investments primarily take place in the manufacturing industry.
>
> (Haugbølle, 2014)

Hence, the knowledge flows to construction projects are predominantly linked to knowledge generated by manufacturing companies who either incorporate this knowledge into their materials and products or provide various types of information services through, for example, information offices or wholesalers. In recent years, Silvan (a major wholesaler) has developed an extensive information service through its website. This web service has proven to be highly successful and attractive for both end-users and contractors with more than 11 million unique users per year; this is twice the number of inhabitants of the country (Figure 15.2).

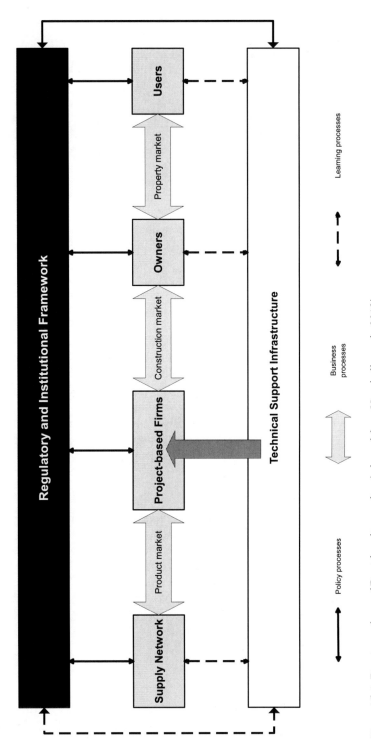

Figure 15.1 Dominant focus of Danish policy studies (adapted from Haugbølle et al., 2012).

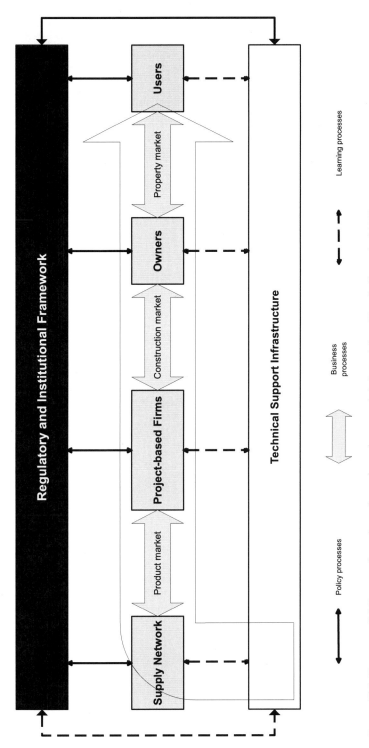

Figure 15.2 Dominant R&D stems from the industrial supply network (adapted from Haugbølle et al., 2012).

252   *Haugbølle*

Table 15.1 Technological frame: knowledge as flows

	Knowledge as flow
Goals	Improve generation and application of knowledge
Key problems	Knowledge as flow
Problem-solving strategies	Remove obstacles impeding the flow
Current theories	Resource-based theories
Users' practice	Fragmented construction sector
Exemplary artefacts	Web-based knowledge centres

Table 15.1 provides an overview of the main elements in the technological frame considering knowledge as flows.

Metaphor 2: knowledge as capabilities

The second metaphor relates to knowledge about knowledge; it considers knowledge as a capability by an organisation to adopt knowledge. Armed with such a metaphor, solutions tend to focus on managerial efforts to improve the adoption of new knowledge in an organisation.

This perspective has a number of important strains of thinkers gathered under such headings as knowledge management and business intelligence. Knowledge management deals with collecting, developing, distributing and applying knowledge within organisations. One of the most important and seminal concepts in this context is 'absorptive capacity'. Cohen and Levinthal (1990) define this ability to absorb knowledge in the following way:

> The ability of a firm to recognize the value of new, external information, assimilate it, and apply it to commercial ends is critical to its innovative capabilities. We label this capability a firm's absorptive capacity and suggest that it is largely a function of the firm's level of prior related knowledge.
>
> (Cohen and Levinthal, 1990)

The development of a firm's absorptive capacity depends on both the individual's and organisation's ability to adopt knowledge, which is both history dependent and path dependent. Cohen and Levinthal (1990) further point out that the firm's ability to adopt new knowledge depends on the communicative structures inside and outside the firm, including the nature and distribution of expertise within the organisation:

> Absorptive capacity refers not only to the acquisition or assimilation of information by an organization but also to the organization's ability to exploit it. Therefore, an organization's absorptive capacity does not simply depend on the organization's direct interface with the external environment. It also depends on transfers of knowledge across and within subunits that may be quite removed from the original point of entry. Thus, to understand the

Four metaphors on knowledge and change 253

sources of a firm's absorptive capacity, we focus on the structure of communication between the external environment and the organization, as well as among the subunits of the organization, and also on the character and distribution of expertise within the organization.

(Cohen and Levinthal, 1990)

In project-based industries such as the construction sector, the absorptive capacity is not solely linked to the firm as such, but as well to the often many projects being managed by the firm. Hence, the interplay between the firm and the individual projects is crucial for understanding and dealing with the problem of knowledge in construction as pointed out by for example Winch (1998; 2010). Projects are not only vital for both new construction and refurbishment but the very organising principle of construction firms. The uniqueness of projects is often used as an explanation for why it is difficult to share knowledge and create change in construction. Despite the widely-held assumption of projects being both unique and isolated events, projects are not isolated islands as Engwall (2003) points out. Instead projects are embedded in an organisational context and with a historicity marked by previous projects, parallel courses of events and ideas about the post-project future. Following Engwall (2003), knowledge in projects is not isolated or necessarily unique, but is tied up in experience from previous projects as well as project-independent factors such as general business policies that go beyond the individual project. In this way, previous experience and general policies are setting the scene for future projects in a path dependency.

With regard to the use of the concept of absorptive capacity in project-based firms in construction, these firms will need to develop their capacity in at least four dimensions: upstream and downstream in the value chain as well as towards regulatory frameworks and knowledge institutions (Figure 15.3).

In most Danish policy studies, there is a strong rhetoric on the project-based nature of construction and the barriers to knowledge-sharing and change created by this very nature. However, especially the capabilities and interaction within each organisation have often been neglected in Danish policy studies, even though this is central to a firm's ability to adopt new knowledge. Hence, a range of possible interventions related to the contingencies influencing the interior process dynamics of a project tend to be ignored. While these contingencies may be overlooked in policy studies, some of the major consulting and contracting firms are addressing these in their internal training and course activities on management as well as development activities. Hence, in recent years, extensive efforts have also been put into developing and applying Building Information Modelling (BIM) in particular among consultants, while some of the major Danish contractors have preferred to establish VDC (virtual design and construction) labs as physical units and applying a range of tools, including BIM.

Management and learning within and across projects is only one side of the knowledge problem in construction. Another side of the problem is related to the type of knowledge that circulates in construction. One of the most important and seminal works on knowledge types was done by Nonaka and Takeuchi (1995),

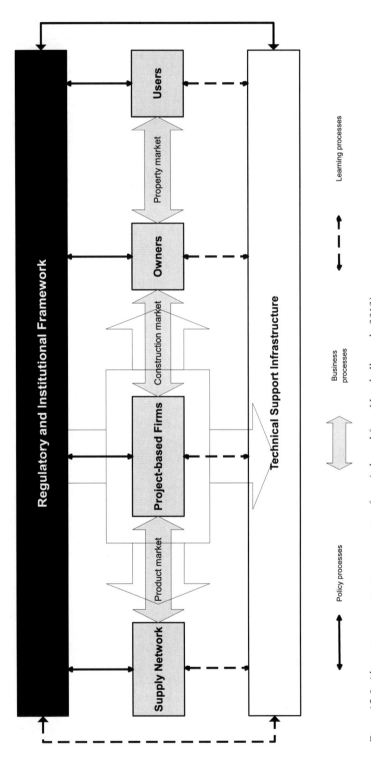

Figure 15.3 Absorptive capacity in construction firms (adapted from Haugbølle et al., 2012).

who studied knowledge and innovation in Japanese companies and developed the so-called SECI model[1] (also referenced by the Bougrain chapter in this book). The SECI model analyses the interaction and transformations between tacit and explicit knowledge through four transformative processes: socialisation, externalisation, combination, and internalisation (see Figure 11.1). The SECI model emphasises that both forms of knowledge and the conversion between them are important issues that construction professionals need to reflect upon, as both explicit and tacit knowledge are strongly prevalent among the different actors of construction.

Several firms and trades are typically involved in the construction process implying numerous and repeated shifts between tacit knowledge and explicit knowledge. Being able to handle these shifts to accumulate, disseminate and apply knowledge in practice is probably one of the construction industry's main challenges with regard to the problem of knowledge. As pointed out bluntly by some Danish practitioners: first, the client articulates his needs for and requirements of a new building or refurbishment project. These may be available in a more or less explicit or tacit form. The consultant then turns these needs and requirements into formal documents such as drawings and specifications. This explicit knowledge is then adopted and adapted into tacit knowledge, which is the primary knowledge base among the craftsmen in construction companies. Finally, the appropriateness of the services delivered by the building is assessed by the end-users against their combined tacit and explicit knowledge of the activities taking place in the building. Table 15.2 provides an overview of the main elements in the technological frame considering knowledge as capabilities.

Metaphor 3: knowledge as practices

The third metaphor views knowledge as an expression of practices. In such optics, the solutions will tend to focus on understanding how practices develop and initiate a diverse set of training and learning activities.

The perhaps most prominent proponent of this perspective is Etienne Wenger who developed the term 'communities of practice' in the early 1990s following his previous work with Jean Lave. Communities of practice have in recent years received considerable attention in the context of organisational development

Table 15.2 Technological frame: knowledge as capabilities

	Knowledge as capability
Goals	Increase innovation and improve business
Key problems	Knowledge as capabilities
Problem-solving strategies	Improve the absorptive capacity
Current theories	Knowledge management
Users' practice	Path-dependency and historicity
Exemplary artefacts	BIM
	VDC

256   *Haugbølle*

and efforts to improve knowledge-sharing in organisations. Wenger (1998) defines learning as an active participation in the practices of a community and forming of identity in relation to these communities. He further states that 'a social theory of learning must therefore integrate the components necessary to characterize social participation as a process of learning and of knowing'. These components include the following:

1   Meaning: 'a way of talking about our (changing) ability – individually and collectively – to experience our life and the world as meaningful'.
2   Practice: 'a way of talking about the shared historical and social resources, frameworks, and perspectives that can sustain mutual engagement in action'.
3   Community: 'a way of talking about the social configurations in which our enterprises are defined as worth pursuing and our participation is recognizable as competence'.
4   Identity: 'a way of talking about how learning changes who we are and creates personal histories of becoming in the context of our communities'.

(Wenger, 1998)

Communities of practice have many different forms, but share three key characteristics related to mutual engagement, a joint enterprise and a shared repertoire (Wenger, 1998):

- *mutual engagement* is a source of coherence of a community. Whatever enables engagement is an essential component of any practice. What makes engagement in practice possible and productive is as much a matter of diversity as a matter of homogeneity. Participants are connected to each other in diverse and complex ways through a shared practice;
- *joint enterprise* is the second source of coherence. It is the result of a collective process of negotiation reflecting the full complexity of mutual engagement. The joint enterprise is defined in the very process by the participants as a response to their situation. It is not a stated goal, but creates relations of mutual accountability among the participants;
- *shared repertoire* is the third source of coherence. The elements of the repertoire can be very heterogeneous. The repertoire is a set of shared resources that reflects a history of mutual engagement and remains inherently ambiguous, which can be reutilised to generate new meanings and effects.

The construction industry with its diverse and fragmented industry structure, many different professions, work gangs, etc., implies that many different communities of practice exist (Figure 15.4). Learning and knowledge-sharing across communities of practice, as is the case with the many trades and professions within construction, present challenges because of the very characteristics that are also their strengths. The internal autonomy of the communities, their

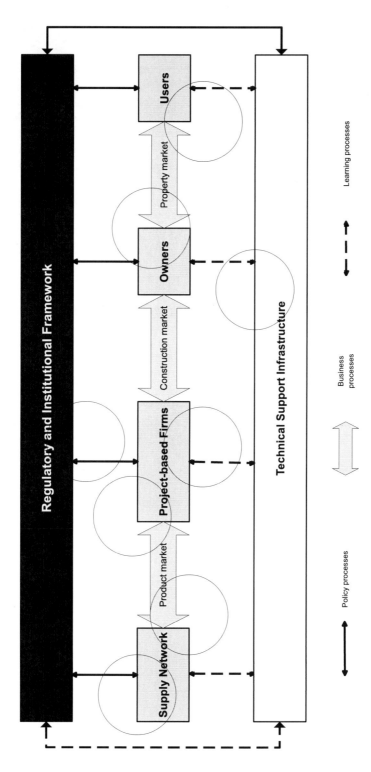

Figure 15.4 Multiple communities of practice in construction (adapted from Haugbølle et al., 2012).

258 *Haugbølle*

often-informal nature and, not least, their differences in prevalence of knowledge types are some of these characteristics.

Looking at the Danish construction industry, this theoretical perspective has received less attention in policy and development activities. One notable exception is the recent completion by the Danish Association of Construction Clients of the i2p initiative ('idea to program') that aimed at improving the initial stages from the very first idea to the formulation of the client brief (i2p, 2015). This programme worked explicitly with the concept of community of practices as the organising principle for creating new innovative networks among the parties of the construction industry.

Despite their relatively overlooked position, communities of practice have consciously, or less consciously, underlined many of the educational activities that have been initiated, such as the initiative of the competence-building programme BygSol (Christensen, 2008). Other initiatives like Bricklayers in Motion (see for example Bertelsen, 2011) have also been targeted at developing vocational training and improving the competences of craftsmen and unskilled labour. Another example of supporting the communities of practice among certain groups of building professionals is a praised initiative that aimed to establish a flexible and moveable training platform on occupational health and safety. Recognising that craftsmen seldom participate in course activities and prefer to adopt new knowledge by on-the-job training, the Safety and Health Preventive Service Bus was established as part of the collective agreement between the parties of the construction sector (Byggeriets Arbejdsmiljøbus, 2016). A similar initiative was taken by a private philanthropic foundation to establish a moveable lab for appropriate construction principles and solutions with regard to bathrooms (BvB and GI, 2015; 2016). Hence, the foundation had a truck specially designed that could be driven to educational institutions or construction sites and be unfolded on site to display various exhibitions, video installations and other learning resources. Table 15.3 provides an overview of the main elements in the technological frame considering knowledge as practices.

Table 15.3 Technological frame: knowledge as practices

	Knowledge as practice
Goals	Social participation as a process of learning and of knowing
Key problems	Knowledge as practice
Problem-solving strategies	How practices develop and initiate a diverse set of training and learning activities
Current theories	Communities of practice
Users' practice	Many different CoP due to many professions, fragmented industry structure, work gangs, etc.
Exemplary artefacts	Educational truck OSH service bus

Metaphor 4: knowledge as formative

The fourth metaphor perceives knowledge as formative: 'social shaping'. Such a perspective focuses on understanding the effects of knowledge along with material objects on producing and reproducing change and stability in, for example, business structures, organisations and work practices. The solutions tend to be about increasing reflexivity among actors.

This perspective draws on studies on sociology of scientific knowledge (SSK) and the later offspring labelled science and technology studies (STS) with its focus on contingency and interpretative flexibility. The perspective has been strongly inspired by thinkers such as:

- Kuhn (1996) on his analysis of paradigm shifts in science;
- Berger and Luckmann (1966) on the dialectic relation between agency and structure;
- Bloor (1976) with his formulation of the so-called strong programme of SSK emphasising four crucial principles of causality, impartiality, symmetry and reflexivity.

Detailed historical accounts of the origin of the constructivist field of science and technology can be found in for example Bijker et al. (1987), Bijker and Law (1992) and Pickering (1992). Focusing on the formative effects of knowledge implies that the previously fixed entities of the different elements of the construction industry and their interrelation may very well change and are at the centre of scrutiny. Hence, Figure 15.5 illustrates how the elements and interrelations may, in principle, change.

Although a constructivist perspective on social and technological developments has increasingly gained momentum in the past 30–40 years among researchers, the perspective is less commonly known or applied in policy and business circles in construction. Some notable works within construction include for example studies by Harty (2005) on the boundedness of innovation with regard to information and communication technologies. Others have studied the use of knowledge through various media such as drawings and what effects knowledge produces, including the meaning and structural constraints of the actors (see for example Whyte et al., 2007). Others, like Dubois and Gadde (2002), have studied construction as a loosely coupled system. With regard to Danish construction, research on knowledge includes studies on the moral and ethical problems inherent in the consulting engineers' profession of knowledge production (Munch, 2005) and the cultural organisation in construction and its role in the application of knowledge (Thuesen, 2006). Table 15.4 provides an overview of the main elements in the technological frame considering knowledge as formative.

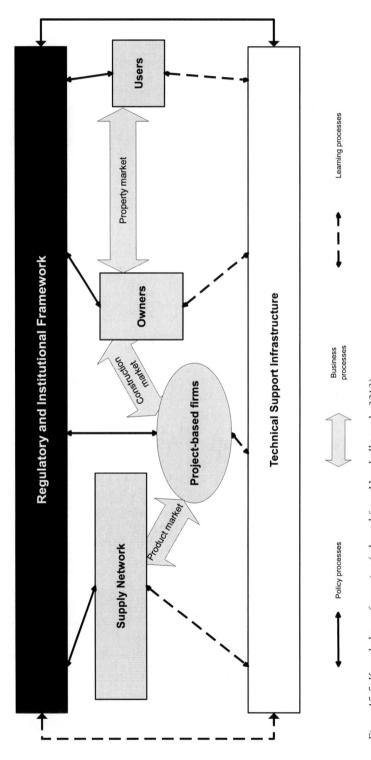

Figure 15.5 Knowledge as formative (adapted from Haugbølle et al., 2012).

Four metaphors on knowledge and change 261

Table 15.4 Technological frame: knowledge as formative

	Knowledge as formative
Goals	Recognise contingency and interpretative flexibility
Key problems	Knowledge as formative
Problem-solving strategies	Increase reflexivity
Current theories	Constructivism/relativism
Users' practice	Changing boundaries and relations
Exemplary artefacts	Deliberating studies

Conclusion: metaphors on knowledge and change

Based on a case study on the problem of knowledge in Danish construction, this chapter provided a brief exploration of research, policy studies and initiatives on knowledge and learning in construction. Although the context laid out is national, the problems addressed are likely to be general problems faced by actors of the built environment in other national contexts. While this chapter may be more explorative than conclusive in character, it is the hope that it will inspire future research on a more integrative and holistic understanding of the problem of knowledge.

This chapter attempts to demonstrate that there is no single way of integrating information into our construction practices. Rather an abundance of perspectives and initiatives is required and practised. This chapter explored four different metaphors for knowledge labelled 'knowledge as flows', 'knowledge as capabilities', 'knowledge as practices' and 'knowledge as formative'. It is the contention that these metaphors of knowledge articulate information integration differently as a problem and thus become controlling of the corresponding solutions to the 'problem of knowledge in construction'.

The four metaphors are not necessarily congruent with each other and may even be in contradiction to each other. The metaphors can help show how different perspectives are enacted in politics and practices to provide different solutions to the problem of knowledge. Whether the ambition is to remove obstacles, build bridges between isolated islands of knowledge, govern new procedures or deal with unintended consequences, it must be kept in mind that any one solution may well only solve part of the problem and cause new ones.

Note

1 The SECI model of knowledge creation explains how tacit and explicit knowledge are converted into organisational knowledge through four modes of knowledge conversion: socialisation, externalisation, combination and internalisation (SECI).

References

Alsted Research, 2003. *Kvalitativ Undersøgelse af Byggeriets Udførende Virksomheders Læring og Behov for Videnformidling for Fonden Realdania (Qualitative Investigation of Learning*

262 Haugbølle

and Needs of Knowledge Dissemination in Construction Companies. For the Foundation Realdania). Copenhagen: Alsted Research.

Bang, H., 1997. *Byggesektoren og Teknologisk Service. (The Construction Sector and Technological Service)*, Copenhagen: Boligministeriet.

Berger, P. L. and Luckmann, T., 1966 (reprinted 1991). *The Social Construction of Reality: A Treatise in the Sociology of Knowledge*, London: Penguin Books.

Bertelsen, N. H., 2011. *Murerfaget i Bevægelse. Vejledning 3 – Kommunikation: Udvikling, metoder, resultater og erfaringer. (The Masonry Trade in Movement. Guideline no 3 – Communication: Development, Methods, Results and Lessons Learned)*. SBi 2011:08. Hørsholm: Statens Byggeforskningsinstitut.

Bijker, W. E., 1995. *Of Bicycles, Bakelites, and Bulbs. Toward a Theory of Sociotechnical Change*. Cambridge: MIT Press.

Bijker, W. E. and Law, J., 1992. *Shaping Technology/Building Society – Studies in Sociotechnical Change*. Cambridge: MIT Press.

Bijker, W. E., Hughes, T. and Pinch, T., 1987. *The Social Construction of Technological Systems*. Cambridge: MIT Press.

Bloor, D., 1976. The strong programme in the sociology of knowledge. In: Knowledge and social imagery. 2nd edn. Chicago and London: The University of Chicago Press, pp. 3–23.

Boligministeriet, 1993. *Byggesektorens Forskningsaktiviteter (Construction Research Activities)*, Copenhagen: Boligministeriet.

Byggeriets Arbejdsmiljøbus, 2016. *BAM-BUS – Byggeriets Arbejdsmiljøbus (BAM-BUS – the OSH Bus of Construction)*, Available at: www.bam-bus.dk/1-56-english---intro. html [Accessed 8 October 2016].

Buildings Performance Institute Europe, 2014. *Renovation Strategies of Selected EU Countries. A Status Report on Compliance with Article 4 of the Energy Efficiency Directive*, Brussels: BPIE. Available at: http://bpie.eu/uploads/lib/document/attachment/86/Renovation_Strategies_EU_BPIE_2014.pdf [Accessed 13 September 2016].

BvB and GI, 2015. Gode Vådrum – Metoder og Løsninger (Good Bathrooms – Methods and Solutions). Available at: http://gi.dk/Publikationer/vaadrum2015.pdf [Accessed 8 October 2016].

BvB and GI, 2016. Gode Vådrum – Metoder og Løsninger (Good Bathrooms – Methods and Solutions). Available at: www.godevaadrum.dk/udstilling/Sider/default.aspx [Accessed 8 October 2016].

Carlsen, M., Dræbye, T. and Kirkeskov, J., 2005. *Byggeviden – Oplæg til Strategi og Handingsplan (Knowledge in Construction – Proposal on Strategy and Action Plan)*, Hørsholm: Statens Byggeforskningsinstitut. Available at: www.sbi.dk/download/byggeviden/byggeviden_strategi_og_handlingsplan.pdf [Accessed 9 October 2016].

Christensen, R. M., 2008. Development practically speaking – Learning processes in the Danish construction industry. Aalborg: Department of Production, Aalborg University. PhD thesis.

Christoffersen, A. K., 2000. *Temagruppe 9 – Videngrundlag. Projekt Hus Slutrapport, Oktober 2000 (Theme Group 9 – Knowledge. Project House Final Report, October 2000)*, Copenhagen: By- og Boligministeriet.

Christoffersen, A. K. and Bertelsen, S., 1990. *Byggesektorens F&U. Forskning og Udvikling i Byggesektoren. Situationen ved 80'ernes Slutning (Construction R&D. Research and Development in the Construction Sector. State-of-the-art at the End of 1980s)*, Copenhagen: Foreningen af Rådgivende Ingeniører.

Four metaphors on knowledge and change 263

Cohen, W. M. and Levinthal, D. A., 1990. Absorptive capacity: A new perspective on learning and innovation, *Administrative Science Quarterly*, 35(1), 128–152.

Det Offentlige Forskningsudvalg for Byer og Byggeri, 2000. *Forskning i Byer og Byggeri (Research in Towns and Construction)*, Copenhagen: By- og Boligministeriet & Forskningsministeriet.

Dræbye, T., 1997. *Teknologisk Byggeviden. Videnbrug, Videnformidling og Videnproduktion – En Kortlægning og Vurdering (Technological Knowledge on Construction. Use, Dissemination and Production of Knowledge – Mapping and Evaluation)*, Copenhagen: Byggeriets Udviklingsråd.

Dubois A. and Gadde L., 2002. The construction industry as a loosely coupled system: Implications for productivity and innovation, *Construction Management and Economics*, 20(7), 621–631.

Engwall, M., 2003. No project is an island: Linking projects to history and context. *Research Policy*, 32 (5), 789–808.

European Commission, 2011. *Communication from the Commission to the European Parliament, the Council, the European Economic and Social Committee and the Committee of the Regions. A Roadmap for moving to a competitive low carbon economy in 2050. COM(2011) 112 final.* Brussels: European Commission. Available at: www.europarl.europa.eu/meetdocs/2009_2014/documents/com/com_com(2011)0112_/com_com(2011)0112_en.pdf [Accessed 13 September 2016].

Flyvbjerg, B., 2006. Five misunderstandings about case-study research, *Qualitative Inquiry*, 12 (2), 219–245.

Gann, D. M. and Salter, A. J., 2000. Innovation in project-based, service-enhanced firms: The construction of complex products and systems, *Research Policy*, 29(7/8), 955–972.

Harty, C., 2005. Innovation in construction: A sociology of technology approach. *Building Research & Information*, 33(6), 512–522.

Haugbølle, K., 2014. Denmark – building/housing R&D investments. In: K. D. Hampson, J. A. Kraatz and A. X. Sanchez (eds). *Construction R&D Investment and Impact in the Global Construction Industry*. London, Routledge, pp. 81–97.

Haugbølle, K. 2016. Changing Construction: Perspectives on Knowledge and Learning. In: A. Saari and P. Huovinen (eds). *WBC16 Proceedings: Volume III. Building Up Business Operations and Their Logic. Shaping Materials and Technologies. Vol. III.* Tampere: Tampere University of Technology. Department of Civil Engineering, pp. 57–68.

Haugbølle, K. and Clausen, L., 2002. *Kortlægning af Bygge/Boligforskningen i Danmark (Mapping of Housing/Construction Research in Denmark)*, Hørsholm: Statens Byggeforskningsinstitut.

Haugbølle, K. and Storgaard, K., 2014. *Renovering og Vidensystemet (Refurbishment and the Knowledge System)*, Copenhagen: Statens Byggeforskningsinstitut, Aalborg Universitet.

Haugbølle, K., Forman, M. and Gottlieb, S. C., 2012. Driving sustainable innovation through procurement of complex products and systems in construction. In: *CIB Proceedings: Joint CIB International Symposium of W055, W065, W089, W118, TG76, TG78, TG81 and TG84, Montreal, 26–29 June*, pp. 444–455.

i2p, 2015. *Den Gode Proces (The Good Process)*, Copenhagen: Bygherreforeningen. Available at: http://i2p.dk/proces [Accessed 8 October 2016].

IEA – International Energy Agency, 2013. *Technology Roadmap. Energy efficient building envelopes*. Paris: IEA Publications. Available at: www.iea.org/publications/free publications/publication/TechnologyRoadmapEnergyEfficientBuildingEnvelopes.pdf [Accessed 19 Sep 2016].

264   Haugbølle

Koordinations- og initiativgruppen for viden i byggeriet, 2009. *Nytænkning i Byggeriet. Forskning Skaber Værdi (Innovation in Construction. Research Adds Value)*, Copenhagen: Erhvervs- og Byggestyrelsen.

Kuhn, T. S., 1996. *The Structure of Scientific Revolutions*, 3rd edn, Chicago: University of Chicago Press.

Munch, B., 2005. *Moral og Videnproduktion. Fragmenter af Praktisk Moral i den Tekniske Rådgivning (Morality and Knowledge Production. Fragments of Practice Moral In Technical Consulting)*, Lyngby: BYG-DTU.

Nonaka, I. and Takeuchi, H., 1995. *The Knowledge Creating Company: How Japanese Companies Create the Dynamics of Innovation*, New York: Oxford University Press.

Oudshoorn, N. and Pinch, T., 2003. *How Users Matter. The Co-Construction of Users and Technologies*, Cambridge: The MIT Press.

Pickering, A., 1992. *Science as Practice and Culture*, Chicago: University of Chicago Press.

Pryke, S. D., 2004. Analysing construction project coalitions: exploring the application of social network analysis, *Construction Management and Economics*, 22(8), 787–797.

Tænketank for Bygningsrenovering, 2013. Fokus på Bygningsrenovering. Syv Initiativer fra Byggebranchen. (Focus on Refurbishment. Seven Initiatives of the Building Sector), Copenhagen: Grundejernes Investeringsfond.

Thuesen, C., 2006. *Anvendelse af den Rette Viden – et Studie af Byggeriets Kulturelle Organisering (Application of the Appropriate Knowledge – a Study on the Cultural Organisation of Construction)*, Lyngby: BYG-DTU.

Udvalget vedrørende byggeforskning i Danmark, 2002. *Byggeriet i Vidensamfundet – Analyse og Anbefalinger fra Udvalget Vedrørende Byggeforskning i Danmark (Construction in the Knowledge Society – Analysis and Recommendations from the Committee on Construction Research in Denmark)*, Copenhagen: Erhvervs- og Boligstyrelsen.

Wenger, E., 1998. *Communities of Practice: Learning, Meaning, and Identity*, Cambridge: Cambridge University Press.

Whyte, J. K., Ewenstein, B., Hales, M. and Tidd, J., 2007. Visual practices and the objects used in design, *Building Research & Information*, 35(1), 18–27.

Winch, G., 1998. Zephyrs of creative destruction: Understanding the management of innovation. *Building Research & Information*, 26(5), 268–279.

Winch, G. M., 2010. *Managing Construction Projects*, 2nd edn, Chichester, UK: Wiley-Blackwell.

16 Contrasting aspects of information integration

Adriana X. Sanchez, Geoffrey London and Keith D. Hampson

Introduction

This book has provided an introduction to the potential value and challenges behind achieving a more integrated built environment and industry. In particular, each chapter has explored different industry sectors where integrating information across organisational boundaries and software domains can help improve the management of human, economic and ecological resources. They have done so by providing context and examples from across the world that showcase the processes involved as well as their potential added value. However, while there are significant gains to be reaped from achieving higher levels of integration of information, there are also many caveats that need to be considered in order to fully harness and leverage this value. The first chapter of this book highlighted that the ideals of the following chapters were implicitly directed towards delivering 'a more efficient and sustainable built environment industry that is able to produce higher quality built assets for global societies' (Chapter 1). This concluding chapter aims to highlight some contrasting aspects of integrated information systems and processes that need to be considered if we are to achieve this objective.

The digital versus the human

In 2014, the United Kingdom (UK) Government released a report titled 'Digital Built Britain'[1] which starts by stating: 'the next stage in the digital revolution has begun. Having transformed retailing, publishing, travel and financial services, digital technology is changing the way we plan, build, maintain and use our social and economic infrastructure' (HM Government, 2015). 'Digital Built Britain' frames a more integrated digital built environment as a keystone of the UK's economic future, asserting that technology and integrated digital technologies, such as Building Information Modelling (BIM) in particular, can transform both the lives and the world of their residents. Their vision includes data-enabled collaborative working across built assets' life-cycles, real-time remote sensing and monitoring of asset and network performance, automated control systems, use of 3D printing and smart factory automation, open data and the internet of services.

266 Sanchez et al.

They, however, also address the 'human' factor by highlighting issues such as the required industry cultural change as well as the need to balance 'hard business elements' such as systems and strategies with social issues that include values, skills, staff and style. This balance, they propose, will 'enable the transition to a more effective service orientated and outcomes based industry, defined by more efficient and effective service organisations' (HM Government, 2015). Australia's more succinct *National Digital Engineering[2] Policy Principles* also mention the potential benefits of technologies such as higher efficiency, value for money, productivity and innovation. These are balanced against more process-oriented issues such as standards and protocol harmonisation across government organisations and life-cycle phases, collaborative efforts to drive best practice across industry and government, capability building and the collection and active incorporation of lessons learned (Australian Government, 2016).

Bougrain (Chapter 11) exemplified this point within the building energy performance sector multi-asset case studies. This chapter concluded that turning digital data into actionable information requires contextual awareness due to the strong organisational impact that socio-technical systems such as BIM have on procurement processes. In this way, leveraging more integrated information systems presents both technical and human challenges. In one of their case studies, knowledge and skills management was a key factor hindering the successful implementation of the new systems which led to a new strategy that included organisational and staffing changes.

Also within the building energy performance sector, Forman (Chapter 10) described a new quality assurance system which was designed in Denmark as a project delivery strategy to handle information integration across life-cycle phases and between service providers and public client organisations. This chapter emphasised the need to determine how knowledge is to be created for the new, more integrated practices and how socio-technical aspects will affect them. Until now, a significant fraction of the research carried out around more integrated information systems has focused on technical aspects of creating and managing digital information. However, as highlighted by Meistad et al. (Chapter 12) the 'people' aspects are often not as thoroughly researched and developed. Nevertheless, 'people are information integrators, independently of the physical information carriers'. Examples of this are found in cases such as the Sydney Opera House in Australia, which has recently started implementing its BIM4FM interface. In this case, key lessons from the implementation journey include having a culture of engagement, developing and nurturing a new set of skills beyond technical expertise and the implementation of soft infrastructure such as an information systems support group formed by 'super users' who can help less experienced team members (Linning et al., 2016).

This means that further research and emphasis is needed on the social aspects along with the technical aspects of implementing more integrated information systems in the built environment. This will require developing a better understanding of how knowledge is transferred and transformed, how organisational processes can be adapted and how different perspectives can be combined to produce cohesive and coordinated delivery and implementation strategies.

Normalisation versus standardisation

> Integration of information is a particularly challenging problem, especially in the face of massive data sources that have been collected by different individuals and institutions in parallel, and rarely specifically for the issue of interest.
>
> (Barrett et al., 2011)

Leveraging and coping with the amount of data and information being created everyday by our increasingly 'smarter' built environment and the communities that inhabit or use it has been one of the main topics discussed throughout this book. On the one hand, there is almost a limitless number of opportunities to leverage this growing 'datasphere' in order to better plan, develop and manage our cities. On the other hand, there is a need to strike a balance between data production and the cost of its processing or analysis requirements. The answer to this dilemma may lie in a mix of standardisation methodologies as well as process and technological innovations. These approaches could ensure the integrity of the data is not lost while eliminating redundancies and growing its usability for decision-making; see for example the case of Amsterdam studied by Taylor et al. (2016).

As proposed by Kraatz et al. (Chapter 2), having access to more integrated systems that allow government and industry knowledge, information and data to be linked and used in a connected way can help deliver a more sustainable and productive social housing sector. This is rooted in the need to track long-term benefits across different levels of governance. Addressing this need poses significant challenges in terms of managing complex arrays of information from multiple sources that are not coordinated or developed with this use in mind. Dealing with this issue will require innovative approaches that provide management strategies 'that meet the underlying needs and values of the affected populations' (Kraatz et al.).

Sanchez et al. (Chapter 3) built on this argument within the context of urban resilience policy. This chapter proposed that due to the complexity of the urban environment and the level of uncertainty that its dynamic nature entails, policy processes need to become more context-aware, adaptive and integrated. Part of the solution to this challenge may lie in information and communication technologies (ICT) and emerging socio-technical systems such as open data portals and urban information models. For these approaches to be effective, standardisation and transparency processes need to be developed and implemented. Standards have a wide-range of positive effects both at the macro-economic level and the individual level (Linning et al., 2016). In particular, they provide a common point of reference that allows data users to integrate different datasets to build a new understanding of complex issues, cause and effect processes and potential impacts of decisions made.

268 *Sanchez et al.*

The development and use of open standards in particular has the potential to reduce interoperability problems significantly while accommodating disparate needs and data uses. Santos and Aguiar Costa (Chapter 6) highlighted how standards in general, and open ones in particular, can help reduce the level of expertise and resource investment needed for data sharing and building analyses within a BIM context. They also help reduce the likelihood of not being able to re-use information embedded in the models throughout an asset's life-cycle. In the case of BIM-based life-cycle assessment (LCA), interoperability, and therefore the use of common standards, is not only required within one organisation or stakeholder group but also across value chains, including manufacturers, designers and LCA developers. At the precinct scale, Newton et al. (Chapter 7), however, also suggest that while open formats are desirable, they often rely on limited funding and struggle 'to keep pace with the rapid innovation that is possible in the commercial sphere'. Nevertheless, 'robust open standards can break this reliance on exclusive proprietary software systems and serve to encourage wider commercial software innovation based on vendor-neutral data formats'.

Abanda and Tah (Chapter 4) dealt with the issue of normalisation, proposing that big data technologies can help cope with the large volume and speed of data production, in terms of its variety, veracity and value. Existing and new platforms need to be further developed for effective and distributed storage and processing of these large datasets. They also highlighted the central role of interoperability and linked data technologies in leveraging the full power of the Internet of Things (IoT) as well as its integration with other technological developments such as BIM and City Information Modelling (CIM).

How societies are to deal with the large amounts of data and information rapidly being produced is at least as important as how this information is produced in the first place. If these two issues are not dealt with hand in hand, the road to maximising the benefits of more integrated and smart built environments will likely be paved by inefficiencies and frustration.

User-driven versus process-driven

> Theories of the built environment have tended to be oriented to process – how it is created and supplied – and/or product – how it functions once it has come into existence.
>
> (Vischer, 2008)

A significant fraction of the work done in information integration systems and the built environment field has been process-oriented. While the development of processes that help improve the effectiveness and efficiency of information integration approaches is important, focusing only on this area of research places the 'how' above the 'why', perhaps limiting potential benefits to be achieved. This lack of focus on communication and alignment of expectations has often led to costly variations and disputes (Hardin and McCool, 2015; Baharuddin et al., 2013). Yet, this continues to be an issue within the built environment industry and now becomes more relevant due to the introduction of systems and approaches that are highly user-dependent. Forman (Chapter 10) for example highlighted how, in Denmark, 'there has not been a tradition of operations and maintenance employees being involved in the design of new buildings'. This leads to design and construction programmes only rarely considering the needs of operational staff who will need to develop corresponding capabilities to manage them. In addition to operational staff, the case studies presented by Bougrain (Chapter 11) showed that the performance of the final product is often dependent on both end-user and staff behaviour and understanding of the system itself.

It is clear that a balance between process- and user-driven approaches and research is needed to take full advantage of more integrated systems. This is especially true for the refurbishment industry. Bröchner and Sezer (Chapter 9) for example highlighted some of the challenges faced by existing buildings aiming to use BIM for refurbishment projects. Some of these refer to the technical challenges of integrating BIM and ICT tools with enterprise systems as well as the lack of automated scan-to-BIM tools and access to data in interoperable formats that can help reduce the entry cost for small projects and companies. Some of the challenges relate to the lack of established processes that facilitate the integration across life-cycle phases from design to operations, and some related to user-centred challenges. The latter include motivations and cultural changes that facilitate information transfer across the value chain. For example, if clients cannot see immediate benefits to them from investing in more integrated information approaches, they are unlikely to proceed. Therefore, in order to maximise the benefits for some project stakeholders, it is important to understand how value can be better communicated and distributed across the entire chain.

Technical solutions such as BIM and advanced ICT generally must therefore be promoted hand-in-hand with strategies and appropriate leadership to overcome process barriers. In this sense, and as highlighted by Støre-Valen et al. (Chapter 13), the built asset 'ought to be considered a strategic means to reach the owner's objective'. This leads to the need to 'translate the requirements and values of the owners and their customers into a design that includes suitable flexible spaces'. Clients are also taking a stronger leadership role across the globe in a bid to improve whole-of-life outcomes for their asset portfolios and the industry as a whole. Sanchez et al. (Chapter 14) briefly explored three types of client strategies that have been observed in Australia leading to more integrated information management approaches. They all had in common the fact that value is sought from early design phases with the operational phase in mind. Performance-driven strategies in particular are becoming more common in market-driven economies as clients become more aware of the potential gains over the life-cycle of their asset portfolio.

Nevertheless, the UK Government's 'Soft Landings' initiatives have recently highlighted the lack of commercial mechanisms that facilitate greater focus on performance requirements (HM Government, 2015). Some of the barriers for the effective implementation of such performance requirements may also lie in the

270 *Sanchez et al.*

speed of development of information technologies and the ability to communicate expectations through a common language. Christensen et al. (2009) proposed that effective communication between suppliers of added value and clients depends on three conditions:

- the client must understand and be able to communicate the attributes that need to be delivered, including the tolerance levels;
- these attributes must have metrics and technology to measure them that are available, reliable and unambiguous;
- the supplier of added value must understand the relationships between the attributes procured and the performance of the system in which they will be used; this means that the supplier must understand the effects of variations on the system's performance.

To address this, Sanchez et al. (Chapter 14) proposed an online platform that allows users to record, share and benchmark value metrics associated with benefits that can be achieved through the implementation of BIM tools and processes. The use of such benchmarking systems can help clients and service providers better articulate their performance expectations throughout the different life-cycle phases and monitor progress towards organisational goals linked to BIM implementation strategies. This approach could also be applied to other integrated information technologies such as smart transport and asset management systems. Meistad et al. (Chapter 12) add to this that there is a need to understand better how integrated design process (IDP) and integrated design and delivery solutions (IDDS) can facilitate team-building and high-performance attitudes.

Useful data versus actionable information and knowledge

> The design, construction, and operation of high-performing facilities depends on the ability of planners and designers to predict the future performance of a facility with reasonable accuracy and granularity, and tailor the performance to support the facility users' business and operational requirements and activities.
>
> (Kim et al., 2012)

The production, storage, sharing of and access to useful data about the elements that form our built environment and how they interact with the communities that inhabit and use them is often mentioned as a limitation for urban research and better asset management. Lupica Spagnolo (Chapter 8) explored the use of ICT web-enabled platforms to integrate information more effectively from the early design phases into the operational phase and optimise asset management and maintenance. This would allow storing data that can be used to produce more reliable maintenance plans based on useful information such as reference service-life (RSL). It could also be used for sharing life-cycle knowledge currently not being transferred across building practices. In this case, the integrated platform is meant to provide useful information stored in an interoperable, identifiable and reliable format which makes it actionable across life-cycle phases. This would in turn allow users to make safer, more sustainable and more aesthetically pleasing choices that also optimise the service-life of the asset.

However, useful information is not limited to characteristics of the built asset and its components. Real-time data collection of indicators that allow managers to take action in a timely manner related to staff safety can be crucial in ensuring that the industry is safe to work in and remains competitive. Yi et al. (Chapter 5) highlighted the significant challenges the construction industry faces in this respect and presented two case studies that aimed to help address this issue. One of the options they discussed is an early-warning system based on heat stress indicators that would allow managers to conduct more effective risk assessments and maintain the safety of their workers.

Nevertheless, one of the challenges of integrating information for decision-making is to have processes that ensure that the data gathered matches the question that needs to be answered rather than just gathering data for 'data's sake' (Barrett et al., 2011).

> The separation and flow of data is the key as it allows the information integration process to perform iterative refinement, which is critical because that enables the specifics of the integrated information to be determined by the context of the problem.
>
> (Barrett et al., 2011)

Bougrain (Chapter 11) explored this issue within the context of energy performance contracts. Here, sensors can monitor a number of asset usage and user behaviour metrics that can help unravel the mystery of why some buildings perform worse than the models predicted. However, these sensors can be expensive to set up and maintain as well as require operating staff to acquire special capabilities. Additionally, while it is important to have access to useful data, the main issue is to produce 'actionable information'. This requires relevant stakeholders being able to access and interpret data in order to make better-informed decisions. This in turn entails contextual awareness which can be facilitated by tools and strategies that produce the required knowledge transformations.

Haugbølle (Chapter 15) explored the problem of knowledge within the context of the Danish construction industry. This chapter concluded that to maximise the value of information integration approaches, it is necessary to develop and apply a variety of perspectives and initiatives. These can include strategies that aim to streamline data and information production and flows, those that aim to improve the acquisition of knowledge and a firm's absorptive capacity, strategies that focus on training and knowledge development, and those that focus on increasing reflexivity.

Concluding remarks

There are a number of contrasting aspects of integrating information across organisational boundaries and software domains that will need to be considered in order to fully harness its potential gains for the built environment industry. This chapter explored some of these issues in terms of their challenges and opportunities based on discussions presented in different sections of this book. Nevertheless, these are merely a sample and more work needs to be done to really understand the importance of other aspects discussed here and elsewhere in the literature such as the issue of scalability versus accessibility. Bröchner and Sezer briefly touched on this topic by highlighting the urgency of downscaling ICT systems and methods 'from large new construction settings to refurbishment projects', while also making them affordable and accessible to this sector of the industry. A different interpretation on these contrasting aspects that was briefly discussed by Newton et al. and still needs to be further explored refers to the accessibility and scalability of information when projects move from object- and building-based models to precinct and metropolitan-scale settings.

Other aspects that were not discussed in detail in this book but may bear significant weight when assessing the effects of more integrated information systems relate to the balance between security, securitisation and democratisation of data and information. Among other things, Sanchez et al. (Chapter 3) discussed open data portals as vehicles for more active citizen participation in urban policy. While the democratisation of information and decision-making processes may be an important factor in improving built environment strategies, this process needs to be balanced against security and financial concerns of data production and ownership. The same is applicable to smaller scales. One of the often-mentioned benefits of implementing BIM is increased transparency and visibility of information flows. That often-elusive *single source of truth* is the BIM-nirvana that many asset and project managers are seeking. However, information and data security is one of the main concerns frequently mentioned by users, especially among firms with high levels of BIM engagement (Sanchez and Joske, 2016). The issue of data ownership and financial value is another concern that will likely rise to prominence as BIM and Digital Engineering processes become more common during the operational phase of built assets.

This book has presented just a taste of the complexities, challenges and opportunities of a growing field of interdisciplinary research, development and practice of information integration. There is therefore much work remaining to capture and leverage the lessons learned from successes and failures encountered on the path to a more integrated built environment industry.

Notes

1 This report was produced by a project chaired by Mark Bew, who was also Chair of the BIM Task Group responsible for delivering the UK Government's BIM Program known as BIM Level 2 (Digital Built Britain, 2016; HM Government, 2015).

2 'Digital Engineering may be defined as the convergence of emerging technologies such as Building Information Modelling (BIM), Geographic Information Systems (GIS) and related systems to derive better business, project and asset management outcomes' (Australian Government, 2016).

References

Australian Government, 2016. *National Digital Engineering Policy Principles*, Canberra: Australian Government.

Baharuddin, H. E. A., Wilkinson, S. and Costello, S. B., 2013. *Evaluating Early Stakeholder Engagement (ESE) as a Process for Innovation*, paper presented at CIB World Building Congress, Brisbane, Australia, 5–9 May.

Barrett, C. L., Eubank, S., Marathe, A., Marathe, M. V., Pan, Z. and Swarup, S., 2011. Information Integration to Support Model-Based Policy. *Innovation Journal*, 16(1), 1–16.

Christensen, C. M., Verlinden, M. and Westerman, G., 2009. Disruption, disintegration and the dissipation of differentiability. In: R. A. Burgelman, C. M. Christensen and S. C. Wheelwright (eds) *Strategic Management of Technology and Innovation*. 5th edn. New York: McGraw-Hill Irwin, pp. 363–387.

Digital Built Britain, 2016. *Digital Built Britain: About Us*. Available at: www.digital-built-britain.com/about [Accessed 12 December 2016].

Hardin, B. and McCool, D., 2015. *BIM and Construction Management: Proven Tools, Methods, and Workflows*. 2nd edn. Indianapolis, IN: Wiley.

HM Government, 2015. *Digital Built Britain. Level 3 Building Information Modelling – Strategic Plan*, London: HM Government.

Kim, T. W., Kavousian, A., Fischer, M. and Rajagopal, R., 2012. *Improving Facility Performance Prediction by Formalizing an Activity-Space-Performance Model*, Stanford, CA: Stanford University.

Linning, C., Sanchez, A. X. and Hampson, K. D., 2016. Tips with hindsight. In: A. X. Sanchez, K. D. Hampson and S. Vaux (eds) *Delivering Value with BIM: A Whole-of-life Approach*. London: Routledge, pp. 81–100.

Sanchez, A. X. and Joske, W., 2016. Enablers dictionary. In: A. X. Sanchez, K. D. Hampson and S. Vaux (eds) *Delivering Value with BIM: A Whole-of-life Approach*. London: Routledge, pp. 205–297.

Taylor, L., Richter, C., Jameson, S. and Perez del Pulgar, C., 2016. *Customers, Users or Citizens? Inclusion, Spatial Data and Governance in the Smart City*, Amsterdam: University of Amsterdam, Maps4Society Final Project Report.

Vischer, J. C., 2008. Towards a user-centred theory of the built environment. *Building Research and Information*, 36(3), 231–240.

Index

3D 48, 72, 203; 3D printing 1, 239, 265

Accessibility 43, 136, 139, 146, 272
Actionable information 178–191, 246, 266, 270, 271
Adaptive action 36, 42, 44, 58, 213
Adaptive capacity 38
Added value 62, 155, 214, 219, 227, 265, 270
Advanced engineering 1, 71, 72, 85, 127, 159, 209, 230
Affordability 7, 14–17, 20, 22, 29, 31, 32, 272
Air-conditioning 134, 188
Air quality 56, 60, 67
American Institute of Architects (AIA) 93
American national institute of standards and technology See NIST
American society of civil engineers 40
Amsterdam 267
Animations 72
Architecture, Engineering and Construction (AEC) 91–96, 105, 106, 134, 195, 199–201, 204, 206
Array of things 54, 66
As-built BIM See Building Information Modelling
Asset and facility management 135, 145, 146, 187, 233
Augmented reality 80
AURIN 30
Australia 5, 14–19, 23–32, 38, 39, 43–46, 55, 73, 75, 91, 111, 122, 215, 231–236, 240, 266, 269

Automation 71–86, 98, 217, 240, 265, 269
Awareness 35, 71, 91, 97, 106, 157, 180, 191, 194, 224, 228, 266, 271

Barcelona 52, 59–61, 65
Barcode 54, 57
Barriers 4, 40, 41, 62, 154, 180, 190, 192, 206, 216, 218, 235, 236, 238–240, 253, 269
BAS 217
Behaviour 53, 56, 64, 65, 72, 81–86, 114, 128, 145, 178, 180, 185, 238, 269, 271
Benchmarking 4, 115, 151, 224, 232–241, 270
Benefit 15, 21, 23, 60, 61, 71–74, 81, 85, 105, 155, 157, 174, 178, 201, 206, 215–219, 227, 232–238, 266, 268, 272
Best practice 23, 127, 128, 266
Big data 43, 62, 64, 65, 106, 217, 268
BIMCS 240
BIM-LCA 92, 97–99, 101, 105, 106
Brazil 235
BREEAM 115, 156
Brisbane 33, 34, 67, 87, 88, 108, 207, 234, 241–243, 273
Brownfield 113, 124, 125, 127
Bto 166, 167, 169, 171–173
Building Information Modelling (BIM) 1, 4, 45, 46, 64, 65, 71–86, 91, 92, 96–99, 101–106, 111–117, 122–129, 134, 135, 140, 145, 152–158, 187–192, 199–201, 203, 206, 215–219, 222–227, 232–240, 253, 255, 265–272; As-built BIM 152–154, 158, 159

Index 275

CAD 46, 90

Canada 13–15, 27, 31, 137, 138

Capabilities 6, 47, 55, 64, 124, 127, 154, 157, 176, 205, 252–255, 261, 266–271

Carbon 52, 53, 113, 114, 117, 122, 125–127, 164

Case studies 3, 15, 23, 31, 52, 59–65, 80–85, 115, 118, 124, 127, 133, 155, 164, 165, 175–183, 189, 191, 199–206, 216, 222–225, 236, 240, 247, 271

Centralised 44, 182, 184–186, 190–192

Centrelink 29

Certification 115, 156–158, 180

Chile 235

China 73

Cisco 54

CityGML 46, 117

City Information Modelling (CIM) 64, 65, 268

Clash 74, 77, 79, 133

Classification 44, 122, 126, 136–138, 140, 144

Client 6, 16, 151–162, 167, 170–179, 186, 196, 201, 206, 220, 227, 232–234, 239, 270

Climate change 37, 38, 41–45, 81, 92, 117, 185

Closede-Circuit Television (CCTV) 53, 60, 62

Cloud-Based 57, 127

Co-creation 218, 223, 227, 228

Collaboration 2, 5, 47, 91, 96, 111, 127, 129, 134, 145, 146, 163, 170, 172, 196, 202–206, 214–233, 237, 240, 265, 266

Commissioning 160–162, 164–173, 175, 176

Communities 7, 16, 17, 22, 23, 25, 29, 30, 41, 45, 127, 161, 163, 164, 170, 172, 235, 255–258

Competence 175, 178, 183, 188, 190, 214–216, 219, 220, 222–224, 228

Competitive advantage 2, 3, 228

Complex problems 1, 23, 26, 35, 37, 38, 44, 46, 101, 111, 180

Compliance 46, 74, 75, 135, 168

Conflict 40, 63, 77, 133, 167, 187, 214

Construction and operation 45, 91, 95, 122, 163, 183, 234, 247, 253

Construction management 72, 213; Construction Information Management (CIM) 134, 140; Construction phase 95, 201, 213, 217, 220, 222, 228; Construction research 70, 248; Construction workers 24, 70, 72, 73, 81, 82, 84–86

Consultant 7, 164–176, 225, 249, 253, 255

Context-aware 73, 85, 267

Contract 3, 6, 27, 162–192, 197, 201, 216, 232–234, 271

Contractor 6, 75, 135, 150–157, 187–190, 198–200, 214–220, 234, 253

Control 16, 56, 59, 60, 63, 80, 82, 85, 135, 172, 182–186, 190–195, 217, 265

Cooling 82, 128, 130, 188

Coordination 3, 4, 7, 36, 41–46, 135, 151, 161–178, 182, 190–197, 201, 216

Copenhagen 161, 164–168, 172, 173

Cost 2, 21, 23, 31, 39, 47, 57, 73, 91, 95, 127, 134, 135, 138, 146, 151, 154, 174, 178–187, 206, 223, 269; Cost-Benefit 27, 29, 202; Cost-effective 21, 53, 85, 223

Crane 72, 77, 78, 138

CRC for Construction Innovation 98, 111, 236

Critical infrastructure 39, 40

Croatia 179

Crowdsourced 4, 237, 239–241

CSTB 136, 192

Culture 5, 39, 70, 152, 200, 266

Customer 2–4, 151, 199, 213, 216, 218, 223, 269

Customisation 2, 31, 76, 127

Cyberattacks 37, 40

Danish 14, 160–168, 175, 177, 246–249, 253–271

Data analysis 44, 57; Database 31, 40, 45, 61, 63, 75, 97–99, 105, 106, 116, 118, 120, 122, 129, 134–146, 157, 165, 187, 233; Data capture 21, 71, 83, 84, 97, 135, 145, 151, 183, 203, 236, 271; Data dictionary 119–123; Data exchange 83,

276 Index

84, 133, 134; Data formats 117, 120, 268; Data-information-knowledge-wisdom (DIKW) 181; Data integration 4, 23, 31, 54, 61, 62, 232; Data management 53, 80; Data mining 86, 106; Data model 118, 119, 124, 126; Data needs 29, 43; Data ownership 4, 272; Data portal 4, 42–44, 47, 65, 267, 272; Data processing 44, 64, 84, 85; Data protocols 47; Data schema 118–123; Data security 272; Data sharing 31, 101, 133, 144, 268; Data standard 46, 47; Data storage 62, 64, 65, 135, 136; Data strategies 43; Data users 267, 268

Decentralisation 20, 42, 70, 151 *See also* Distributed

Decision-support 73, 80, 85

Decommissioning 153, 231 *See also* end-of-life

Design, construction, and operation 5, 116, 150, 196, 245, 270; Design-bid-build 153, 154, 187

Design alternatives 72, 153; Design analysis 135

Designer 6, 74, 95–99, 103–106, 133–136, 145, 199, 200, 218; Design practice 172, 173; Design team 162–173, 197, 200, 213, 214

Design phase 98, 133, 136, 153, 160–176, 196–200, 206, 225–228, 269, 270; Design process 134, 161–175, 196, 199–201, 206, 218, 270; Detailed designing 200, 204

Detection 55–58, 73–76, 82, 86, 116, 117

Developers 99, 105, 106, 122, 128, 130, 140, 268

Devices 52–63, 84, 203, 216, 217, 227, 228

Digital built environment 1, 3, 116, 240, 265; Digital Built Britain 47, 265, 272; Digital engineering 241, 272, 273

Digital platform 42, 46, 111, 115, 218, 222, 227, 240

Digital prototyping 117

Disabilities 14, 19, 30, 86

Disputes 24, 268

Distributed 41, 56, 64, 125, 128, 268

Documentation 74, 152, 165–174, 226, 232

Drivers 3, 14, 129, 156, 199, 201, 237

Duplication 98, 133

Durability management 133, 135, 136

Dwelling 5, 16, 17, 31, 52, 128, 153, 155, 182–190

Dynamo 68

Early design 200, 227, 269, 270; Early involvement 194, 195; Early planning 195, 196, 214–228

Early-warning 71, 80–85, 271

Eco-Efficiency 52, 61, 126

Education 5, 14, 16, 23, 30, 81, 125, 175, 197, 222–228, 245, 258

Electricity 61, 94

Emergencies 22, 81, 167

Emission 45, 52, 53, 56, 59, 62, 91, 96, 97, 99, 113, 127, 128, 182, 183, 185; Embodied carbon 95, 125

Enabler 14, 56, 72, 80, 178, 206, 214–219, 240, 241

End-of-life 95, 136

End-user 3, 5, 6, 116, 129, 140, 175, 194, 195, 213–228, 249, 255

Energy analysis 95, 106; Energy audits 190; Energy data 62, 95, 178, 183, 190, 191; Energy demand 53, 96, 128, 164; Energy efficiency 59–61, 64, 95, 96, 127, 128, 146, 179, 183, 191, 195, 196; Energy performance 3, 150, 160–190, 197, 215, 217, 266, 271; Energy Performance Contracting (EPC) 178–186, 188, 190–192, 201, 202; Energy saving 56–62, 92, 178, 184–186; Energy usage 53, 95, 97, 156, 178, 184, 185

Energy systems 128, 165, 183, 184, 189

Engagement 23, 116, 127, 218, 233, 256, 266, 272

Engineers 6, 71, 77, 116, 129, 161, 162, 167, 172, 176, 197, 198, 231, 259

England 30

Environmental 3, 5, 7, 14, 26, 52–62, 84, 91–93, 95–99, 105, 106, 116, 125, 128, 144, 156, 158, 183, 196, 197, 205, 206, 223, 225

Error 63, 136, 144, 190, 192, 236, 240
Estimated Service Life (ESL) 135, 138, 139
Europe 13, 17, 91, 92, 97, 137, 161, 179, 180, 245, 247
Evaluation 23, 27, 29, 33, 72, 82, 86, 92, 136, 138, 157, 180, 226, 236, 240, 263
Execution 139, 180, 195–198, 201, 203–205, 215, 223
Experimental 29, 57, 82, 140

Fabrication 201, 202, 215
Façade 183, 187, 188, 222–227
Facilities 5, 55, 60, 77, 106, 116–126, 135, 150, 152–170, 206, 213–219, 227–234, 245, 270; Facility management 135, 145, 148, 178, 183–189, 199, 202, 206, 225, 233
Failure 40, 41, 81, 156
Falling-prevention 76; Falls 70, 73, 74, 89
Fatalities 70, 74
Fatigue 81, 84, 85
Feedback 5, 37–46, 55, 171, 182
File 98, 101, 122, 145, 158
Fire 43, 146
Fiscal 14, 21–24, 31
Flexibility 123, 163, 219, 224, 259, 261; Flexible 36, 42, 123, 151, 176, 216, 223, 258, 269
Forecast 35, 60, 96, 111, 135, 152, 164
Format 4, 43, 44, 47, 61–64, 99–107, 114, 117, 120, 126, 145, 146, 238, 240, 268, 269, 271
Formwork 99
Fragmentation 39, 40, 111, 176, 218, 236, 252, 256, 258
France 3, 15, 91, 136, 180, 183, 186, 192, 193
Frequency 37, 52, 72
Front-end 199
Functional data 75
Functionalities 7, 43, 45, 122, 124, 176, 178, 186, 190, 219, 220, 238, 240; Functional requirement 166, 169, 173; Functional-spatial 140, 143
Future performance 58, 71, 270; Future research 7, 64, 86, 105, 191, 195, 240, 261; Future risk 38, 45

Gains 4, 128, 146, 206, 233–235, 239, 265, 269, 272
Game 72, 237
Gap 2, 4, 25, 58, 76, 82, 135, 150, 160, 161, 164, 184, 236
Garbage 57
General contractor 188–190, 199
Geographic information systems (GIS) 1, 32, 43–46, 72, 85, 86, 112–117, 126, 127
Geolocated 118, 123
Geometric 46, 54, 74, 76, 98, 101, 122, 144, 152, 153, 157, 158, 168, 203
Geospatial 115, 117
Geotechnical 151
Germany 14, 45, 91, 92, 137, 157
Global warming 53, 96
GML 117, 130
GNSS 117
Goods and services 6, 154
Governance 4, 35, 37–47, 125–127, 204, 205, 215, 267
Government 1, 13–23, 31–43, 59–64, 81, 95, 114, 115, 120–129, 153, 165, 231, 238, 239, 249, 266, 267
GPS 86
Granularity 124, 270
Green building 127, 225
Greenhouse gases 45, 52, 53, 56, 59, 96, 130, 182, 183, 226
Green infrastructure 125
Green procurement 157
Green star 114, 127, 157
Greyfields 113, 124
Greywater 127
Grouping 123, 137
GSA 134, 135
GSM 84
Guideline 13, 80, 160, 167, 226, 228, 233

Handover 168, 176, 197, 204, 228
Hardware 55, 58, 203
Harmonisation 97, 111, 144, 266
Hazards 38, 41, 58, 71–73, 76, 81, 82, 85, 86, 151
Heating 63, 128, 134, 183, 188
Heat stress 81–85, 271
Heritage 125, 153, 233

278 *Index*

Hierarchy 118, 137, 144, 145, 181
High-performance 161, 198, 270
High schools 182–186, 189–192
Historical analyses 29
Hospital 153, 214, 216, 222–228, 233, 234
Housing organisations 15, 16; Housing provider 17, 19, 23, 25, 29; Housing sector 3, 15–17, 19, 188, 267; Housing stock 16, 17, 182, 186, 188, 190, 191
Humidity 60, 61, 81, 82, 84
HVAC 134, 143, 188

Information and Communications Technology (ICT) 3, 47, 59, 61, 71, 72, 80–86, 111, 133–136, 140–157, 215–218, 232–241, 247, 267–272
IDDS 196, 198, 208, 270
Identifier 54, 119
IDP 196, 198, 270
IFC 46, 99, 101–106, 118, 123, 126, 134–137, 145, 146
Implementation 3–6, 37–39, 41, 47, 59, 73, 80, 91, 95, 118, 126, 179, 180, 192, 204, 206, 215, 217, 232–234, 236, 238, 240, 266–270
Incentives 39, 40, 44, 157, 188, 201, 227
Index 29, 84, 163
India 91
Indicators 22–27, 29, 30, 47, 70, 71, 81, 84, 114, 127, 151, 154, 179, 197, 220, 235, 271
Indoor climate 225
Information integrators 198, 266
Information management 38, 53, 65, 85, 134, 135, 140, 146, 233, 238, 269; Information requirements 41, 98, 164; Information service 249; Information sharing 2, 3, 134, 187, 190, 191, 232; Information systems 4, 61, 150, 181, 185, 207, 265, 266, 272; Information technologies 4, 7, 41, 53, 63, 65, 80, 150, 157, 195, 206, 270
Infrastructure 5, 6, 16, 20, 37–45, 53, 91, 96, 112–127, 134, 163, 164, 170, 172, 234–248, 265, 266
Injuries 70, 71, 73, 81, 86

Innovations 1, 6, 39, 43, 116, 239, 267
Input 55, 73–82, 92, 95, 135, 136, 154, 162, 166, 168, 170–172, 176, 179, 182, 204, 205, 215, 216, 239, 240
Institutional investment 17, 24, 27
Integration platforms 46; Integration process 92, 271; Integration systems 268
Integrators 198, 266
Integrity 55, 267
Intelligent buildings 52
Intelligent waste containers 57
Interactive 43, 72, 145, 228, 237
Interdependencies 1, 26, 41, 114, 134
Interdisciplinary 200, 272
Interfaces 43, 45–47, 54, 134, 200, 201, 215, 228, 233, 252, 266
International Organization for Standardization (ISO) 84, 91–93, 99, 103, 106, 111, 112, 134, 135, 138–140
Internet of Things (IoT) 4, 52, 54–57, 59–65, 217, 222, 227, 268
Interoperability 2–5, 39–47, 61–65, 91, 92, 98–106, 112, 133–146, 154, 192, 206, 215, 219, 232, 233, 240, 268–271
Inter-organisational 3, 80, 236
Interventions 22, 23, 42, 57, 82, 135, 136, 138, 139, 144, 253
In-use 135–144, 199, 202, 214
Inventories 2, 47, 55, 58, 92–98, 180, 181
Investment 3, 7, 16, 17, 21–29, 40, 91–96, 101, 125, 150–154, 178–180, 183, 184, 190, 195, 198, 206, 236, 238, 249, 268
Israel 14
Italy 137, 138, 140, 143, 179
Item 58, 101–103, 144
Iterative 190, 271

Knowledge acquisition 62; Knowledge management 55, 252, 255; Knowledge production 259; Knowledge sharing 162–164, 169, 245, 248, 253, 256
Knowledge-based 73, 232
Knowledge boundaries 163, 164, 176
Knowledge systems 246, 247, 263
Knowledge transformations 271

Labelling 54, 115
Labour 6, 14, 81, 237, 258
Landscape 5, 116, 129, 248
Land use 94, 119, 120, 124
Language 63, 119, 123, 130, 136–146, 191, 205, 270
Layer 21, 25, 54, 61, 83–85, 143
Layout 76, 77
Leadership 128, 156, 214, 216, 232, 236, 239, 269
Learning process 161, 173, 175, 176, 248
Legacy 39, 40, 233
Legal 23, 38, 44, 118, 119, 161, 169, 179, 188, 232, 236
Legislation 160, 173, 179, 180, 225
Level of accuracy 124, 203
Level of detail 153, 171
Level of development 98, 101, 102
Library 103, 106, 122, 129, 145, 163
Life-cycle costs (LCC) 95, 224–226; Life-cycle energy analysis (LCEA) 95, 96; Life-cyle assessment (LCA) 92–106, 136, 268
Limitations 41, 57, 64, 96, 139, 191, 240, 268, 270
Linked data 64, 268
Lisbon 98
Literacy 247
Liveability 1, 111, 114, 127
Local authorities 13, 17, 21, 23, 39, 56, 57, 114, 126
Location 43, 53, 72–86, 116, 124
Logistics 2, 3, 44, 55, 57, 58, 157, 198
Longitudinal 29, 30, 162
Low carbon 46, 96, 111, 127, 128; Low energy 207, 215
Low incomes 13, 14, 17, 29

Machine learning 64, 106
Macroeconomic 21, 23, 267
Maintenance 6, 7, 16, 20, 40, 58, 73, 133–136, 146, 150, 153, 169, 175, 180–188, 203, 206, 217, 223, 224, 234, 269, 270
Management system 3, 42, 55, 57, 58–63, 80, 126, 146, 178, 206, 231, 232, 270

Managers 1, 6, 7, 71, 72, 116, 129, 135, 151–156, 178, 185–191, 206, 219, 225, 228, 271, 272
Mandatory 44, 179, 236, 249
Manual effort 73, 81, 240
Manufacturer 6, 75, 97–106, 135, 140, 145, 146, 268; Manufacturing 6, 53, 95, 151, 196, 249
Materials 1, 2, 6, 53, 57, 74, 81, 86, 97–99, 101, 135–157, 168, 173, 194, 201, 215, 222–227, 249
Maturity 200, 216, 219, 222
Measurement 27, 71, 124, 138, 151, 152, 158, 179–181, 184, 203, 215, 224, 235, 236
Mechanical 106, 134, 146
Media 47, 53, 234, 259
Melbourne 41, 124, 127, 128
Metadata 30, 124
Metrics 21, 30, 236, 238–241, 270, 271
Metropolitan 39, 42, 114, 120, 272
Milan 133, 136, 138, 140, 144
Mitigation 36, 38, 53, 96
Mobile 21, 43, 55, 75–84, 155; Mobility 14, 59, 70
Model 3, 17, 26, 41, 43–46, 64, 71, 72, 74–76, 85, 96, 98, 101, 105, 114, 119, 122–126, 133, 134, 138, 139, 152–160, 179, 184, 194, 200–205, 216, 233, 234, 255, 261
Modular 75, 78
Monitoring 3, 5, 52, 56, 59–61, 71, 80, 127, 150–156, 178–190, 197, 238, 239, 265
Municipal 13, 38, 42, 56, 57, 161, 164, 165
Mutopia 115, 127, 128

NATHERS 128
National Institute of Standards and Technology (NIST) 218, 235
Natspec 101, 106, 239
Neighbourhood 14, 23–27, 57, 62, 112, 113, 129
Netherlands 13–16, 29, 137
Network 2, 5, 8, 17, 22, 30–42, 45–47, 52–64, 73, 85, 86, 111, 116, 157, 181, 215, 220, 231, 232, 236, 258, 265

280 *Index*

New buildings 17, 125, 162, 175, 176, 224, 234, 245, 247, 255, 269; New construction 4, 118, 151–154, 156, 157, 160, 183, 253, 272
Noise 59, 86, 146, 151, 152
Non-proprietary 101
Normalisation 267, 268
Norway 3, 38, 41, 91, 199–203, 214–216, 220–228
Not-for-profit 13, 15, 16, 20, 21, 29, 31, 106, 117, 239

Object 38, 46, 54, 55, 63, 72–76, 91, 97–106, 112, 114, 121, 122, 129, 134–138, 140, 144, 145, 152, 158, 162–175, 203, 246, 259, 272; Object-oriented 73, 101, 115, 122, 127, 152, 153
Occupancy 30, 53, 56, 127, 176, 178, 180, 185
Office 25,107, 108, 125, 150, 156, 157, 159, 184, 187, 188, 195, 199, 214, 216, 225, 228, 247, 249
Offshore 199
Offsite 6, 53, 215
Omniclass 102, 103, 138
Online 165, 185, 186, 236, 239, 270
On-Site 57, 74, 75, 91, 101, 143, 144, 155, 157
Open access 43, 236, 238
Open BIM 126
Open data 4, 42–44, 46, 47, 64, 65, 120, 123, 130, 265, 267, 272
Open formats 99, 268; Open standard 3, 47, 106, 114, 117, 118, 268
Open government 43, 59
Operations 3–6, 45, 53, 60, 71, 73, 84, 86, 91, 95, 115, 116, 126, 135, 139, 145, 153, 157, 160–191, 196, 197, 203, 206, 213–234, 269, 270; Operational phase 161, 165, 166, 192, 214, 216, 217, 228, 231, 269, 270, 272
Operational conditions 174; Operational cost 57, 122, 125; Operational energy 56, 128; Operational log 166, 169, 170, 172–174; Operational services 223, 224; Operational staff 57, 58, 175, 269

Operator 27, 71, 72, 185, 188, 195, 197, 206, 232
Optimisation 7, 57, 106, 135, 146, 222, 224
OSCAR 195, 207, 216, 220, 222, 223, 228
Oslo 225
Output 21, 91, 92, 97, 99, 111, 115, 127, 140, 144, 150, 151, 179, 182, 204, 205, 247
Outsourcing 175, 176
Owner 6, 43, 58, 125–129, 154, 183, 194–198, 204–206, 213–223, 227–234, 245, 269
Ownership 4, 7, 18, 47, 55, 118, 124, 125, 186, 204, 219, 223, 272

Parameters 62, 74, 75, 80, 84, 97, 123, 165, 180, 184, 205, 236
Parametric 74, 75, 98
Paris 53, 96
Parking 56, 59, 60
Participation 14, 24, 140, 166, 171, 173, 226, 238, 239, 256, 258, 272
Partnering 16, 194; Partnership 2, 14, 17, 20, 162, 180, 183, 214
Path-dependency 255
Performance assessment 115, 128, 135, 235; Performance-based 160–162, 175, 179; Performance characteristics 103, 125, 144; Performance contracting 3, 178, 179, 183, 186, 188, 271; Performance data 125, 235, 236, 241; Performance-driven 234, 269; Performance evaluation 29, 240; Performance indicators 25, 114, 151, 154, 179, 235; Performance measurement 179, 215; Performance metrics 238, 239; Performance requirements 165, 175, 269; Performance specifications 143, 167, 169–171
Perth 5, 233
Phone 55, 82, 84, 155, 183
Photogrammetry 117
Photovoltaic 128
Pilot 22, 57, 155, 234, 238
Planners 41, 45, 113, 116, 135, 218, 270

Index 281

Planning phase 195, 196, 214–231;
  Planning processes 38, 72, 219;
  Planning systems 2, 199
Platform 21, 23, 40–53, 60, 71–76,
  111–118, 127–140, 145, 146, 154, 155,
  162, 185, 186, 195, 215–222, 227,
  237–240, 258, 268–271
Point cloud 152, 158
Policy 4, 16, 23, 25–27, 30, 31, 36–47, 95,
  117, 167, 192, 225, 232, 248, 249,
  253–272
Policy-makers 41, 245; Policy-making 47,
  232
Political 23, 38, 55, 183, 234–236
Pollution 41, 56, 58–60, 156, 176
Portal 4, 42–44, 47, 65, 127, 145, 155,
  267, 272
PPP 183, 188
Precinct 46, 112–130, 268, 272
Precinct Information Modelling (PIM)
  See Precinct
Precision 151, 153, 186
Predict 1, 37, 57, 58, 64, 71, 84, 205, 270
Predictive 43, 106, 127
Prefabrication 200
Prevention 71–74, 76, 81
Priorities 7, 41, 46, 154, 160, 161, 173,
  174, 176, 195, 227, 238, 240
Privacy 63
Private 5, 13–38, 44, 45, 60, 129, 145,
  162, 178–183, 195, 232, 245, 249, 258
Process documentation 169, 170, 174
Process-oriented 266, 268
Procurement 7, 180–182, 197–202, 216,
  232, 234, 238, 240, 247, 266
Product development 164
Production 2, 25, 36, 39, 46, 47, 55, 139,
  157, 196, 199–201, 248, 259, 267, 268,
  270–272
Productivity 4, 13–15, 21–26, 46, 71–75,
  80, 81, 86, 111, 114, 151, 191–198,
  215–219, 227, 235–241, 248, 266
Productivity commission 13, 236, 241
Project management 134, 194, 195, 197,
  198, 202–204, 206, 272; Project
  planning 73, 154
Project success 181, 197

Project team 187, 199, 227
Property 16, 58, 103, 106, 119, 124, 125,
  129, 154, 161, 166–173, 204, 248
Proprietary 117, 126, 268
Protocol 47, 56, 64, 111, 125, 134, 266
Prototype 75, 80–83, 117, 129, 134, 140
Public client 179, 232, 234, 239, 240, 266;
  Public owners 207, 215, 222, 228;
  Public sector 13, 21, 91, 180
Public housing 13–16
Public policies 43, 192
Public procurement 140, 157, 183, 188,
  231, 235, 240
Public transport 61, 113
Public works 31, 183

Quality 52–63, 97, 135, 146, 151,
  157–162, 167, 179–189, 195–198, 201,
  206, 215, 228, 234, 239, 265, 266
Quality assurance 160–162, 175, 176, 197,
  234, 266
Query 45, 63–65, 124, 237

Radio frequency identification (RFID)
  52–72, 85, 86, 118
Rail 234
Ratings 40, 71, 85, 114, 115, 127, 156
Real-time 53–63, 71, 72, 83, 84, 111, 265,
  271
Recyclable 58, 157; Recycling 57, 95, 127
Redundancy 40, 140, 146, 182, 190, 192,
  267
Reference service life (RSL) 135, 136,
  138–140, 270
Refurbishment 4, 150–158, 183, 184,
  189–191, 245, 247, 253, 255, 269, 272
Regeneration 13, 124
Regional 25, 37, 46, 114–119, 183, 184,
  232, 235
Regulation 16, 22, 27, 74, 135, 151, 160,
  161, 168, 198, 217; Regulatory 116,
  248, 253
Reliability 63, 124, 135; Reliable 5, 44, 85,
  91, 97, 101, 117, 133–136, 146,
  178–181, 187, 190, 270, 271
Remediation 125
Remote sensing 265

282 *Index*

Renewable 53, 56, 58, 59, 61, 96, 127, 128
Renovation 150, 153–158, 179, 180, 182, 183, 189–192, 245
Rental 13–17, 22, 25
Reporting 25, 96, 127
Repository 18, 43, 44, 47, 119, 126
Representation 46, 91, 101, 112, 125, 126, 134, 157, 169, 171, 186
Residential 16, 122, 124, 125, 128–130, 138, 143
Resilience 4, 7, 35–47, 114, 117, 267; Resilient 1, 35–37, 41, 42, 45, 47, 111, 240
Resources 6, 7, 31–39, 46, 47, 55–62, 91–93, 125, 133–146, 192, 202, 206, 220, 232, 236, 256, 258, 265
Retrofit 119, 124, 126, 157, 179
Return on investment 23, 24, 27, 241; Returns 17, 24, 27, 236
Reuse 101, 125, 146, 206, 213, 268
Revenue 16, 22, 23
Rework 133, 134
RIBA 133, 204, 208
Risk 3, 17, 38–45, 71–86, 152, 153, 171, 174–183, 188, 191, 206, 234, 271
Road 3, 41, 46, 56, 118, 144, 268
Robotics 1, 86, 239
Rova 57
Rules 74–76, 101, 198, 247; Rule-based 74, 77, 85

Safety 4, 7, 23, 29, 43, 70–76, 80–86, 146, 151, 258, 271
Savings 56–62, 92, 179, 184–189, 231, 234
Scaffolding 71, 73–80, 85
Scanning 86, 117, 153, 158, 202, 203
Scan-to-BIM 198, 199, 202, 203, 269
Schedule 43, 57, 58, 71, 82, 151–157, 167, 195, 197, 234; Scheduling 73, 134–136
Schema 101, 106, 118, 119, 123, 129, 170
Security 14, 25, 29, 35, 63, 146, 272
Semantic 98, 118, 157
Sensitivity 42, 44
Sensor 52–63, 73, 84–86, 118, 178, 184, 186, 217, 222, 271
Server 55, 83, 84, 122, 145, 238

Shipbuilding 199
Simulation 72, 73, 78, 79, 91, 96, 98, 127, 135, 168–170, 176, 184, 188
Singapore 52, 59–61, 65, 70
Small and medium-sized enterprises (SMEs) 215, 216
Smart 41, 45, 52–61, 64, 65, 72, 82–86, 112, 117, 118, 178, 216, 217, 227–230, 244, 265–273
Social capital 22–24
Social housing 3, 7, 13–17, 21–31, 186–192, 267
Socio-technical 4, 41, 45, 47, 177, 232, 266, 267
Software 4–7, 37, 40–44, 55, 57, 63, 82, 84, 96–99, 105, 106, 115–135, 145, 146, 154–157, 187–192, 203, 206, 217, 232, 237, 240, 265, 268, 272
Solar 81, 128
Spatial data 111, 114, 117; Spatial information 72, 85, 113
Specification 75, 77, 101, 106, 134, 143, 147, 151, 167–171, 179, 182, 224, 232, 255
Stakeholder 7, 8, 43, 124, 127, 133, 145, 194, 195, 222, 231, 233, 268
Standard 3, 4, 40–47, 56, 75, 92, 93, 97, 101–106, 111–123, 129, 134–138, 144, 155–170, 176, 179, 201–207, 220, 228–240, 266–268
Standardisation 46, 47, 63, 92, 136, 267
Storage 4, 57, 62–65, 77, 128, 135, 136, 145, 268, 270
Storm water 125
Structural 3, 6, 74, 102, 133, 134, 151, 155, 259
Subcontractor 6, 154, 155, 157, 180, 201, 218, 234
Suburbs 113, 124
Supplier 2, 6, 59, 154, 164, 191, 204, 217–227, 234, 270; Supply 2, 6, 16, 17, 19, 31, 60, 96, 111, 126, 181, 231, 235; Supply chain 1–3, 111, 150, 157, 179, 236
Surveyor 188, 233
Sustainability 7, 32, 36, 40, 47, 52–65, 91–95, 112, 114, 127, 135, 136, 146,

150, 156–158, 196, 197, 205, 206, 216, 219, 223–231, 248
Sustainable building 53, 157, 196, 215, 225; Sustainable design 127
Sustainable built environment national research centre (sbenrc) 5, 23, 27, 31, 74, 236
Sustainable development 55, 192, 204
Sweden 5, 137, 153, 155, 157, 158
Sydney 120, 124, 126, 233, 241, 266
Systems support 207, 266
Systems thinking 22, 204

Tactical 2, 197, 206, 225, 227
Technical solutions 39, 145, 162, 170, 171, 184, 214, 269
Technical staff 167, 170–173
Technical systems 4, 41, 45, 161, 162, 167–170, 174, 176, 177, 232, 266, 267
Temperature 60, 61, 81, 82, 84, 96, 184, 186, 217, 225
Temporal 37, 58, 114, 163
Tender 151, 167, 179–183, 187–191
Terminology 137, 217
Thermal modelling 184, 188
Tool 27, 28, 43, 46, 65, 74, 93, 98–100, 105, 127, 129, 134–146, 155–157, 170–173, 186, 192, 200, 203, 206, 218, 226, 236–240
Tracking 21, 57, 58, 101, 135
Traffic 56, 60, 77, 78, 80, 113, 118
Training 23, 70–72, 81, 86, 166, 185–187, 247, 253, 255, 258, 271
Transformation 1, 7, 19, 20, 111, 150, 206, 255, 271
Transition 41, 111, 126, 162, 185, 266
Translation 43, 162, 164, 216, 225, 269
Transmission 57, 83, 84, 191
Transparency 44, 47, 98, 267, 272
Transport 43, 56, 61, 74, 113–117, 127, 132, 139, 234–239, 270; Transportation 14, 57, 58, 70, 99, 100, 234
Trees 43, 58, 144
Trend 1, 20, 26, 35, 37, 46, 47, 55, 65, 81, 105, 106, 222
Trust 27, 63, 182, 191, 192, 200, 224, 239
Tunnels 117, 118

Turnover 185, 186, 192
Typology 121, 122, 164, 183

Uncertainty 26, 37, 38, 40–42, 151, 163, 174, 187, 249, 267
Underground 59, 82
Uniclass 137, 138
Uniformat 102
Unique 22, 31, 54, 119, 124, 134, 136, 145, 146, 155, 199, 236, 249, 253
Universal 15, 134, 137
Upstream 2, 93, 253
Urban design 113, 114, 127, 129; Urban development 14–16, 64, 111, 114, 124; Urban environment 40, 55, 267; Urban forest 58; Urban governance 37, 38, 42; Urban information 4, 44–47, 111, 267; Urban planning 39, 45, 46, 129
Urbanisation 35, 117
Usability 46, 156, 194, 217, 224, 267
Utilities 116, 118, 125

Validation 47, 139, 184, 236, 241
Value 2–9, 23–34, 50, 62, 64, 71, 84, 90, 107, 108, 126–129, 146–164, 177–183, 188, 189, 194–199, 205–244, 252, 253, 264–273; Value chain 2, 3, 5, 7, 146, 253, 269; Value creation 199, 206, 213, 214, 216, 218, 219, 222, 224; Value management 216, 222–225, 227
Variables 75, 84, 103, 139
Variations 15, 133, 135, 268, 270
Vehicle 1, 56, 58, 77, 78, 138, 140, 144, 272
Vendor-neutral 117, 268
Ventilation 82, 134, 139, 167, 188, 225
Veracity 62, 64, 268
Verifiability 181
Verification 78, 79, 101, 180, 184
Virtual design and construction (VDC) 241, 253, 255
Visualisation 43, 71–73, 76–78, 85, 86, 115, 127, 187, 225
Volunteer 29, 30

Walkability 113
Warning 71, 72, 78, 80, 82–85, 271

284 *Index*

Waste 17, 40, 52–60, 97, 114, 122, 127, 130, 150–157, 178, 220
Wastewater 125, 127
Water 38, 39, 59, 60, 82, 85, 97, 99, 100, 114, 116, 122, 125, 127, 128, 130
Wearable 84
Web-based 42, 71, 122, 136, 252; Web-enabled 127, 135, 270
Well-being 16, 23, 25, 27, 30
Wikipedia 237
Wireless 55, 58, 59, 80, 83, 86

Worker 70–74, 81–86, 151, 187, 197, 216, 225, 228, 271
Workflow 227
Workforce 23, 24, 73, 86, 247
Workplace 70, 74, 76, 82, 113, 213, 223

XML 63, 106, 130, 146
Xquery 63

Zonal 121, 122
Zone 5, 62, 118–120, 122, 124